T0309564

to the case of partial differential equations.

In Section 5.2 we consider an important case of the equation

$$\frac{du}{dt} = a_0(u, x) + \varepsilon a_1(x, u),$$

where a_0 is a vector function depending on an unknown vector u and a spatial variable x, and $a_1(x, u)$ is a partial differential operator acting on u. A characteristic property of these problems is the fact that if the operator a_1 in the right-hand side is removed, an ordinary differential equation with spatial variable x as a parameter results.

Further, there are considered some problems included here because of their importance and owing to the authors' tastes. In particular, we discuss the known Whitman method and an "algorithmic" part of the Maslov method for solving many-dimensional problems of geometric optics.

In conclusion, notice that we have not discussed boundary problems. However, while for ordinary differential equations a "general" solution is sought, the boundary problems can be considered as being not to principal; and for partial differential equations the situation differs completely and too much is still obscure.

the uniform normal form of the system is introduced. Arguments given are well-known to specialists (perhaps only the notion of a 0-point is not very customary); however, we decided to formulate them in a compact form. In this Section (in Section 3.4) the computational aspect of obtaining some partial solutions, which in mathematical literature are connected with the so-called "center manifold theorem" or the Lyapunov–Schmidt reduction [14], is discussed. Similar problems of bifurcational type arise, e.g., when attempting to find a periodic solution corresponding to the Hopf theorem.

Examples 3.11 and 3.12 illustrate the uniform normal form. In Section 3.13 the Hopf theorem is considered. In Sections 3.13–3.16 more general computational aspects of bifurcation theory are discussed.

In Chapter 4 we consider problems connected with reconstruction. They are conditionally divided into two types. In problems of the first type it is supposed that the operator is already reduced to the normal form and the question arises which operator should be considered as the leading one in similar situations. Until Chapter 5 this question does not arise since the previously developed methods were sufficient. The known problems of this type in the theory of linear ordinary differential equations are problems with a nilpotent leading operator. In the theory of linear equations a method to solve these problems was developed: the so-called shearing transformation [46], whose characteristic feature is the degeneracy of the corresponding change of variables as the parameter tends to 0. The goal of the shearing transformation is the decreasing of a number of Jordan cells in the new leading operator as compared with the initial one.

The linear formulation of nonlinear perturbation problems enables us to generalize the method to certain nonlinear problems using results of Chapter 2.

In problems of the second type there are singular surfaces in whose vicinity certain coefficients of the initial operators or of the changes of variables operators lose their smoothness (boundedness). In reconstruction problems of this type it is again possible to make use of algebraic methods. It should be noted in particular that this possibility arises because in the vicinity of singular surfaces the leading linear first order operator becomes degenerate (e.g., it often becomes nilpotent). Sections 4.3 and 4.4 are devoted to problems of this kind.

An additional problem of "matching" asymptotics constructed far from and near a singular surface arises. Since the matching process is very cumbersome, it is advisable to develop the most economic matching method. Section 4.5 is devoted to this problem. The rest of the chapter contains examples.

Chapter 5 deals with equations involving partial derivatives. Partial differential equations are in principle more complicated and possibilities for "theoretization" abruptly diminish. That is why Chapter 5 is somewhat eclectic. In Section 5.1 we introduce, in a convenient form, a language of functional derivatives necessary to extend the formalism developed earlier

The book requires only a standard mathematical background for engineers and does not require reference to the special literature.

Let us review the contents of the book. Chapter 1 contains in a convenient form an exposition of the matrix perturbation theory that is the model for constructing the general theory. The uses of Sections 1.4 and 1.5 will arise later. Section 1.4 contains a model for general formulations of perturbation theory (see Chapter 2) and Section 1.5 is needed only for reconstruction problems and contains an account of the well-known "shearing transformation" generalized, in what follows, to the nonlinear case.

In Chapter 2, main definitions are introduced and almost all "theoretical material" is gathered. In Sections 2.1 and 2.2 we give a general formulation of perturbation theory problems in the linear case and a notion of the canonical form of an operator. Section 2.3 deals with generalizations of notions of an eigenfunction and an eigenvalue for first order operators, notions which serve as the basis for the computational apparatus of Section 2.3. Exposition of the Jordan case, following the notion of an extended basis, can be omitted at the first reading. In Section 2.4 we elucidate how one should deal with eigenfunctions to reduce an operator to the canonical form. At the first reading one should note the sentences clarifying formula (2.4.1) and the simple variant of the proof of Theorem 2.4.3, then turn to formulas (2.4.7) and (2.4.8). It is worthwhile to look through Section 2.6, where the relation with N.N. Bogolyubov's ideas is established, and then turn to problems from the beginning of Chapter 3.

Section 2.5 supplies the background necessary for reconstruction problems as contains both a theorem on the normal form of an operator with respect to a nilpotent one and its generalization. This theorem provides a generalization of a shearing transformation and the basic idea of the algebraic reconstruction needed. Section 2.5 is the most "mathematicized" part of our survey.

Section 2.7 is devoted to problems of motion near a stationary manifold. This problem (called "the reduction of information problem") is often encountered in different problems of physics. One example is the problem of deduction of hydrodynamical equations from the Boltzmann equation; i.e., Hilbert and Chapman-Enskog methods. In Section 2.8 the case of Hamiltonian systems is briefly discussed.

Chapter 3 contains examples of differing complexity. Their purpose is indicated in the text. Nevertheless, notice that we advise looking through Examples 3.1 and 3.2 as soon as possible since they embody notions introduced earlier. In Example 3.9 the search for eigenfunctions is first discussed. In the preceding problems the leading part is linear and eigenfunctions were determined without any difficulty. Besides, in this example we discuss a certain specific technique of solving problems with a boundary layer. Section 3.10 is devoted to a (theoretical) discussion of some problems connected with a system with a parameter. The nature of a solution might steeply change for "resonance" values of a parameter. A notion of

Due to the immense diversity of problems where perturbation theory is used, the authors did not even try to present a rigorous mathematical theory. We tried to find, as has been said earlier, only a suitable formalism, having in mind that proofs of validity of the results it obtains should be carried out separately in each concrete problem (in the framework of the uniform formal approach).

While reading the "theoretical" part one should consider it as a formal recipe to be performed under appropriate conditions; in particular, all functions are supposed to be sufficiently smooth (if it is not stated to the contrary), i.e., bounded in the given domain along with their derivatives of the needed orders.

It ought to be said that even under the formal approach, the authors could not fulfill the above program completely because of its vastness. The reader will easily note quite a number of drawbacks and open questions. However, we hope that the progress we have obtained can justify the appearance of this book.

The book is arranged as follows. In Chapter 1, matrix perturbation theory is presented, first for a diagonal X_0, then for an arbitrary (Jordan) case; and a shearing transformation of Turritin [46] is given (in a form convenient for our purposes) as a special case of a reconstruction. In Chapter 2, results of Chapter 1 are generalized to the case of nonlinear equations, and a generalization of a shearing transformation (considered also as a special case of a reconstruction) to the case of polynomial perturbations is given. In Chapter 3, we give quite a few examples of solution of (most "regular") problems and consider them in detail. Chapter 4 deals with problems of reconstruction and contains several examples. In Chapter 5, problems involving partial derivatives are discussed.

The reader who wishes to get as quickly as possible to the core of the proposed algebraic approach may read first only those parts of Chapters 1 and 2 which refer to the diagonal X_0 and can then pass immediately to Examples 3.1 and 3.2. But for patient consecutive study of the "theory" of Chapters 1 and 2, we advise reference to corresponding examples of Chapter 3 as often as possible (the reason for each example is usually given) since examples constitute an essential part of the book. We do not advise reading Chapter 4 without previous detailed study of Chapters 1–3.

Another comment about examples is appropriate here. Though examples are necessary to understand the "theory," they are given with purely illustrative intentions. Most of them are already known problems solved here otherwise; the authors did not try to obtain here new results regarding completeness of the study or its elegance. That is why we did not try to give an exhaustive bibliography, restricting ourselves to citing the literature which is directly involved.

We hope that, though somewhat overloaded with "theory," the book will be helpful to those who are interested in applications.

izing the Poincaré–Bogolyubov–Krylov–Mitropolsky notion of a normal form; and to give a "standard" method for calculating the asymptotics (without, in particular, having to guess their form).

In the basis of this formalism lie the following considerations:

1. Instead of a *nonlinear* system we study a *linear* first order operator in partial derivatives for which the initial system is a characteristic one. The corresponding *linear* problem arises for a perturbed operator of the form $X = X_0 + \varepsilon X_1$, where X_0 and X_1 are first order operators such that "all is known" about X_0.
   When X_0 and X_1 are matrix operators, we deal with the well-studied classical theory consisting of reducing the matrix of X to the normal (Jordan) form under a transformation close to the identity.

2. The linear algebraic formulation of the theory mentioned above, the main tool being an appropriate generalization of notions of eigenvectors (eigenfunctions) and eigenvalues (from the viewpoint of mechanics these eigenfunctions can be considered as a generalization of action-angle coordinates), is extended to the "general" case of nonlinear equations. It provides an effective computational apparatus and a better understanding of the structure of a solution.

(b) *To propose an effective approach to reconstruction problems(singular perturbation problems) and a satisfactory matching procedure.*

A systematic linear algebraic approach enables us to define a reconstruction problem as a problem such that in a vicinity of the singular manifold the standard procedure of reduction to normal form fails (singularities arise) and the operator should be reconstructed in the form $X = \nu(\varepsilon)(\tilde{X}_0 + \delta(\varepsilon)\tilde{X}_1 + \cdots)$ under a singular (not close to the identity) transformation.

The standard scheme is applicable to the new operator $\tilde{X} = \tilde{X}_0 + \delta(\varepsilon)\tilde{X}_1$. Methods of search for such a transformation are developed both from the algebraic and the "trajectory" viewpoints.

The matching problem is considered as a problem of establishing relationships between constants of integration entering both asymptotics using the existence of the unique asymptotic development in the "overlapping" domain, where both asymptotics fit.

(c) *To discuss a possibility of its minimization, taking into account a quite large volume of computations in considered problems.*

(d) *To show possible ways of extending the proposed formalism to equations involving partial derivatives.*

Preface

Many books have already been written about the perturbation theory of differential equations with a small parameter. Therefore, we would like to give some reasons why the reader should bother with still another book on this topic.

Speaking for the present only about ordinary differential equations and their applications, we notice that methods of solutions are so numerous and diverse that this part of applied mathematics appears as an aggregate of poorly connected methods. The majority of these methods require some previous guessing of a structure of the desired asymptotics.

The Poincaré method of normal forms and the Bogolyubov–Krylov–Mitropolsky averaging methods, well known in the literature, should be mentioned specifically in connection with what will follow. These methods do not assume an immediate search for solutions in some special form, but make use of changes of variables close to the identity transformation which bring the initial system to a certain normal form. Applicability of these methods is restricted by special forms of the initial systems.

We may further notice that the known solved problems can be divided (however, rather conditionally) into two types: "regular," with the asymptotics sufficiently simple to be constructed in the whole domain of variation of variables; and "singular," ones whose characteristic feature is the presence of singular manifolds on which the qualitative behavior of a trajectory changes so that one has to construct different asymptotics in various subdomains and "match" the developments that have been obtained. The widely known example of "singular" problems, which henceforth will be called "reconstruction problems," are problems with turning points.

In "singular" problems two questions arise: how to find the asymptotics near the singular manifold or equations which define it, and how to "match" these asymptotics with the one found "far from" the singular manifold. We have not found in the literature general approaches to the solution of these problems. (The first question is rigorously put only in the works devoted to linear equations; in particular, we note the so-called shearing transformation.) The character of the asymptotics is usually guessed and the "matching" procedure is a collection of a not-too-clear recipe, see, e.g., [38].

All this makes it possible to explain our goals:

(a) *To develop, making use of a change of variables, a formalism general-*

V.N. Bogaevski
(deceased)

A. Povzner
Institute of the Physics of the Earth
Academy of Science—USSR
Bolshaya Gruzinskaya 10
Moscow, USSR

Mathematics Subject Classification Codes: 58Fxx, 34Dxx, 76Dxx

Library of Congress Cataloging-in-Publication Data

Bogaevskiĭ, V. N. (Vladamir Nikolaevich)
[Algebraicheski metody v nelineinoi teori vozmushcheniĭ.
English]
Algebraic methods in nonlinear perturbation theory / Bogaevski, A.
Povzner.
p. cm. — (Applied mathematical sciences)
Translation of: Algebraicheski metody v nelineinoi teori vozmushcheniĭ.
Includes bibliographical references and index.
1. Perturbation (Mathematics) 2. Nonlinear Theories.
I. Povzner, A. I͡A. (Aleksandr I͡Akovlevich) II. Title. III. Series: Applied
mathematical sciences (Springer-Verlag New York Inc.)
QA1.A647
[QA871]
510 s—dc20
[515.'35] 90-19608

Printed on acid-free paper.

Photocomposed pages prepared using LaTeX.
Printed and bound by R.R. Donnelley & Sons, Harrisonburg, VA.
Printed in the United States of America.

9 8 7 6 5 4 3 2 1

ISBN 0-387-97491-1 Springer-Verlag New York Berlin Heidelberg
ISBN 3-540-97491-1 Springer-Verlag Berlin Heidelberg New York

V.N. Bogaevski A. Povzner

Algebraic Methods in Nonlinear Perturbation Theory

With 17 Illustrations

Springer-Verlag
New York Berlin Heidelberg London
Paris Tokyo Hong Kong Barcelona

Applied Mathematical Sciences

1. *John:* Partial Differential Equations, 4th ed.
2. *Sirovich:* Techniques of Asymptotic Analysis.
3. *Hale:* Theory of Functional Differential Equations, 2nd ed.
4. *Percus:* Combinatorial Methods.
5. *von Mises/Friedrichs:* Fluid Dynamics.
6. *Freiberger/Grenander:* A Short Course in Computational Probability and Statistics.
7. *Pipkin:* Lectures on Viscoelasticity Theory.
8. *Giacaglia:* Perturbation Methods in Non-Linear Systems.
9. *Friedrichs:* Spectral Theory of Operators in Hilbert Space.
10. *Stroud:* Numerical Quadrature and Solution of Ordinary Differential Equations.
11. *Wolovich:* Linear Multivariable Systems.
12. *Berkovitz:* Optimal Control Theory.
13. *Bluman/Cole:* Similarity Methods for Differential Equations.
14. *Yoshizawa:* Stability Theory and the Existence of Periodic Solution and Almost Periodic Solutions.
15. *Braun:* Differential Equations and Their Applications, 3rd ed.
16. *Lefschetz:* Applications of Algebraic Topology.
17. *Collatz/Wetterling:* Optimization Problems.
18. *Grenander:* Pattern Synthesis: Lectures in Pattern Theory, Vol I.
19. *Marsden/McCracken:* Hopf Bifurcation and Its Applications.
20. *Driver:* Ordinary and Delay Differential Equations.
21. *Courant/Friedrichs:* Supersonic Flow and Shock Waves.
22. *Rouche/Habets/Laloy:* Stability Theory by Liapunov's Direct Method.
23. *Lamperti:* Stochastic Processes: A Survey of the Mathematical Theory.
24. *Grenander:* Pattern Analysis: Lectures in Pattern Theory, Vol. II.
25. *Davies:* Integral Transforms and Their Applications, 2nd ed.
26. *Kushner/Clark:* Stochastic Approximation Methods for Constrained and Unconstrained Systems.
27. *de Boor:* A Practical Guide to Splines.
28. *Keilson:* Markov Chain Models—Rarity and Exponentiality.
29. *de Veubeke:* A Course in Elasticity.
30. *Sniatycki:* Geometric Quantization and Quantum Mechanics.
31. *Reid:* Sturmian Theory for Ordinary Differential Equations.
32. *Meis/Markowitz:* Numerical Solution of Partial Differential Equations.
33. *Grenander:* Regular Structures: Lectures in Pattern Theory, Vol. III.
34. *Kevorkian/Cole:* Perturbation methods in Applied Mathematics.
35. *Carr:* Applications of Centre Manifold Theory.
36. *Bengtsson/Ghil/Källén:* Dynamic Meteorology: Data Assimilation Methods.
37. *Saperstone:* Semidynamical Systems in Infinite Dimensional Spaces.
38. *Lichtenberg/Lieberman:* Regular and Stochastic Motion.
39. *Piccini/Stampacchia/Vidossich:* Ordinary Differential Equations in R^n.
40. *Naylor/Sell:* Linear Operator Theory in Engineering and Science.
41. *Sparrow:* The Lorenz Equations: Bifurcations, Chaos, and Strange Attractors.
42. *Guckenheimer/Holmes:* Nonlinear Oscillations, Dynamical Systems and Bifurcations of Vector Fields.
43. *Ockendon/Tayler:* Inviscid Fluid Flows.
44. *Pazy:* Semigroups of Linear Operators and Applications to Partial Differential Equations.
45. *Glashoff/Gustafson:* Linear Optimization and Approximation: An Introduction to the Theoretical Analysis and Numerical Treatment of Semi-Infinite Programs.
46. *Wilcox:* Scattering Theory for Diffraction Gratings.
47. *Hale et al.:* An Introduction to Infinite Dimensional Dynamical Systems—Geometric Theory.
48. *Murray:* Asymptotic Analysis.
49. *Ladyzhenskaya:* The Boundary-Value Problems of Mathematical Physics.
50. *Wilcox:* Sound Propagation in Stratified Fluids.
51. *Golubitsky/Schaeffer:* Bifurcation and Groups in Bifurcation Theory, Vol. I.

(continued following index)

Applied Mathematical Sciences
Volume 88

Contents

1

Matrix Perturbation Theory

1.1 Perturbation Theory for a Linear Operator

$$X = X_0 + \varepsilon X_1 + \varepsilon^2 X_2 + \cdots \qquad (1.1.1)$$

which acts in the n-dimensional (complex) vector space R.

Here ε is a small parameter and matrices of operators, X_i, where $i = 0$, 1, ..., are defined in the fixed basis $e_1, \ldots, e_n$ and do not depend on ε.

The perturbation theory of the operator (1.1.1) consists of methods of obtaining asymptotic (as $\varepsilon \to 0$) expansion of eigenvalues and coordinates of vectors of the Jordan basis of X in the basis $e_1, \ldots, e_n$. This is usually done by a direct search for the mentioned values under certain assumptions on the form of the asymptotics. For example, suppose the "leading" operator X_0 possesses an eigenbasis $e_1, \ldots, e_n$ (from the very beginning we will assume that the matrix of X is defined using this basis) and corresponding eigenvalues $\lambda_1, \ldots, \lambda_n$ are mutually distinct. Then eigenvalues $\lambda_1^*, \ldots, \lambda_n^*$ and eigenvectors $e_1^*, \ldots, e_n^*$ of X are sought in the form

$$\lambda_j^* = \lambda_j + \varepsilon\lambda_j^{(1)} + \varepsilon^2\lambda_j^{(2)} + \cdots, \qquad e_j^* = e_j + \varepsilon e_j^{(1)} + \varepsilon^2 e_j^{(2)} + \cdots$$

where

$$e_j^{(k)} = \sum_{1\le\nu\le n} a_{j\nu}^{(k)} e_\nu \qquad \text{for } j = 1, 2, \ldots, n.$$

The process of the consecutive (with respect to the powers of e) search for coefficients is easily performed. In this process all the coefficients are defined via arithmetic operations so that the only transcendent problem is that of the initial reduction of the matrix of X_0 to the diagonal form.

The situation becomes more complicated when some of $\lambda_1, \ldots, \lambda_n$ coincide. In this case we suppose that $\lambda_{j_s}^*$ and $e_{j_s}^*$ are in the form

$$\lambda_{j_s}^* = \lambda_{j_s} + \varepsilon\lambda_{j_s}^{(1)} + \varepsilon^2\lambda_{j_2}^{(2)} + \cdots, \qquad e_{j_s}^* = \sum_{1\le r\le m} b_{j_s} e_{j_r} + \varepsilon_{j_r} e_{j_r}^{(1)} + \varepsilon^2 e_{j_s}^{(2)} + \cdots$$

where

$$e_{j_s}^{(k)} = \sum_{1\le\nu\le n} \alpha_{j_s\nu}^{(k)} e_\nu \qquad \text{for } s_{j_s} = 1, 2, \ldots, m_j,$$

e_{j_r} corresponds to λ_j, and m_j is the multiplicity of λ_j.

Unlike the previous case, the problem of determining the corrections to the eigenvalues (even the first corrections $\lambda_{j_s}^{(1)}$) turns out to be transcedent: it is necessary to "split" eigenvalues.

Still more difficult is the case when the matrix of the leading operator possesses Jordan cells. One has to consider expansions involving fractional powers of ε.

Due to the importance of the matrix theory which serves as a model for the consideration to follow we discuss it below in complete generality in an equivalent formulation, convenient for our purposes: find an operator $C(\varepsilon)$ such that the matrix of the operator

$$M = C^{-1}XC \tag{1.1.2}$$

is in Jordan form (in the same basis $e_1, \ldots, e_n$ for which the matrix is X is defined).

We will start from the assumption that C is of the form

$$C = E + eC_1 + e^2C_2 + \cdots \tag{1.1.3}$$

where E is the identity operator and the matrices of the operators C_i in the basis $e_1, \ldots, e_n$ do not depend on ε. We will try to find the simplest form of the matrix of

$$M = X_0 + eM_1 + e^2M_2 + \cdots \tag{1.1.4}$$

which may be obtained by an appropriate choice of C. (We assume that the matrix of X_0 is already in Jordan form.)

Remark 1.1.1 From the above it is clear that, generally speaking, it will not be possible to obtain the Jordan form for the matrix of M by making use of (1.1.3). However, we will see later how we should modify (1.1.3) naturally (in a sense) to achieve this goal.

First we give some formulas and definitions which will often be used.

1.2 Main Formulas

Let us write the operator (1.1.3) in the form

$$C = e^S = E + S + \frac{1}{2!}S^2 + \cdots, \tag{1.2.1}$$

where $S = \varepsilon S_1 + \varepsilon^2 S_2 + \cdots$ and the matrices of the operators S_j, $j = 1, 2, \ldots$, in the basis $e_1, \ldots, e_n$ do not depend on ε. Later we will see the usefulness of such a form of S.

Put $M = e^{-S}Xe^{S}$. The following *Hausdorff formula* holds:

$$M = X + [X,S] + \frac{1}{2!}[[X,S],S] + \cdots, \tag{1.2.2}$$

where brackets indicate the *commutator* of two operators:

$$[A,B] = AB - BA.$$

The validity of (1.2.2) may be justified as follows. Put $M(t) = e^{-tS}Xe^{tS}$, where t is a parameter; clearly $M(1) = M$, $M(0) = X$. Then

$$\begin{aligned}\frac{dM(t)}{dt} &= -Se^{-tS}Xe^{tS} + e^{-tS}Xe^{tS}S = -SM(t) + M(t)S \\ &= [M(t),S], \\ \frac{d^2M(t)}{dt^2} &= [[M(t),S],S], \ldots .\end{aligned}$$

Making use of these formulas, and expanding $M(t)$ in a Taylor series at $t_0 = 0$ we get the Hausdorff formula at $t = 1$. Note also this simple but important property of the commutator:

$$C^{-1}[A,B]C = [C^{-1}AC, C^{-1}BC].$$

Let us substitute in (1.2.2) the series (1.1.1), (1.2.1) for X and S, respectively and collect summands with the same powers of ε. We will obtain the following formulas for operators M_j, where $M_0 = X_0$ and $j = 1, 2, \ldots$:

$$\begin{aligned}M_1 &= [X_0, S_1] + X_0, \\ M_2 &= [X_0, S_2] + X_2 + [X_1, S_1] + \tfrac{1}{2}[[X_0, S_1], S_1], \ldots .\end{aligned} \tag{1.2.3}$$

These formulas are recursive:

$$M_j = [X_0, S_j] + Y_j, \tag{1.2.3'}$$

where $Y_1 = X_1$ and Y_j for $j \geq 1$ are known if $S_1, \ldots, S_{j-1}$ are known. It is clear now that our first concern will be mainly with the case $j = 1$ in (1.2.3).

Let us introduce some more definitions. The operator Y' will be referred to as the *derivative of the operator* Y *with respect to the operator* X_0 if

$$Y' = [X_0, Y]. \tag{1.2.4}$$

In what follows, the operator with respect to which the differentiation is performed will not be mentioned explicitly. This will not cause a misunderstanding since the expression Y' will be used only for differentiation with respect to the leading operator X_0. Note also a simple rule for differentiation of the commutator:

$$[A,B]' = [A',B] + [A,B'] \tag{1.2.5}$$

(the Jacobi identity). Further, we say that the *height* of Y equals μ if the μ-th derivative of Y vanishes while the $(\mu-1)$-th does not. If no derivative of Y is zero, then Y will be referred to as *unbounded in height*. Clearly, the operator of height 1 commutes with X_0. An operator Y representable in the form $Y = Z'$, where Z is a linear operator, will be called *integrable*, and *nonintegrable* otherwise. Finally, an operator will be referred to as *diagonal* if it possesses a basic system of eigenvectors.

1.3 Diagonal Leading Operator

This special case is of particular significance in spite of its simplicity. It is the only case studied in this section.

Let $\lambda_1, \ldots, \lambda_k$ for $k \leq n$ be (all) distinct eigenvalues of X_0 and $m_1, \ldots, m_k$ their multiplicities; i.e., $m_1 + \cdots + m_k = n$. In the eigenbasis the matrix of X_0 is of the form

$$X_0 = \begin{Vmatrix} \Lambda_1 & & 0 \\ & \ddots & \\ 0 & & \Lambda_k \end{Vmatrix},$$

where $\Lambda_s = \lambda_s \xi_s$ and ξ_s is the square $m_s \times m_s$ unit matrix for $s = 1, \ldots, k$. Let us split the matrix of Y (in the same basis) into cells y_{pq} corresponding to cells Λ_s of the matrix $\mathcal{X}_0$ so that the cell y_{pq} is situated on the intersection of rows and columns whose numbers coincide with the numbers of rows and columns, respectively, of the cell Λ_q; i.e., the cell y_{pq} has m_p rows and m_q columns. We will write $\mathcal{Y} = \|y_{pq}\|$, where $p, q = 1, 2, \ldots, k$. Let us compute $Y' = [X_0, Y]$. The matrix $\mathcal{Y}'$ of this operator in a similar cell notation is of the form $\mathcal{Y}' = \|(\lambda_p - \lambda_q) y_{pq}\|$. From here we deduce several simple corollaries.

(1) *The operator Y commutes with X_0 if and only if its matrix in the canonical basis of X_0 is quasidiagonal; i.e.,*

$$\mathcal{Y} = \begin{Vmatrix} y_{11} & & 0 \\ & \ddots & \\ 0 & & y_{kk} \end{Vmatrix},$$

where y_{ss}, $s = 1, \ldots, k$, are arbitrary square cells of order m.

The invariant reformulation of (1) is the following statement.

Let R_s be the linear span of the eigenvectors of X_0 corresponding to the eigenvalue λ_s so that R_s is a X_0-invariant subspace, i.e., $x \in R_s$ implies $X_0 x \in R_s$ and R is the direct sum of subspaces R_s, where $s = 1, \ldots, k$. Then

(1′) *The necessary and sufficient condition of permutability of Y and X_0 is Y-invariance of R_s.*

(2) *When X_0 is diagonal, any operator is either unbounded in height or is of height 1, i.e., commutes with X_0. In fact, $Y'' = 0$ implies $Y' = 0$.*

(3) *The operator Y is integrable if and only if diagonal cells y_{ss} of its matrix are zero. If Y is integrable, then a solution of $Y = Z'$ is given as follows: out-of-diagonal cells of the matrix $\|z_{pq}\|$ are equal to*

$$z_{pq} = \frac{1}{\lambda_p - \lambda_q} y_{pq} \qquad \text{for } p \neq q \tag{1.3.1}$$

while diagonal ones are arbitrary. (Z is defined up to a summand permutable with X_0.)

(4) *Any operator is uniquely representable in the form of the sum of an integrable operator and a height 1 operator (permutable with X_0).*

Since Corollary (4) has an invariant formulation, we give a method to decompose Y into permutable and integrable parts which does not require basic eigenvectors of X_0.

Suppose the matrices of X_0 and Y are defined with respect to a common basis. Let us compute $Y', Y'', \ldots, Y^{(n^2)}$. Since there are no more than n^2 linearly independent $n \times n$ matrices there exist numbers $\omega_0, \ldots, \omega_{n^2}$ not all equal to zero and such that

$$\omega_0 Y + \omega_1 Y' + \cdots + \omega_{n^2} Y^{(n^2)} = 0. \tag{1.3.2}$$

This equation will be called the *commutation relation.* (It holds for any X_0.) For a diagonal X_0 we may assume without loss of generality that either $\omega_0 \neq 0$ or $\omega_1 \neq 0$. In fact, if $\omega_0 = \omega_1 = 0$, then by Corollary (2) we may make $\omega_1 \neq 0$ by integrating (1.3.2) as many times as necessary and relabeling the coefficients ω. If $\omega_0 \neq 0$, then Y is integrable; i.e.,

$$Y = \left\{ -\frac{1}{\omega_0}(\omega_1 Y + \omega_2 Y' + \cdots + \omega_{n^2} Y^{(n^2-1)}) \right\}'.$$

If $\omega_0 = 0$ and $\omega_1 \neq 0$, then, by (1.3.2),

$$\bar{Y} = Y + \frac{1}{\omega_1}(\omega_2 Y' + \cdots + \omega_{n^2} Y^{(n^2-1)}),$$

and

$$Y - \bar{Y} = \left\{ -\frac{1}{\omega_1}(\omega_2 Y + \cdots + \omega_{n^2} Y^{(n^2-1)}) \right\}$$

is integrable.

Now, turning to the formula for M_1 (see (1.2.3) and (1.2.4)), we get $M_1 = S_1' + X_1$. We may represent X_1 as the sum of an integrable operator and an operator permutable with X_0 and eliminate the integrable part of M_1 by the choice of S_1. Under such a choice of S_1 the operator M_1 commutes with X_0, i.e., $M_1' = 0$. Quite similarly, making use of the recursive formulas (1.2.3′), we may get $M_2' = 0$, $M_3' = 0, \ldots$ so that

$$M' = 0. \tag{1.3.3}$$

The operator M_1 is uniquely defined. Operators $M_2, M_3, \ldots$ are not uniquely defined due to the ambiguity in the choice of the summand permutable with X_0 in operators $S_1, S_2, \ldots$.

Remark 1.3.1 Formulas for the operators M_i, where $M_i' = 0$, constructed above may be written in the form [similar to (1.2.3)]

$$M_2 = [M_1, \bar{S}_1] + N_1,$$
$$M_3 = [M_1, \bar{S}_2] + N_2 + [N_1, \bar{S}_1] + \tfrac{1}{2}[[M_1, \bar{S}_1], \bar{S}_1], \ldots,$$

where $\bar{S}_j$, $j = 1, 2, \ldots$, are permutable with X_0 and parts of S_j and operators N_j, $j = 1, 2, \ldots$, are uniquely defined. (We skip the proof.)

We have therefore arrived at the notion of the normal form of (1.1.1) with respect to X_0 in the case of a diagonal X_0: the operator $M = e^{-S}Xe^{S} = X_0 + \varepsilon M_1 + \varepsilon^2 M_2 + \cdots$ will be referred to as the *normal form* of X (with respect to X_0) if $M_1, M_2, \ldots$ commute with X_0. For the perturbation theory problem in this case we understand the problem of reducing X to the normal form (with accuracy up to any needed power of ε).

By Corollary (1), the equality (1.3.3) shows that the problem of reducing the matrix of X to Jordan form breaks into several (k) problems of lesser dimension in subspaces k_s, where $s = 1, \ldots, k$. Let us consider two extreme cases:

1. All eigenvalues of the leading operator X_0 are distinct ($k = n$; the simple spectrum). Then the matrix of the operator M is diagonal in the basis $e_1, \ldots, e_n$ and the problem is solved.

2. All eigenvalues of X_0 coincide ($k = 1$; the spectrum is completely degenerate). Then any operator commutes with X_0 and $M_1 = X_1$. The problem is reduced to the study of $X_1 + \varepsilon X_2 + \cdots$, as was evident beforehand.

In the intermediate case ($1 < k < n$; the spectrum is partially degenerate) the spectrum of the leading operator X_0 in R_s for $m_s \neq 0$ is completely degenerate and the operator M_1 becomes the *new leading operator* on R_s (see also Remark 1.3.1). Thus, having studied only the leading operator we can obtain only the above-mentioned reduction to problems of lesser dimension.

Remark 1.3.2 We call attention to the last circumstance since it models the prevailing situation when working with some problems of nonlinear differential equations with a small parameter. In most of these problems degeneracy of the spectrum is typical.

The reduction of the matrix of the new leading operator M_1 to diagonal form is a new transcendent problem mentioned in Section 1.1.

1.4 General Case. The Normal Form of the Matrix of the Operator M

In this section we will give only formulations of main statements for an arbitrary X_0. The proofs may be found in the next chapter which contains more general propositions, those presented here included as a special case.

Let, as before, $\lambda_1, \ldots, \lambda_k$ be all different eigenvalues of the leading operator X_0 and let $m_1, \ldots, m_k$ be their respective multiplicities; $m_1 + \cdots + m_k = n$. As is known, R_n splits into the direct sum of subspaces R_s of dimension m_s for $j = 1, \ldots, k$ defined by the condition

$$\text{if} \quad e \in R_s, \quad \text{then} \quad (X_0 - \lambda_s E)^p e = 0, \tag{1.4.1}$$

where $p = p(e) \geq 1$ is an integer. (This statement may be considered as a "part" of the theorem on reduction of a matrix of an operator to the Jordan form. It remains only to pick in each R_s an appropriate, Jordan, basis.) As we have just shown in the case of a diagonal X_0, i.e., $p = 1$, by an appropriate choice of S the reduction problem for X splits into problems onto subspaces R_s. The same reduction holds as well in the general case and the permutability property (1.3.3) is replaced by a more general condition of boundedness in height,

$$M^{(\mu)} = 0 \qquad \text{for } \mu \geq 1. \tag{1.4.2}$$

Respectively, in the general case of an arbitrary X_0, by a *normal form of* X (with respect to X_0) we will mean, by definition, an operator $M = e^{-S} X e^{S} = X_0 + \varepsilon M_1 + \varepsilon^2 M_2 + \cdots$ such that $M_1, M_2, \ldots$ are bounded in height.

The possibility of such a reduction is justified by the following lemmas.

Lemma 1.4.1 *The operator Y is bounded in height if and only if R_s, where $s = 1, 2, \ldots, k$, are Y-invariant subspaces (see proof in Theorem 2.4.1).*

Lemma 1.4.2 *The operator S, which gives (1.4.2), does exist.*

Proof of this statement is easily obtained by making use of the commutation relations (1.3.2); the only differences in the arguments as compared with Section 1.3 are that it is no longer possible to arrive at the same conclusion about coefficients ω_0 and ω_1, and the first index of the nonzero coefficient, ω_μ, is μ, which is, generally speaking, greater than 1. For a more "economic" method of constructing S [similar to (1.3.1)] see Theorem 2.4.3.

We can proceed now with the study of the general case of considering problems onto R_s. As is known [and transparently from (1.4.1)], X_0 is representable on R_s in the form of the sum of a diagonal operator and a *nilpotent* one, i.e., an operator whose sufficiently large powers vanish. [Equality (1.4.1) shows that $X_0 - \lambda_s E$ is nilpotent in R_s.] The diagonal part of X_0 on R_s does not affect further simplification of X, as we have already seen, and we may without loss of generality put $\lambda_s = 0$. Thus we come to the study of the case when the leading operator X_0 in (1.1.1) is nilpotent. It is well known that the space R_n, where the nilpotent operator X_0 acts, splits into the direct sum of subspaces R_s of dimension q_s for $s = 1, 2, \ldots, r$ such that $\sum q_s = n$, and so that in each of R_s a (Jordan) basis $e^{(s)}, \ldots, e^{(s)}$ exists satisfying

$$X_0 e_1^{(s)} = 0, \quad X_0 e_2^{(s)} = e_1^{(s)}, \quad \ldots, \quad X_0 e_{q_s}^{(s)} = e_{q_0 - 1}^{(s)}.$$

In the basis

$$e_1^{(1)}, \ldots, e_{q_1}^{(1)}, e_1^{(2)}, \ldots, e_{q_s}^{(2)}, \ldots, e_{q_r}^{(r)},$$

the matrix of X_0 possesses the Jordan canonical form

$$\mathcal{X}_0 = \left\| \begin{matrix} H_{q_1} & & 0 \\ & \ddots & \\ 0 & & H_{q_r} \end{matrix} \right\|, \tag{1.4.3}$$

where H_{q_s}, for $s = 1, \ldots, r$, is a Jordan cell of order q_s with eigenvalue zero. (In what follows we assume $q_1 \geq \cdots \geq q_r$.) Let us suppose that the basis $e_1, \ldots, e_n$ in which the matrix of X is defined is a canonical Jordan basis of X_0. For nilpotent X_0 any operator is bounded in height (as is evident since $Y^{(\mu)}$ is a linear combination of operators $X_0^\alpha Y X_0^\beta$, where $\alpha + \beta = \mu$, and it is not difficult to establish that the maximal possible height equals $2 \max_s q_s - 1$). The main goal of the constructions to follow is to represent an arbitrary operator Y in the form

$$Y = \bar{Y} + Z', \tag{1.4.4}$$

where $\bar{Y}$ is an operator of the minimal possible height. Such a representation is surely not unique: if, for instance, the maximal height of Y is $2 \max_s q_s - 1$, then the same is true of $\bar{Y}$ for any Z. Nevertheless, as we will see later, *minimal height* is an invariant notion.

The invariant formulation of the problem is as follows. Linear operators acting in R_n span the linear space T, where the operator $\mathcal{D}$ of differentiation, $\mathcal{D}Y = Y'$ for $Y \in T$, is a linear nilpotent operator. Therefore T is the direct sum of subspaces $T^{(\rho)}$ each having a (Jordan) basis

$$Y_{\mu_p}^{(\rho)}, \; Y_{\mu_p - 1}^{(\rho)}, \ldots, Y_1^{(\rho)} \tag{1.4.5}$$

satisfying $\mathcal{D}Y_\nu^{(\rho)} = Y_{\nu-1}^{(\rho)}$, $\mathcal{D}Y_1^{(\rho)} = 0$. (Here ν is the height of $Y_\nu^{(\rho)}$.) The operator $Y_\nu^{(\rho)}$ for $\nu < \mu_\rho$ is evidently integrable, and $Y_{\mu_\rho}^{(\rho)}$ is not integrable.

Splitting the operator Y with respect to the basis (1.4.5), we can clearly obtain

$$\bar{Y} = \sum_\rho c_\rho Y_{\mu_\rho}^{(\rho)}, \tag{1.4.6}$$

where the c are constants. The form (1.4.6) of $\bar{Y}$ is referred to as *normal.*

It is possible to choose S so that M takes the normal form (1.4.6). The matrix of $\bar{Y}$ [in the normal form (1.4.6)] in the basis $\{e_1^{(s)}, \ldots, e_{q_s}^{(s)}\}$ depends, of course, on the choice of bases (1.4.5), whereas the height of Y is an invariant. In the sequel we give a concrete method for construction of a basis in T. This basis is used hereafter.

Each operator $Y \in T$ may be considered as a sum of operators $Y^{(q_i q_j)}$ that map R_i into R_j in agreement with the splitting of the matrix of Y into cells $\|a_{qp}\| = A_{q_j q_i}$ corresponding to cells H_{q_i}, H_{q_j} in $\mathcal{X}_0$.

Thus

$$Y^{(q_i q_j)} e_p^{(i)} = \sum_{1 \le q \le q_j} a_{pq} e_q^{(j)} \quad \text{for } p = 1, \ldots, q, \tag{1.4.7}$$

and

$$\mathcal{D}Y^{(q_i q_j)} e_p^{(i)} = \sum_{1 \le q \le q_j} (a_{p,q+1} - a_{p-1,q}) e_q^{(j)}, \quad \text{where } a_{p,q_j+1} = a_{0,q} = 0. \tag{1.4.8}$$

Denote by $Y^{(q_i q_j)}(\alpha)$ the "diagonal" operator such that $a_{pq} = 0$ for $q - p \neq \alpha$, where $-(q_i - 1) \le \alpha \le q_j - 1$, and note that $\mathcal{D}Y^{(q_i q_j)}(\alpha)$ is an operator of the form $Y^{(q_i q_j)}(\alpha - 1)$.

Making use of (1.4.8) it is easy to verify the following three statements.

1. For $-(q_i - 1) \le \alpha \le \max(q_j - q_i, 0) - 1$, the "diagonal" operator is integrable.
2. For
$$\max(q_j - q_i, 0) \le \alpha \le q_j - 1, \tag{1.4.9}$$
a "diagonal" operator is integrable if and only if the sum Σ of its matrix elements is zero.
3. Among operators $Y^{(q_i q_j)}(\alpha)$ for α as in (1.4.9) and $\Sigma = 1$ there is one, $Y_\mu^{(q_i q_j)}(\alpha)$, of minimal height μ, where
$$\mu = 2\alpha + q_i - q_j + 1. \tag{1.4.10}$$

The following statements are less elementary.

Lemma 1.4.3 *Operators $Y^{(q_i q_j)}(\alpha)$ constitute a complete set of generating elements $Y_{\mu_j}^\beta$ introduced in (1.4.5).*

Lemma 1.4.4 *Elements $a_{p,p+\alpha}$ of the matrix of $Y_\mu^{(q_i q_j)}(\alpha)$ are positive.*

Proofs of these lemmas as well as those of statements (1), (2) and (3) will be given later; in Chapter 2 we consider a much more general problem of reducing a (nonlinear) operator to the normal form with respect to a nilpotent X_0.

1.5 Nilpotent Leading Operator. The Reconstruction Problem

Let us assume that the operators X_j in (1.1.1), where $j = 1, 2, \ldots$, are in normal form. In this case one cannot "simplify" X by using an operator C of the form (1.1.3) or (1.2.1) which tends to the identity as $\varepsilon \to 0$. On the other hand, the simple example

$$\mathcal{X} = \begin{pmatrix} 0 & 1 \\ 0 & 0 \end{pmatrix} + \varepsilon \begin{pmatrix} 0 & 0 \\ 1 & 0 \end{pmatrix}$$

shows that for a nilpotent leading operator the parameter ε is introduced in a sense incorrectly; in this example the leading and the perturbation operators are "equivalent" for the problem does not actually contain a small parameter since X is equivalent to

$$\begin{pmatrix} \sqrt{\varepsilon} & 0 \\ 0 & -\sqrt{\varepsilon} \end{pmatrix}.$$

In what follows we aim to extricate in a "natural" way the summands in the perturbation which are "equivalent" to the leading operator and which, when added to it, will constitute the "proper" leading operator. This procedure of finding a "correct" leading operator will constitute the meaning of the notion "reconstruction." Let Y be an operator of height μ whose matrix is in the normal form; i.e., Y is a linear combination of operators $Y_\mu^{(q_i q_j)}(\alpha)$ with the common μ defined in Section 1.4. Consider the operator

$$X_0 + \Delta Y, \tag{1.5.1}$$

where Δ is a small parameter.

Making use of the operator $P(\delta)$, where δ is also a small parameter, whose matrix $\mathcal{P}(\delta)$ in the chosen basis is of the form

$$\mathcal{P}(\delta) = \begin{Vmatrix} \mathcal{D}_{q_1} & & 0 \\ & \ddots & \\ 0 & & \mathcal{D}_{q_r} \end{Vmatrix}, \tag{1.5.2}$$

where

$$\mathcal{D}_{q_i} = \begin{Vmatrix} \delta^{1-q_i/2} & & 0 \\ & \delta^{2-q_i/2} & \\ & & \ddots & \\ 0 & & & \delta^{q_i/2} \end{Vmatrix},$$

we construct the operator

$$P(\delta)^{-1}(X_0 + \Delta Y)P(\delta) \tag{1.5.3}$$

equivalent to (1.5.1). The transformation with the operator $P(\delta)$ defined by (1.5.2) is a trivial modification of the so-called *shearing transformation* [46]. It is easy to see [with the help of (1.4.10) and (1.5.2)] that the operator (1.5.3) is

$$\delta X_0 + \delta^{(1-\mu)/2}\Delta Y \tag{1.5.3'}$$

where only the "diagonality" of the cells of the matrix has been used. The specific form of the elements of this matrix will be needed later in Theorem 1.5.1. Now, if

$$\delta = \Delta^{2/1+\mu} \tag{1.5.4}$$

so that $\Delta = \delta^{(1+\mu)/2}$, then the operator (1.5.3′) takes the form

$$\delta(X_0 + Y). \tag{1.5.3''}$$

Thus operators X_0 and Y are "equivalent." Taking (1.5.4) into account we may say that the operator ΔY is *added* to X_0 and that its *weight* is $\Delta^{2/(1+\mu)}$. The greater the height μ, the greater the weight of Y. Now consider the case $Y = Y_1 + Y_2$, where Y_1, Y_2 are in the normal form with heights μ_1 and μ_2 where $\mu_1 > \mu_2$. Then, putting

$$\delta = \Delta^{2/1+\mu_1},$$

we get

$$P^{-1}(X_0 + \Delta Y)P = \delta\left\{X_0 + Y_1 + \Delta^{(\mu_1+\mu_2)/(1+\mu_1)}Y_2\right\}.$$

The operator $X_0 + Y_1$ becomes the new leading operator with a new small parameter,

$$\Delta^{(\mu_1-\mu_2)/(1+\mu_1)}.$$

What has been said above may render unnecessary the investigation of different variants of reconstruction of (1.1.1) depending on heights of operators constituting X. We note that the new leading operator clearly depends only on a finite number of terms, the latter being of the form $X_0 + \bar{Y}$, where $\bar{Y}$ is an operator in the normal form of any height and the new perturbation operator depends on some power of ε.

The problem of finding a new leading operator is independent of perturbation theory.

This problem being resolved, we find ourselves in a situation analogous to the one just considered, and one may wonder if this process will terminate. But before we discuss this question, let us consider the same problem from the viewpoint of the theory of differential equations.

Let us consider $dx/dt = (X_0 + \Delta Y)x$, where X_0 is assumed to be in the Jordan form. If t varies over a finite interval, then it is possible to obtain a solution in the form $x = x_0 + \Delta u_1 + \Delta^2 u_2 + \cdots$. However, the larger $|t|$ is, the greater is the number of terms needed to obtain any given accuracy. Therefore, the initial variables of the problem, $x = (x_1, \ldots, x_n)$, are inconvenient when t is large and we should find some new variables better suited for computations in this case.

Suppose that to the Jordan cell H_{q_i} in (1.4.3) variables $x^i = (x^i_1, \ldots, x^i_{q_i})$ correspond. The leading system $dx/dt = X_0 x$ can be rewritten in the form

$$\frac{dx^i}{dt} = H_{q_i} x^i.$$

Consider a solution of a single equation when $|t| \gg 1$. It is convenient to set $t = \delta\tau$ and assume that $\tau \sim 1$, $\delta \gg 1$. It is easy to see that x^i_k is proportional to δ^{k-1}. Hence it is natural to introduce new variables

$$x^i_k = \delta^{k+\mu_i} y^i_k$$

(the reason for introducing the additional multiple δ^{μ_i+1} is that the transformation

$$\bar{y}^i_k = \delta^{\mu_i+1} y^i_k$$

does not change the leading equation; i.e., the operator corresponding to this transformation commutes with X_0). In the new variables, the system takes the form

$$\frac{dy^i_k}{d\tau} = y^i_{k-1} + \Delta \sum_{s,\rho} a^{is}_{k\rho} y^s_\rho \delta^{\mu_s - \mu_i - k + \rho + 1}.$$

First, suppose that all numbers $\Delta\delta^{\mu_1 - \mu_i - k + \rho + 1}$ are "small." Then it is possible to rewrite the system in the form $dy/d\tau = X_0 y + Hy$, where H is a "small" operator as compared with X_0. Let us make more precise what we mean by this. Put $y = e^{\tau X_0} z$. (The search for solutions in such a form is an operator generalization of the variation of parameters method. In quantum mechanics this trick is known under the name of "passage to the mixed representation" and is widely used.) Then

$$\frac{dz}{d\tau} = e^{-\tau X_0} H e^{\tau X_0} z.$$

Since H is small, then for $\tau \sim 1$ (so that a solution of the leading equation may be considered as finite) we solve the equation for z by iterations [in variables y and assuming that initial values are of order $O(1)$]. Then z changes slowly; i.e., for the computation of y we may make use of perturbation theory. This is the meaning of the phrase "the operator H is small as compared with X_0." If at least one of the numbers

$$\Delta\delta^{\mu_s-\mu_i-k+\rho+1}$$

is not small, then the perturbation theory is not applicable; the behavior of a solution of the complete equation differs essentially from those of the leading problem. Thus the order of t such that the perturbation theory is still applicable is bounded by $\bar{\delta}(\Delta)$, i.e., by the smallest of those $\delta \gg 1$ for which

$$\Delta\delta^{\mu_s-\mu_i-k+\rho+1} = O(1).$$

For $|t| \sim \delta$ the leading equation should be the new one; this may be obtained by putting $\delta = \bar{\delta}(\Delta)$.

In these considerations Y has not been reduced to the normal form, which shows that the process of reconstruction is not unique. The process described above is preferable because of its algebraic (invariant) nature. Now let us return to the starting point. Consider the reconstructed problem with the new leading operator $X_0 + \bar{Y}$. The following cases will be encountered.

1. *The operator $X_0 + \bar{Y}$ has at least two different eigenvalues.* Then we obtain a new reduction problem.
2. *The spectrum of $X_0 + \bar{Y}$ is completely degenerate but $X_0 + \bar{Y}$ is not nilpotent.* Then we get the process similar to that in the case of diagonal X_0; i.e., from X we extract the diagonal operator of the "main" order.
3. *The operator $X_0 + \bar{Y}$ is nilpotent.* In this case we start a new reconstruction process and the question is whether the procedure ever stops. The answer is affirmative. Each reconstruction yields a certain step toward the solution, as the following theorem shows.

Theorem 1.5.1 *If $X_0 + \bar{Y}$ is nilpotent and $p_1 \geq \cdots \geq p_t$ are dimensions of its invariant subspaces, then the first of the p_s's different from the corresponding number q_s, where $q_1 \geq \cdots \geq q_r$, satisfies $p_s > q_s$, and if $\bar{Y}$ is nonzero then such a number exists.*

PROOF

If theorem is false, then

$$(X_0 + \bar{Y}_0)^{q_1} = 0, \tag{1.5.5}$$

where X_0 and $\bar{Y}$ are matrices of operators X_0, $\bar{Y}$ in the canonical basis of X_0 (cf. Section 1.4).

In what follows we consider every matrix $A(\dim A = \dim X_0)$ as a block-matrix $(A_{q_iq_j})$, where $A_{q_iq_j}$ corresponds to blocks (cells) H_{q_i}, H_{q_j} of X_o [cf. (4.3) as in §3 and §4].

We call $A_{q_iq_j}$ a senior block if $\max(q_i, q_j) = q_1$.

If it could be shown that from (1.5.5) follows that senior blocks of $\bar{Y}$ are zeros, then one would have to consider matrices of lesser dimension, which would finish the proof.

Let us enumerate diagonals of cells $A_{q_iq_j}$, starting from the upper right-hand corner and call a diagonal free if its elements are zeros, together with all elements of preceding diagonals. We note that cells $\bar{Y}_{q_iq_j}$ possess no less than $\max(q_i, q_j) - 1$ free diagonals and so have a "triangular" form (cf. Section 1.4).

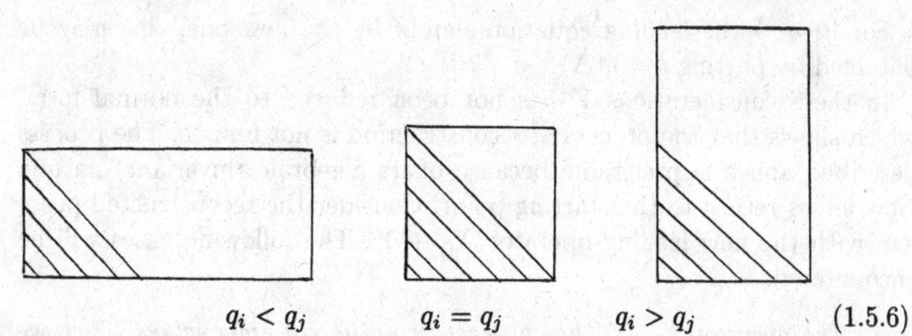

$$q_i < q_j \qquad\qquad q_i = q_j \qquad\qquad q_i > q_j \tag{1.5.6}$$

[In (1.5.6) diagonals which are not necessarily free are shown schematically].

Let us consider now the left-hand side of (1.5.5) as a sum of diverse matrices

$$\begin{gathered} X_0^{\alpha_1}\bar{Y}^{\beta_1}\cdots X_0^{\alpha_s}\bar{Y}^{\beta_s}, \\ \alpha_1 + \cdots + \alpha_s + \beta_1 + \cdots + \beta_s = q_1, \quad \beta_1 + \cdots + \beta_s \geq 1 \\ (X_0^{q_1} = 0). \end{gathered} \tag{1.5.7}$$

Supposing that senior cells of $\bar{Y}$ have no less than $q_1 - 1 + k$ $(k \geq 0)$ free diagonals we want to show that senior cells of matrix (1.5.7) possess no less than $q_1 - 1 + k - (\alpha_1 + \cdots + \alpha_s)$ free diagonals. But first, we shall prove that our theorem follows from this estimate.

Indeed, consider a fixed senior cell in the left-hand side of (1.5.5) and its $(k+1)$-th diagonal. Using representation of this cell as a sum of matrices (1.5.7), we see that in nonlinear with respect to $\bar{Y}$ summands this diagonal is free because $\beta_1 + \cdots + \beta_s \geq 2$, and this gives $\alpha_1 + \cdots + \alpha_s \leq q_1 - 2$. Let us consider now the (linear with respect to $\bar{Y}$) terms $X_0^{\alpha_1}\bar{Y}X_0^{\alpha_2}$, $\alpha_1 + \alpha_s \leq q_1 - 1$.

One obtains the cell $(X_0A)_{q_iq_j}$ from $A_{q_iq_j}$ by moving its lines one step upwards, and one obtaining cell $(AX_0)_{q_iq_j}$ by moving the $A_{q_iq_j}$ columns one step to the right.

This shows that elements of $(k+1)$-th diagonal are sums of some elements of (q_1+k)-th diagonal of a corresponding senior cell $\bar{Y}$. The appearing equations have only a trivial solution $(C_{q_iq_j\alpha} = 0)$ as it follows from Lemma 4.4

that elements of $(q_1 + k)$-th diagonal are positive (disregarding to common factor $C_{q_i q_j \alpha}$) and the beforementioned construction shows that at least in the summand $X_0^{q_1-1}\bar{Y} + \bar{Y}X_0^{q_s-1}$ the $(k+1)$-th diagonal can not be free. Because $k \geq 0$ is arbitrary one gets that senior cells of $\bar{Y}$ have to be zeros.

Now we have to prove that our estimate is correct. The fundamental property of "triangular" cells (1.5.6) is as follows: if $A_{q_i q_j}$, $B_{q_i q_j}$ are two cells and one of them has the form (1.5.6) and the second one has no less than ν free diagonals, then the product $A_{q_i q_j} B_{q_i q_j}$ has no less than ν free diagonals. From this we conclude that if m_{ij} is a number of free diagonals of $A_{q_i q_j}$, then: (a) the cells of i-th block-line of matrix $A\bar{Y}$ have no less than $\min_j m_{ij}$ free diagonals and (b) the cells of the j-th block-column of $\bar{Y}A$ have no less than $\min_i m_{ij}$ free diagonals.

Considering now the matrix (1.5.7) as a product $(\ldots(((A_1 A_2)A_3)A_4)\ldots)$ where A_p denotes X_0 or $\bar{Y}$ we conclude that our estimate of the number of free diagonals is correct for the senior block-line. Indeed each multiplication by $\bar{Y}$ (when one $\bar{Y}$ had already occurred) does not change our estimate, and each multiplication by X_0 diminishes its maximum by one. For the senior block-column, one obtains the estimate in a similar way, considering (1.5.7) in the form $(\ldots(B_4(B_3(B_2 B_1)))\ldots)$ where B_p is $\bar{Y}$ or X_0.

The theorem is proved.

2

Systems of Ordinary Differential Equations with a Small Parameter

In this chapter we construct an analogue of the matrix perturbation theory for systems of the form

$$\varepsilon^{\alpha}\frac{dx_i}{dt} = a_{0i}(x) + \varepsilon a_{1i}(x) + \varepsilon^2 a_{2i}(x) + \cdots = f_i(x;\varepsilon),$$

where $x = (x_1, \ldots, x_n)$, $i = 1, \ldots, n$, or, in vector notation,

$$\varepsilon^{\alpha}\frac{dx}{dt} = a_0(x) + \varepsilon a_1(x) + \varepsilon^2 a_2(x) + \cdots = f(x;\varepsilon).$$

We will consider this system in a given domain D and suppose that the right-hand sides of the system satisfy certain quite restrictive requirements stated below. Examples given in the following chapters show that in problems of interest these requirements are often satisfied in practice. We will also see that right-hand sides of the systems themselves and the formulation of the problems usually motivate the choice of D.

2.1 Passage to the Linear Problem. Change of Variables Operator

The contents of this section do not depend on the presence of a small parameter, which will be omitted here. Consider the system

$$\frac{dx}{dt} = f(x) \tag{2.1.1}$$

and, connected with it, the first order linear partial differential operator

$$X = \sum_{i \le i \le n} f_i(x)\frac{\partial}{\partial x_i} = (f, \nabla) \tag{2.1.2}$$

such that (2.1.1) is its characteristic system. Here and hereafter (unless it is mentioned to the contrary) all functions considered are assumed to be sufficiently smooth and bounded in D together with all of their derivatives.

Let $x'(t,x)$ be a solution of the system (2.1.1) with the initial value $x'(0,x) = x \in D$. (We find it convenient to denote by x the initial values which, in what follows, will be considered as independent variables.) For any $F(x)$ we have, by (2.1.1),

$$\frac{\partial F(X'(t,x))}{\partial t} = \sum_{1 \le i \le n} f(x') \frac{\partial F(x')}{\partial x'_i}. \tag{2.1.3}$$

Put $G(t,x) = F(x'(t,x))$. Then by (2.1.3) and (2.1.2) we have

$$G(0,x) = F(x), \qquad \left.\frac{\partial G}{\partial t}\right|_{t=0} = XF(x), \qquad \left.\frac{\partial^2 G}{\partial t^2}\right|_{t=0} = X^2 F(x), \qquad \ldots$$

Summing the Taylor series for $G(t,x)$ obtained with the help of these formulas, we get

$$F(x') = G(t,x) = e^{tX} F(x). \tag{2.1.4}$$

In particular, if we put $F(x) = x_i$, where $i = 1, \ldots, n$, then

$$x' = e^{tX} x. \tag{2.1.5}$$

Differentiating both parts of (2.1.4) in t we get the following equation for the function $G(t,x)$:

$$\frac{\partial G}{\partial t} = XG, \tag{2.1.6}$$

where $G(0,x) = F(x)$ is the initial value. Thus the integration of the system (2.1.1) is equivalent to the integration of one linear partial differential equation, (2.1.6). This well-known fact enables us to reduce the formal perturbation theory of a *nonlinear* system to the perturbation theory of a first order *linear* operator. Suppose S is a first order operator. Let us substitute x' from (2.1.5) into (2.1.4) and denote tX by S, where t is a parameter. We get

$$F(e^S x) = e^S F(x). \tag{2.1.7}$$

Let us apply the operator e^{-S} to both parts of (2.1.6), where S does not depend on t. Since e^{-S} and $\partial/\partial t$ commute, and due to (2.1.7), we get

$$\frac{\partial H(t,x)}{\partial t} = MH(t,x), \tag{2.1.8}$$

where $H(t,x) = G(t, e^{-S}x)$, $M = e^{-S} X e^S$. Equation (2.1.8) is a first order equation like (2.1.6). In fact, M is an operator of the form (2.1.2), and, due to the Hausdorff formula (1.2.2) and since the commutator of two first order operators is a first order operator, we have

$$[(a,\nabla),(b,\nabla)] = ((a,\nabla)b - (b,\nabla)a, \nabla). \tag{2.1.9}$$

In fact,

$$\left(\sum_{1\le i\le n} a_i(x)\frac{\partial}{\partial x_i}, \sum_{1\le i\le n} b_i(x)\frac{\partial}{\partial x_i}\right)h(x)$$
$$=\sum_{1\le i\le n}\left(\sum_{1\le i\le n}\left(a_j(x)\frac{\partial b_i(x)}{\partial x_j} - b_j(x)\frac{\partial a_i(x)}{\partial x_j}\right)\right)\frac{\partial h(x)}{\partial x_i}.$$

Equation (2.1.8) can be considered as a change of variables

$$x' = e^S x \tag{2.1.10}$$

in (2.1.6), where the old notation, x, is used for the new variables since

$$H(t,x') = H(t,e^S x) = e^S H(t,x) = e^S G(t,e^{-S}x) = G(t,x).$$

If we choose some S and solve (2.1.8) instead of (2.1.6), then the operator e^{-S} should be applied in order to get a corresponding solution of (2.1.6). The operator e^S is naturally referred to as the *change of variables operator.*

2.2 General Formulation of the Perturbation Theory Problem

Let us return to the system with a small parameter. The equation of the form (2.1.6) for it is

$$\varepsilon^\alpha \frac{\partial G}{\partial t} = XG, \tag{2.2.1}$$

where

$$X = X_0 + \varepsilon X_1 + \varepsilon^2 X_2 + \cdots, \quad X_j = (a_j(x), \nabla), \qquad j = 0, 1, \ldots.$$

The operator e^S transforms (2.2.1) into an equation of the form (2.1.8),

$$\varepsilon^\alpha \frac{\partial H}{\partial t} = MH, \tag{2.2.2}$$

where $M = e^{-S}Xe^S$, $S = (s(x;\varepsilon), \nabla)$, and $S = \varepsilon S_1 + \varepsilon^2 S_2 + \cdots$, where the S_j's do not depend on ε.

Now, using formulas and definitions of Section 1.2, we are able to start the construction of an analogue of the matrix perturbation theory.

Remark 2.2.1 Since the system (2.1.1), $\varepsilon^\alpha(dx/dt) = Xx$ is equivalent to the equation

$$\varepsilon^\alpha \frac{dx}{dt} = Mx, \tag{2.2.3}$$

we intend to try to simplify the operator M as much as possible.

The direct analogue of the algebraic perturbation theory is the following requirement on the choice of the change of variables: find S so that

$$M_j^{(\mu_j)} = 0 \qquad \text{for } \mu_j \geq 1 \tag{2.2.4}$$

(see 1.4.2). This requirement will be taken as a definition of the perturbation theory problem and it constitutes a direct generalization of the matrix perturbation theory considered above.

Remark 2.2.2 The hasty thought that we may try to require, for instance, $M_1 = 0$ (since the equation $S_1' + X_1 = 0$ is locally solvable) must immediately be rejected since we need a global (in D) simplification of M. So, in order to obtain in D the property (2.2.4), S must satisfy very restrictive conditions. This remark will clarify to an extent the naturalness of the restrictions introduced below.

Thus, under the general perturbation theory problem we understand by definition, as in the matrix theory, the problem of reducing the operator X in (2.2.1) to a "canonical" form as in (2.2.4). Then there arises the necessity of the study of a new "canonical" operator (system); see (1.1.3), the case of the spectrum with multiplicities.

Remark 2.2.3 Let us accentuate once more the difference of linear systems from the matrix case. The nature of the new problem (for the operator M_1) and means of its investigation in the matrix case are known beforehand. In nonlinear systems such a problem is new in principle (as compared with X_0, which is, surely, always supposed to be known). The perturbation theory of nonlinear systems consists at first of a method of obtaining new equations simplified as compared with the exact problem. Their solution is (in general) not connected any longer with the presence of a small parameter and must be considered as a new problem.

2.3 Canonical Form of First Order Operator X_0

Since we discuss a linear operator, we introduce the notion of eigenvectors and their generalization in the sense of (1.4.1) almost as for matrices. The difference is that it is now not natural to assume $\lambda = \text{const}$. We only need that it must be constant *with respect to* X_0. We will say that a (single-valued in D) function $\varphi(x)$ corresponds to the eigenvalue $\lambda(x)$ of X_0 if for all $x \in D$ we have $X_0\lambda(x) = 0$ and $(X_0 - \lambda(x))^p\varphi(x) = 0$ for sufficiently large integers $p \geq 1$. When $p = 1$, i.e., $X_0\varphi = \lambda\varphi$, the function φ will be referred to as the *eigenfunction* of X_0 corresponding to $\lambda(x)$. An eigenfunction $\omega(x)$ corresponding to the function which is identically zero, i.e., $X_0\omega(x) = 0$, will be referred to an an *invariant* of X_0. Invariants are constants with respect to X_0; i.e., $X_0(\omega F) = \omega X_0 F$, and any function in invariants is an invariant itself. An eigenvalue is an invariant. Thus, invariants and eigenvalues (as invariants) play a twofold role: that of a number and of an eigenfunction. The product of eigenfunctions is an eigenfunction corresponding to the sum of their eigenvalues. So are functions $\varphi(x)$ for $p > 1$, as follows from the binomial formula [where $C_m^k = m!/k!(m-k)!$],

$$(X_0 - \lambda_1 - \lambda_2)^m(\varphi_1\varphi_1) = \sum_{0\leq k\leq m} C_m^k (X_0 - \lambda_1)\varphi_1 (X_0 - \lambda_2)^{m-k}\varphi_2,$$

valid since λ is a constant with respect to X_0.

Let $z_1(x), \ldots, z_n(x)$ be a system of functions such that in D we may pass to new variables $z_1, \ldots, z_n$ without affecting the smoothness of operators being considered. Such a system is referred to as *basic*.

In variables $z_1, \ldots, z_n$, the linear operator Y is of the form

$$Y = (Yz_1)\frac{\partial}{\partial z_1} + \cdots + (Yz_n)\frac{\partial}{\partial z_n},$$

where coefficients are expressed in terms of z; i.e., $Yz_i = \varphi_i(z)$. Now suppose that there is a basic system consisting of eigenfunctions and invariants of X_0; i.e.,

$$\varphi_1(x), \ldots, \varphi_k(x), \quad \omega_1(x), \ldots, \omega_m(x) \qquad \text{for } k + m = n,$$

where $\lambda_i(x) \not\equiv 0$ in D for $i = 1, \ldots, k$. Let us suppose also that the eigenvalues $\lambda_i(x)$ are functions in basic invariants $\lambda_i(x) = \lambda_i'(\omega) = \lambda_i$. Then in basic variables the operator X_0 takes the form

$$X = \lambda_1\varphi_1\frac{\partial}{\partial\varphi_1} + \cdots + \lambda_k\varphi_k\frac{\partial}{\partial\varphi_k}. \tag{2.3.1}$$

An operator reducible to such a form is naturally referred to as a *diagonal operator* and the form (2.3.1) is called *canonical*. (In this form λ_i looks like a genuine constant.)

Remark 2.3.1 If we consider now Equation (2.2.1) in the zeroth approximation,

$$\varepsilon^\alpha \frac{d\omega_i}{dt} = 0 \qquad \text{for } i = 1, \ldots, m,$$
$$\varepsilon^\alpha \frac{d\varphi_j}{dt} = \lambda_j\varphi_j \qquad \text{for } j = 1, \ldots, k,$$

then we get

$$\omega_i(x(t)) = \text{const}, \qquad \varphi_j(x(t)) = \text{const} \cdot \exp\left(\frac{\lambda_j t}{\varepsilon^\alpha}\right),$$

where $\lambda_j = \text{const}$ along the trajectory. The variables ω and φ generalize mechanical action-angle coordinates.

Denote "vectors"

$$(\varphi_1, \ldots, \varphi_k), \qquad \left(\frac{\partial}{\partial \varphi_1}, \ldots, \frac{\partial}{\partial \varphi_k}\right),$$

and the diagonal matrix with eigenvalues $\lambda_1, \ldots, \lambda_k$ by φ, ∇_φ, and Λ. Then (2.3.1) may be written in the form

$$X_0 = (\Lambda\varphi, \nabla_\varphi). \tag{2.3.1'}$$

Before we pass to the general Jordan case, we note the following. We have required λ to be functions in ω only. The case when λ depends on φ is quite possible. For example, two basic functions may correspond to eigenvalues of opposite signs. Then their product is an invariant and may serve as a new eigenfunction. This situation not only destroys the introduced canonical form but, as we will see later, entangles the procedure of the construction of S. Now we will exclude, in a sense, this case by means of the following notion of an extension of an operator.

The system $dz_i/dt = Yz_i$, where $i = 1, \ldots, n$, may be extended trivially by considering how some additional functions $z_{n+1}(z), \ldots, z_{n+p}(z)$ which depend on $z = (z_1, \ldots, z_n)$ vary on its trajectories. For this we must adjoin the equations for z_{n+j}, $j = 1, \ldots, p$, to the initial equations. If we consider all new variables z as independent ones, then the new system arises. However, when initial conditions match, i.e., $z(0) = z_0$, $z_{n+j}|_{t=0} = z_{n+j}(z_0)$, we will not obtain anything new.

Accordingly, let us adjoin to the basic system new functions $z_{n+1}(z), \ldots, z_{n+p}(z)$, and consider the operator

$$Y_p = (Yz_1)\frac{\partial}{\partial z_1} + \cdots + (Yz_n)\frac{\partial}{\partial z_n} + \cdots + (Yz_{n+1}(z))\frac{\partial}{\partial z_{n+1}} + \cdots + (Yz_{n+p}(z))\frac{\partial}{\partial z_{n+p}},$$

where now all variables $z_1, \ldots, z_{n+p}$ are assumed to be independent and Yz_i for $i = 1, \ldots, n+p$ are expressed in terms of $z_1, \ldots, z_{n+p}$ in any "convenient" fashion. Such an operator Y_p will be referred to as *extended operator* Y_p. Evidently, when additional functions $z_{n+1}, \ldots, z_{n+p}$ are invariants of Y, coefficients of the extended operator Y_p coincide with those of Y on the manifold $z_{n+1} = z_{n+1}(z), \ldots, z_{n+p} = z_{n+p}(z)$. The system of functions $z_1, \ldots, z_n, z_{n+1}, \ldots, z_{n+p}$ is referred to as an *extended basic system*. .

Now suppose $\varphi_1(x), \ldots, \varphi_k(x); \omega_1(x), \ldots, \omega_m(x); \lambda_1(x), \ldots, \lambda_k(x)$ is an extended basic system of the diagonal operator X_0 and express the coefficients in terms of these variables: $X_0\varphi_i = \lambda_i\varphi_i$. The extended operator X_0 coincides with (2.3.1), where λ_i no longer depends on φ (nor on ω) and is in the canonical form.

Let us now pass to the Jordan case. Suppose there is a basic system of functions $\varphi_1, \ldots, \varphi_n$ corresponding to eigenvalues $\lambda_1, \ldots, \lambda_n$ of X_0. Let us find for the function φ_i, where $i = 1, \ldots, n$, the minimal number p_i such that $(X_0 - \lambda_i)^{p_i}\varphi_i = 0$ and construct by the standard procedure the series of functions corresponding to λ_i, putting

$$\begin{aligned} \varphi_{i0} &= \varphi_i, \\ X_0\varphi_{i0} &= \lambda_i\varphi_{i0} + \varphi_{i1}, \\ X_0\varphi_{i1} &= \lambda_1\varphi_{i1} + \varphi_{i2}, \\ &\vdots \\ X_0\varphi_{i,p-1} &= \lambda_i\varphi_{i,p-1}. \end{aligned}$$

The system of functions

$$\varphi_{10}, \varphi_{11}, \ldots, \varphi_{1,p_1-1}; \ldots; \varphi_{n0}, \varphi_{n1}, \ldots, \varphi_{n,p_n-1}; \lambda_1, \ldots, \lambda_n \tag{2.3.2}$$

will be referred to as an *extended Jordan basis*. The extended operator X_0 in this system is representable in the form

$$X_0 = (\mathcal{J}\varphi, \nabla_\varphi), \tag{2.3.3}$$

where $\mathcal{J}$ is the Jordan matrix with blocks of size $p_i \times p_i$ for $i = 1, \ldots, n$ and the λ_i are eigenvalues. This form will be called *canonical* and the operator reducible to it in the above sense will be referred to as a *Jordan one*.

In addition, we introduce the notion of a quasilinear operator. If there exists an extended basic system $z_1, \ldots, z_n;\ z_{n+1}, \ldots, z_{n+p}$ such that

$$X_0 = (\Omega z, \nabla_z), \tag{2.3.4}$$

where $\Omega = \|\omega_{ij}(z)\|$ is the matrix whose elements are invariants of X_0, then X_0 is said to be *quasilinear*. For instance, a Jordan operator is quasilinear.

2.4 An Algebraic Formulation of the Perturbation Theory Problem

Suppose the leading operator X_0 is Jordan. Under an algebraic formulation we will understand the following requirements for the choice of S [cf. (1.1.4)]: find S such that M transforms any function corresponding to an eigenvalue of X_0 into a function corresponding to the same eigenvalue; i.e.,

$$(X_0 - \lambda(x))^p\varphi(x) = 0 \quad \text{implies} \quad (X_0 - \lambda(x))^q M\varphi(x) = 0 \tag{2.4.1}$$

for sufficiently large q for any $x \in \mathcal{D}$. The following generalization of Lemma 1.4.1 holds.

Theorem 2.4.1 *If X_0 is in Jordan form, then Y is bounded in height if and only if*

$$(X_0 - \lambda(x))^p \varphi(x) = 0 \quad \textit{implies} \quad (X_0 - \lambda(x))^q Y \varphi(x) = 0 \qquad (2.4.1')$$

for sufficiently large q; i.e., (2.3.3) and (2.4.1) are equivalent.

PROOF

First, note that if $\varphi = \varphi(x)$ corresponds to $\lambda = \lambda(x)$, then so does $(X_0 - \lambda)^\alpha \varphi$, where $\alpha > 0$ is any integer; conversely, if $(X_0 - \lambda)^\alpha \varphi$ corresponds to λ, then so does φ. Suppose (2.4.1′) is satisfied. Let us show that $Y^{(\mu)} = 0$ for a sufficiently large μ. Consider $\hat{X}_0 = X_0 - \lambda E$, where E is the identity operator. Denote by $\hat{Y}'$ the derivative of Y with respect to this operator; i.e., $\hat{Y}' = [\hat{X}_0, Y] = Y' + (Y\lambda)E$. Note that $\hat{Y}^{(\mu)}$ is a linear combination of operators of the form

$$\hat{X}_0^\alpha Y \hat{X}_0^\beta = (X_0 - \lambda E)^\alpha Y (X_0 - \lambda E)^\beta,$$

where $\alpha + \beta = \mu$. Further, it is easy to verify that $\hat{Y}^{(\mu)} = Y^{(\mu)} + \mu(X_0^{\mu-1}(Y\lambda))E$. But by definition λ corresponds to zero and due to (2.4.1′) so does $Y\lambda$. Therefore, $\hat{Y}^{(\mu)} = Y^{(\mu)}$ for a sufficiently large μ.

Further, $(X_0 - \lambda E)^\beta \varphi = 0$ for sufficiently large β by the definition of φ, and $Y(X_0 - \lambda E)^\beta \varphi$ corresponds to λ due to (2.4.1′) for any β. Therefore $(X_0 - \lambda E)^\alpha Y (X_0 - \lambda E)^\beta \varphi = 0$ for sufficiently large $\mu = \alpha + \beta$; i.e., $\hat{Y}^{(\mu)} \varphi = 0$ for any φ. Thus, $Y^{(\mu)} \varphi = 0$ for any φ and sufficiently large μ. Since there is a basic system consisting of functions φ, we have $Y^{(\mu)} = 0$, Now let $Y^{(\mu)} = 0$, for some $\mu \geq 1$. Let us show that (2.4.1′) is satisfied. It suffices to prove the following: if (2.4.1′) is satisfied for an operator Z', then it also holds for Z. (For the operator $Y^{(\mu)} = 0$ this condition is clearly verified.) For any function $h = h(x)$ we have

$$Z'h = X_0 Zh - ZX_0 h = (X_0 - \lambda) Zh - Z(X_0 - \lambda)h - (Z\lambda)h.$$

Note that $Z\lambda$ corresponds to zero; in fact, $Z'\lambda = X_0(Z\lambda)$ and since $Z'\lambda$ corresponds to zero, then so does $Z\lambda$. Let $(X_0 - \lambda)^p \varphi = 0$. Put $h_1 = (X_0 - \lambda)^{p-1} \varphi$. Then $Z'h_1 = (X_0 - \lambda) Zh_1 - (Z\lambda)h_1$. Since $Z'h$ corresponds to λ, h_1 corresponds to λ, and $Z\lambda$ corresponds to zero, we get that $Z'h_1 + (Z\lambda)h_1 = (X_0 - \lambda)Zh_1$ corresponds to λ. Hence Zh_1 corresponds to λ. Further, set $h_2 = (X_0 - \lambda)^{p-2}\varphi$. Then $Z'h_2 = (X_0 - \lambda)Zh_2 - Zh_1 - (Z\lambda)h_2$ and we find similarly that Zh_2 corresponds to λ. Continuing this process we see that $Zh_p = Z\varphi$ corresponds to λ. This completes the proof.

Returning once more to the proof of the first part of the theorem, we note that $Y^{(\mu)} = 0$ holds if (2.4.1′) is true only for basic functions and corresponding eigenvalues, provided that for any basic φ the function $(X_0 - \lambda)^s \varphi$ for $s \geq 1$ is also basic or zero. Since this condition is verified by an extended Jordan basis by its construction, the following theorem holds.

Theorem 2.4.2 *To solve the perturbation problem, it suffices to find S so that the property (2.4.1′) holds only for functions from the extented Jordan basis of X_0.*

The following theorem gives one of the practical methods for constructing S (see the proof).

Theorem 2.4.3 *Suppose Y satisfies the following:*

1. *for any function $\varphi(x)$ from the extended basis of X_0 corresponding to the eigenvalue $\lambda(x)$ we have*

$$Y\varphi(x) = \sum_{1\le k\le m} \psi_k(x), \qquad (2.4.2)$$

where $\psi_k(x)$ corresponds to the eigenvalue $\mu_k(x)$ of X_0;

2. *either $\mu_k(x) \equiv \lambda(x)$ or $\mu_k(x) \neq \lambda(x)$ for all $x \in \mathcal{D}$. Then the equation (in M and S) $M = [X_0, S]+Y$, where M satisfies (2.4.1′), is solvable.*

The complete proof of this simple fact will be given later. It looks somewhat cumbersome because it is stated too generally. Therefore we will present it first for the case when the basis consists only of eigenfunctions and invariants, eigenvalues depend only on invariants of the basis, and in the decomposition (2.4.2), only invariants or eigenfunctions enter.

The equation $M = [X_0, S] + Y$ requires computation of M and S only on basic functions since M and S are first order operators. It is essential that we start by defining their values on invariants. By assumption $Ye_\nu = \psi_0 + \psi_1 + \cdots$, where e_ν and ψ_0 are invariants whereas $X_0\psi_i = \mu_i(e)\psi_i$ and $e = (e_1, \ldots, e_p)$ are basic invariants; $\mu_i(e) \neq 0$ in D. But then $Me_\nu = X_0 Se_\nu + \psi_0 + \psi_1 + \cdots$. Set

$$Se_\nu = -\sum_i \psi_i/\mu_i(e).$$

Then $Me_\nu = \psi_0$; i.e., M transforms an invariant into an invariant.

Now define $M\varphi_\lambda$ when $X_0\varphi_\lambda = \lambda\varphi_\lambda$. We have

$$M\varphi_\lambda = X_0 S\varphi_\lambda - (S\lambda)\varphi_\lambda - \lambda S\varphi_\lambda + Y\varphi_\lambda.$$

But $S\lambda = \sum_\nu (\partial\lambda/\partial e_\nu)Se_\nu$ and formulas for Se_ν imply that $(S\lambda)\varphi_\lambda$ is a sum of eigenfunctions corresponding to the eigenvalues γ such that $\gamma - \lambda \neq 0$ in D. Hence $M\varphi_\lambda$ may be written in the form

$$M\varphi_\lambda = X_0 S\varphi_\lambda - \lambda S\varphi_\lambda + \psi_\lambda + \sum_\gamma \psi_\gamma, \qquad (2.4.3)$$

where the ψ_γ are eigenfunctions and ψ_λ corresponds to λ whereas ψ_γ corresponds to γ so that $\lambda - \gamma \neq 0$ in D. Setting

$$S\varphi_\lambda = \sum_\gamma \frac{\psi_\gamma}{\lambda - \gamma},$$

we get $M\varphi_\lambda = \psi_\lambda$ and in this case the theorem is proved. The complete proof is as follows.

PROOF

First, consider an auxiliary equation

$$X_0 y - \lambda y + \psi = 0, \tag{2.4.4}$$

where y is an unknown function, $\psi = \psi(x)$ corresponding to an eigenvalue $\mu(x)$ of X_0 such that $\mu(x) \neq \lambda(x)$ for $x \in D$ and $\lambda = \lambda(x)$ is an invariant of X_0. Let $p = \min \hat{p}$, where $(X_0 - \mu)^p \psi = 0$. Let us construct the series

$$\begin{aligned} \psi^{(0)} &= \psi, \\ \psi^{(1)} &= X_0 \psi^{(0)} - \mu\psi^0, \\ \psi^{(2)} &= X_0 \psi^{(1)} - \mu\psi^{(1)}, \\ &\vdots \\ X_0 \psi^{(p-1)} &= \mu\psi^{(p-1)}. \end{aligned}$$

Set

$$y = \omega_0 \psi^{(0)} + \omega_1 \psi^{(1)} + \cdots + \omega_{p-1} \omega^{(p-1)}, \tag{2.4.5}$$

where $\omega_0, \ldots, \omega_{p-1}$ are invariants of X_0. Substituting this expression in (2.4.4), we get

$$\begin{aligned} &[\omega_0(\mu - \lambda) + 1]\psi^{(0)} + [\omega_1(\mu - \lambda) + \omega_0]\psi^{(1)} \\ &\quad + \cdots + [\omega_{p-1}(\mu - \lambda) + \omega_{p-2}]\psi^{(p-1)} = 0. \end{aligned}$$

Putting

$$\omega_0 = -\frac{1}{\mu - \lambda}, \quad \omega_1 = \frac{1}{(\mu - \lambda)^2}, \quad \ldots, \quad \omega_{p-1} = \frac{(-1)^p}{(\mu - \lambda)^p}, \tag{2.4.6}$$

we obtain one of the solutions of Equation (2.4.4).

The solution (2.4.5), (2.4.6) is evidently a function corresponding to the eigenvalue $\mu(x)$ of X_0 since $X_0 \lambda = 0$. Now let $\Omega(x)$ be an arbitrary invariant that enters an extended basis of X_0. Then $M\Omega = X_0(S\Omega) + Y\Omega$. Under the hypotheses, $Y\Omega$ is a sum of functions corresponding to eigenvalues of X_0 which either are identically zero (since $\lambda \equiv 0$ for Ω) or vanish nowhere in D. Denote by $(Y\Omega)_0$ the sum of all summands in $Y\Omega$ corresponding to zero eigenvalues. It also corresponds to zero. Putting $M\Omega = (Y\Omega)_0$ we satisfy the requirements for M on Ω. For $S\Omega$ we have the equation $X_0(S\Omega) + Y\Omega - (Y\Omega)_0 = 0$, where $Y\Omega - (Y\Omega)_0$ does not contain summands which correspond to zero. By the linearity of this equation we may find its solution, $S\Omega$, making use of the above auxiliary problem ($\lambda \equiv 0$). It is essential to notice that $S\Omega$, like $Y\Omega$, will satisfy (2.4.2) and (2.4.3); i.e., they are the sum of functions corresponding to eigenvalues of X_0 that do not vanish anywhere in D. But then to the solution just constructed any invariant may be added. We will not deal here with questions of nonuniqueness of S and M.

In particular, this implies that the function $\varphi(x)S\Omega$, where $\varphi(x)$ corresponds to $\lambda(x)$, is representable as a sum of the type (2.4.2) whose summands correspond to eigenvalues that do not coincide with $\lambda(x)$ anywhere in D. This remark will be utilized immediately.

Thus, operators S and M on invariants of the extended basis are known. Now let us find them on the basic eigenfunctions.

Suppose $\varphi = \varphi(x)$ is an eigenfunction with the eigenvalue $\lambda = \lambda(x)$. Then $M\varphi = X_0(S\varphi) - \lambda(S\varphi) - \varphi(S\lambda) + Y\varphi$. Since λ is an invariant, then $S\lambda$ is known, hence so is $-\varphi(S\lambda) + Y\varphi$. By the hypothesis of the theorem the second summand is the sum of functions corresponding to eigenvalues which either do not coincide with λ anywhere in D or coincide with λ identically. According to the remark made above so is $-\varphi(S\lambda)$. Now we construct $M\varphi$ and $S\varphi$ as we have constructed $M\Omega$ and $S\Omega$. Like $Y\varphi$, the function $S\varphi$ will satisfy (2.4.2) and (2.4.3).

After S and M on eigenfunctions are found, we pass to the construction of functions $\varphi(x)$ satisfying $(X_0 - \lambda)^2\varphi = 0$. From the series $X_0\varphi = \lambda\varphi + \varphi^{(1)}$, $X_0\varphi^{(1)} = \lambda\varphi^{(1)}$ we get

$$M\varphi = X_0(S\varphi) - \lambda(S\varphi) - \varphi(S\lambda) - S\varphi^{(1)} - Y\varphi.$$

Since $\varphi^{(1)}$ is an eigenfunction, $-S\varphi^{(1)}$ is known. Due to the above, $-\varphi(S\lambda) - S\varphi^{(1)} + Y\varphi$ satisfies (2.4.2) and (2.4.3) as $Y\varphi$ does. Thus the construction of $M\varphi$, $S\varphi$ is performed by the same scheme as that of $M\Omega$ and $S\Omega$. The case $(X_0 - \lambda)^p\varphi = 0$ for any p is easily dealt with by induction.

Remark 2.4.4 In problems that arise in practice we often encounter infinite series of the form (2.4.2). Theorem 2.4.3 supplies us with a method of reducing operators to the normal form and is a reformulation of the well-known Schroedinger perturbation theory. In examples we will show that this problem may often be solved by other means.

Remark 2.4.5 The existence of the Jordan basis and decompositions with respect to eigenfunctions of X_0 doubtless imposes severe restrictions. But, as we have already mentioned, they are often satisfied. This is hardly so, however, with the requirement (2.4.3). In problems encountered in practice this condition often fails on a submanifold in D which has the evident meaning of the resonance set (small denominators). Problems connected with the construction of asymptotics in such cases are considered in Chapters 4–5. We will not discuss now what is gained by the above procedure in the general case and will only consider the following simple situation.

Consider the case of a diagonal X_0 when M transforms invariants into invariants and eigenfunctions φ_j of the basic system of X_0 into the same functions up to a multiple which is an invariant of X_0. To simplify and avoid "extensions," let us suppose that each invariant is a function of basic invariants ω_i only. Then

$$\begin{aligned} M\omega_i &= (X_0 + \varepsilon M_1 + \cdots)\omega_i = (\varepsilon M_1 + \cdots)\omega_i \qquad (2.4.7) \\ &= \varepsilon q_{1i}(\omega) + \cdots + M\varphi_j = (X_0 + \varepsilon M_1 + \cdots)\varphi_j \\ &= \lambda_j(\omega)\varphi_j + (\varepsilon M_1 + \cdots)\varphi_j \\ &= [\lambda_j(\omega) + \varepsilon h_{1j}(\omega) + \cdots]\varphi_k. \end{aligned}$$

The corresponding system is

$$\varepsilon^\alpha \frac{d\omega_i}{dt} = \varepsilon g_{1i}(\omega) + \cdots$$
$$\varepsilon^\alpha \frac{d\varphi_j}{dt} = [\lambda_j(\omega) + \varepsilon h_{1j}(\omega) + \cdots]\varphi_j,$$

where $i = 1, \ldots, m$; $j = 1, \ldots, k$, and $k + m = n$. The system for the variables ω_i is an analogue of an algebraic problem of lesser dimension (equations for φ_j are similar analogues; they are one dimensional due only to our rigid requirements). If $\omega_i(t)$ are found, then $\varphi_j(t)$ are found by quadratures.

Remark 2.4.6 This example enables us to notice that even this simple case is similar to the presence of a Jordan cell in an algebraic problem notwithstanding the diagonality of X_0: indeed, for $\lambda \neq \text{const}$ we have $M'' = 0$ instead of $M' = 0$ as in the algebraic problem.

Let us give one more proposition, an analogue of an invariant method for construction of S and M by commutation relations [see (1.3.2)].

Theorem 2.4.7 *Let X_0 be quasilinear and Y, defined in the extended basic system $z_1, \ldots, z_n, z_{n+1}, \ldots, z_{n+p}, \omega_{11}(z), \ldots, \omega_{n+p,n+p}(z)$ [see (2.3.4)], be of the form*

$$\sum P_s \frac{\partial}{\partial z_s} + \sum Q_{ij} \frac{\partial}{\partial \omega_{ij}},$$

where P_s and Q_{ij} are polynomials in z whose coefficients are invariants of X_0. Then there exists a commutation relation

$$\omega^{(0)}Y + \omega^{(1)}Y' + \cdots + \omega^{(2)}Y^{(r)} = 0 \qquad \textit{for } r \geq 1$$

where $\omega^{(i)}$ are invariants of X_0, not all identically zero.

Corollary 2.4.8 *If $\omega^{(0)} \equiv \omega^{(1)} \equiv \cdots \equiv \omega^{(\mu-1)} \equiv 0$ and $\omega^{(\mu)} \neq 0$ for $1 \leq \mu \leq r-1$ then in the domain $\omega^{(\mu)} \neq 0$ we may construct S and M exactly as in Lemma 1.4.2; i.e., $M^{(\mu)} = 0$, where $M = [X_0, S] + Y$.*

PROOF

The proof follows from the following simple considerations:

1. Due to quasilinearity of X_0, the derivative of $\sum P_s(\partial/\partial z_s)$ is an operator of the same form (ω_{ij} are independent in the extended system) with coefficients whose degrees do not exceed the maximal degree of P_s;

2. the derivative of $\sum Q_{ij}(\partial/\partial\omega_{ij})$ is an operator of the form

$$\sum P_s^* \frac{\partial}{\partial z_s} + \sum Q_{ij}^* \frac{\partial}{\partial \omega_{ij}},$$

where the maximal degree of the polynomials Q_{ij}^* does not exceed the maximal degree of the Q_{ij}'s. As for the degree of P_s^*, though it is greater, generally speaking, by 1 than that of Q, $\sum P_s^*(\partial/\partial z_s)$ is a first order operator; therefore the maximal degree of coefficients of the second derivative does not surpass that of P. It is worth mentioning the connection of Remark 2.4.6 with these arguments.

2.5 The Normal Form of an Operator with Respect to a Nilpotent X_0. The Reconstruction Problem

Here we give, for the nonlinear case, the generalization of the results of Chapter 1 for a nilpotent X_0. (It is a generalization of a shearing transformation to the case when Yx_i are polynomials in x.)

First, we present two algebraic theorems. Let P_{N-1}, where $N = 1, 2, \ldots$, be the linear space of homogeneous polynomials of degree N in n variables (n is fixed) and X_0 a fixed first order operator that transforms P_0 into P_0 and is nilpotent in P_0 [i.e., for any $z \in P_0$ there is an integer $m(z)$ such that $X_0^{m(z)} z = 0$]. Since $X_0(uv) = uX_0v + vX_0u$, it is easy to verify that X_0 is nilpotent in any P_{N-1}. In P_0, fix a Jordan basis with respect to X_0. Elements of a particular chain of this basis will be indexed symmetrically: $e_{-k}, e_{-k+1}, \ldots, e_{k-1}, e_k$, where $2k$ is an integer and $X_0 e_s = e_{s-1}$, $X_0 e_{-k} = 0$. For simplicity's sake we skip the index that denotes the number of the chain. Each element of P_{N-1} will be considered henceforth as a homogeneous polynomial of degree N in the fixed basic variables (e).

Theorem 2.5.1 *There exists in P_{N-1} a Jordan (with respect to X_0) basis consisting of chains $\hat{e}_r, \hat{e}_{r-1}, \ldots, \hat{e}_{-r+1}, \hat{e}_{-r}$, where $X_0\hat{e}_s = \hat{e}_{s-1}$, $X_0\hat{e}_{-r} = 0$, such that under the transformation [in all chains of the basis(e)]*

$$e_q \to t^q e_q \qquad \text{for } q = k, \ldots, -k, \tag{2.5.1}$$

where t is a parameter, we have

$$\hat{e}_s \to t^s \hat{e}_s, \tag{2.5.1'}$$

where $s = r, \ldots, -r$, for all chains of the basis $(\hat{e})$.

Such bases $(\hat{e})$ will be referred to as *normal.*

Remark 2.5.2 For $N = 1$ the theorem is trivial: the basis (e) is evidently normal.

Proof

First, we give a concrete method for constructing elements of the normal basis and then justify it. Introduce two new first order operators Z_0 and Z_+, where Z_0 is the diagonal operator, $Z_0 e_s = s e_s$, and Z_+ is the nilpotent operator $Z_+ e_s = C_{ks} e_{s+1}$, where $C_{ks} = (k-s)(k+s+1)$ for all chains $e_k, \ldots, e_{-k}$, and $Z_+ e_k = 0$. For symmetry's sake, set $X_0 \equiv Z_-$, where $Z_- e_s = e_{s-1}$, $Z_- e_{-k} = 0$. The operators Z_1, Z_0, and Z_+ clearly satisfy

$$[Z_0, Z_-] = -Z_-, \quad [Z_-, Z_+] = -2Z_0, \quad [Z_0, Z_+] = Z_+. \tag{2.5.2}$$

The proof is based on this formula. Note also that since s is the eigenvalue to which e_s corresponds with respect to Z_0, then each monomial M from P_{N-1}, i.e., a product of elements of (e), corresponds to the eigenvalues S equal to the sum of eigenvalues of its multiples (e) since $Z_0 M = SM$. But clearly M turns into $t^S M$ under the transformation (2.5.1). Therefore the desired basis $(\hat{e})$ is normal if and only if $Z_0 \hat{e}_s = s\hat{e}_s$.

Denote by $P_{r,N-1} \subset P_{N-1}$ the subspace of polynomials p corresponding to the eigenvalue r with respect to Z_0 such that

$$Z_0 p = rp, \quad Z_+ p = 0 \qquad \text{for } p \in P_{r,N-1}. \tag{2.5.3}$$

This means that the space $P_{r,N-1}$ consists of all linear combinations of monomials p that correspond to r and such that $Z_+ p = 0$. As we will see later, $P_{r,N-1} \equiv 0$ when $r < 0$ and $P_{r,N-1} \neq 0$ for $r \geq 0$. Therefore in computations we may restrict ourselves to r such that $0 \leq r \leq r_{\max}$. In each $P_{r,N-1} \not\equiv 0$ choose a basis and for any $\hat{e}$ from this basis let us construct the chain $\hat{e}_r, \hat{e}_{r-1}, \ldots$ in P_{N-1} so that

$$\hat{e}_r \equiv \hat{e}, \quad \hat{e}_{s-1} = Z_- \hat{e}_s, \tag{2.5.3'}$$

where $s = r, r-1, \ldots$. A little later we will see that this chain terminates in $\hat{e}_{-r}$, which means that $Z_-\hat{e}_{-r} = 0$. We claim that elements of all chains $\hat{e}_r$, ..., $\hat{e}_{-r}$ are linearly independent and constitute a normal basis in P_{N-1}. Induction in $s = r, r-1, \ldots$ and (2.5.3), (2.5.2) easily imply that for $p \equiv \hat{e}^* \equiv \hat{e}_r$ we have

$$Z_0\hat{e}_s = s\hat{e}_s, \qquad Z_+\hat{e}_s = C_{rs}\hat{e}_{s+1}, \tag{2.5.4}$$

where $e_{rs} = (r-s)(r+s+1)$. The first formula in (2.5.4) shows that if $(\hat{e})$ is a basis, then this basis is normal. Further, putting $s = -r-1$ we have $Z_+\hat{e}_{-r-1} = 0$, and if $\hat{e}_{-r-1} \neq 0$ then (2.5.4) implies that $\hat{e}_{-r-2} \neq 0 \ldots$ since C_{rs} vanishes only for $s = -r-1$ and $s = r$. But the chain (2.5.3) cannot be infinite. Thus $Z_-\hat{e}_{-r} = 0$. The linear span of $(\hat{e})$ coincides with P_{N-1} since linear combinations of all $\hat{e}_r$'s exhaust all eigenvectors of Z_+ and $\hat{e}_{-r}$'s are generating elements with respect to Z_+.

It remains to demonstrate the linear independence of $(\hat{e})$. Let s_0 be a fixed eigenvalue with respect to Z_0 such that $-r_{\max} \leq s_0 \leq r_{\max}$. From each chain $\hat{e}_r, \ldots, \hat{e}_{-r}$ for $r \geq |s_0|$ we pick the element $\hat{e}_{s_0}$. Suppose that

$$\sum \alpha\hat{e}_{s_0} = 0, \tag{2.5.5}$$

where the α's are constants and sum runs over all picked $\hat{e}_{s_0}$'s.

Let us apply Z_+^γ, where $\gamma \geq 0$, to both parts of (2.5.5) and make use of the fact that $C_{rs} > 0$ for $s < r$ [see (2.5.4)]. For $\gamma = r_{\max} - s_0$ we deduce from (2.5.5) a linear relation between elements of the form $\hat{e}_{r_{\max}}$. Since the latter ones are linearly independent, the corresponding values of α's should be zero. Setting $\gamma = r_{\max} - s_0 - 1$, etc., we get that all α's are zero. Now suppose that elements of $(\hat{e})$ are linearly dependent. Then so must be elements of the form $\hat{e}_{s_0}$ for some s_0 since polynomials corresponding to different eigenvalues are linearly independent; but independence of $\hat{e}_{s_0}$'s was just established. Theorem 2.5.1 is proved.

Denote by T_{N-1} the linear space of the first order operators that map P_0 into P_{N-1}, and define the operator $\mathcal{D}Y = [X_0, Y] = X_0Y - YX_0$ in T_{N-1} (which is manifestly nilpotent).

Theorem 2.5.3 *In T_{N-1}, there is a Jordan basis with respect to D consisting of a series Y_m, Y_{m-1}, ..., Y_{-m+1}, Y_{-m}, where $\mathcal{D}Y_S = Y_{S-1}$, $\mathcal{D}Y_{-m} = 0$, and such that for all series $Y_S \mapsto t^sY_s$, where $s = m, \ldots, -m$, under the transformation (2.5.1).*

Such a basis in T_{N-1} will be called normal.

Proof

We carry out the direct construction of the normal basis. Let $P_0^{(e)} \subset P_0$ be a subspace spanned by the fixed chain e_k, ..., e_{-k}; $P_{N-1}^{(\hat{e})} \subset P_{N-1}$ the subspace spanned by the fixed chain $\hat{e}_r, \ldots, \hat{e}_{-r}$; and let $T_{N-1}^{(e\hat{e})} \subset T_{N-1}$ be the subspace of operators $Y^{(e\hat{e})}$ which map $P_0^{(e)}$ into $P_{N-1}^{(\hat{e})}$. Assign to the operator

$$Y^{(e\hat{e})}e_s = \sum_p a_{sp}\hat{e}_p,$$

where $a_{sp} = \text{const}$, the bilinear form

$$\omega^{(e\hat{e})} = \sum_{s,p} (-1)^s a_{sp} e_s \hat{e}_p,$$

where e_s and $\hat{e}_p$ are considered formally as independent variables. This is a one-to-one correspondence. Then to the operator

$$\mathcal{D}Y^{(e\hat{e})}e_s = \sum_p (a_{s,p+1} - a_{s-1,p})\hat{e}_p,$$

where $a_{-k-1,p} \equiv a_{s,r+1} \equiv 0$, corresponds the form

$$X_0\omega^{(e\hat{e})} = \sum_{s,p} (-1)^s (a_{s,p+1} - a_{s-1,p}) e_{-s}\hat{e}_p.$$

In the space of bilinear forms we will construct the Jordan basis with respect to X_0. Generating elements $\omega_m^{(e\hat{e})}$ may be found from the condition $Z_+\omega_m^{(e\hat{e})} = 0$, where

$$\omega_m^{(e\hat{e})} = \sum_s (-1)^s b_s(m) e_{-s}\hat{e}_{m+s} \qquad \text{for } m \geq 0.$$

We have

$$\begin{aligned} Z_+\omega_m^{(e\hat{e})} \equiv \sum_s (-1)^s \{&(k+s)(k-s+1)b_s(m) \\ &- (r-m-s+1)(r+m+s)b_{s-1}(m)\} e_{s+1}\hat{e}_{m+s} = 0. \end{aligned}$$

Nontrivial solutions $b_s(m)$ of this equation exist only for $|r-k| \leq m \leq r+k$, where $-k \leq s \leq r-m$, and up to a constant multiple are equal to

$$b_s(m) = \frac{(r+m+s)!(k-s)!}{(r-m-s)!(k+s)!}. \tag{2.5.6}$$

This is a special case of the well-known Clebsch–Gordan formulas.

"Diagonal" operators corresponding to $\omega_m^{(e\hat{e})}$,

$$Y_m^{(e\hat{e})} e_s = b_s(m)\hat{e}_{s+m}, \tag{2.5.7}$$

are generating elements of a Jordan basis in $T_{N-1}^{(e\hat{e})}$ with respect to D. [It is not difficult to notice that up to a multiple and reindexing of basic elements operators $Y_m^{(e\hat{e})}$ for $N = 1$ are "diagonal" operators of the minimal height $Y_\mu^{(q_i q_j)}(\alpha)$ mentioned in (1.1.2).] It is easy to verify that this basis is normal if the basis $(\hat{e})$ is normal. Theorem 2.5.3 is proved.

Remark 2.5.4 Coefficients (2.5.6) are positive.

Remark 2.5.5 As is easy to verify, $X_0 \mapsto t^{-1}X_0$ under the transformation (2.5.1). If we consider $X_0 + \Delta Y_S$, where Y_S is an element of a normal basis in T_{N-1}, then for $t = \Delta^{-1/1+S}$, where $S \neq -1$, we have

$$X_0 + \Delta Y_S \mapsto \Delta^{1/1+S}(X_0 + Y_S).$$

Theorems 2.5.1 and 2.5.3 enable us to generalize the shearing transformation to the case of nonlinear perturbations.

Now let the leading operator X_0 be a Jordan nilpotent operator and coefficients of X_j in the basic system of X_0 be polynomials in basic variables.

We want to give a definition of a canonical form with respect to X_0 of an operator $X_0 + \varepsilon X_1 + \varepsilon^2 X_2 + \cdots$. This is equivalent to the following: let us consider an operator

$$M_j = [X_0, S_j] + Y^{(j)}, \tag{2.5.8}$$

where $Y^{(j)}$ is given and

$$Y^{(j)} P_0 \subset \sum_{j \leq m \leq N} P_{m-1}.$$

It is always possible to choose S_j in such a way that M_j is a linear combination of just-constructed generating elements from (2.5.7).

Any sum of M_j's which are such linear combinations will be referred to as *canonical*.

Remark 2.5.6 The height of Y_m in (2.5.8) is $\mu = 2m + 1$.

Thus, the operator $X = X_0 + \varepsilon X_1 + \varepsilon^2 X_2 + \cdots$ is equivalent to

$$M = e^{-S} X e^{S} = X_0 + \varepsilon M_1 + \varepsilon^2 M_2$$

and may be reduced (in any order), under the above conditions, to the canonical form (for each M_j).

In M, fix a summand $X_0 + \Delta Y_m$, where $\Delta = \varepsilon^k$, and in Y_m choose one of the canonical summands of M_k. After applying the transformation (2.5.1) with

$$t = \Delta^{-1/1+m} = \Delta^{-2/1+\mu} \equiv \delta^{-1} \tag{2.5.9}$$

let us express M in the form $M = \delta \tilde{M}$. Generally speaking, the decomposition of $\tilde{M}$ with respect to powers of ε might contain negative powers. Evidently, the summand ΔY_m may be chosen so that negative powers vanish. Such a choice of ΔY_m will be referred to as the *standard* one. The summands in $\tilde{M}$ with zero power of ε are not exhausted in the general case by $X_0 + Y_m$ (see Remark 2.5.3). All summands in $\tilde{M}$ corresponding to the zeroth power of ε may be interpreted as the new leading operator (up to the multiple δ). The transformation (2.5.1), (2.5.9) corresponding to the standard choice of the summand ΔY_m is a generalization of the shearing transformation (1.5.2). Let us emphasize that the canonical form and the shearing transformation are defined only by algebraic properties of the system.

Remark 2.5.7 As in section 1.5, the parameter $\delta = \Delta^{2/1+\mu}$ will be called the "weight" of ΔY_m, where $m = \mu - 1/2$ (see Remark 2.5.4). The standard choice of ΔY_m is then the choice of the summand of highest "weight."

Remark 2.5.8 One of the special features of the nonlinear case ($N > 1$) as compared with the linear one ($N = 1$) is that the number of summands of the given (in particular, maximal) weight might be infinite. In Chapter 3 we give a (formal) example of such a case.

Remark 2.5.9 It seems impossible to generalize the theorem on termination of the reconstruction procedure (Theorem 1.5.1) to the nonlinear case.

Now let us discuss previous results, keeping applications in mind. In the considered case (X_0 is nilpotent, perturbations are polynomial) the perturbation theory problem is *formally solved from the very beginning* since X_j are bounded in height; we try to gain more. So we choose the new leading operator with the help of the generalized shearing transformation and study if it is possible to fulfill again the perturbation procedure (with respect to the new leading operator). The procedure indicated here provides us with a regular method of choosing a new leading operator. Further, it is clear that its practical application is connected, generally speaking, with cumbersome computations which can often be avoided in a roundabout way. Let us consider a simple example. Let $X = X_0 + \varepsilon(Y_m + Y_i^*)$, where X_0 is nilpotent. Y_m is a generating element of a normal basis, Y_i^* is not a generating element of the normal basis, and $i > m$. Clearly, we will single out a new leading operator by (2.5.1) with $t = \varepsilon^{-1/1+i}$ (see Remark 2.5.3) corresponding to the inclusion of Y_i^* into the new leading operator. If we act formally, we must first find S_1 so that $M_1 = Y_m$, i.e., Y_i^* is "killed." Now the term with ε^2 pops up and we must expand it with respect to the normal basis, etc. "Traces" of Y_i^* will appear in all orders with respect to powers of ε and it is intuitively clear that the highest "weight" δ of canonical operators will nevertheless turn out to be $\varepsilon^{1/1+i}$ since the commuting increases the degree and the height of polynomials-coefficients. (Strictly speaking, the equivalence of the two procedures is not established. We may even guess that canonization procedure in order to distinguish the operator of the maximal weight might lead to an infinite process; cf. Remark 2.5.6).

Experience in solving problems shows that for a nilpotent leading operator (and not necessarily polynomial perturbation) it is often comparatively simple to find the highest "weight" of summands of the perturbation. When we know δ, it is advisable to make use of the shearing transformation (2.5.1) with $t = \delta^{-1}$ and take $(X\delta^{-1})|_{\varepsilon=0}$ for the new leading operator. In this way we avoid the reduction to the canonical form. This unjustified trick is often successful and its application will be illustrated below.

In the case of an arbitrary X_0 the following consideration turns out to be helpful. The *weight* of ΔY, where Δ is a small parameter and Y is an operator of height μ with respect to X_0, is $\delta = \Delta^{2/1+\mu}$ [as above, cf. (2.5.5)]. Let $X = X_0 + \varepsilon X_1 + \cdots$ be reducible to the normal form $M = X_0 + \varepsilon M_1 + \cdots$ where $M_j^{(\mu j)} = 0$ and μ_j are minimal (S_1 is chosen for μ_1 to be minimal, then S_2 is chosen for μ_2 to be minimal, etc.). Let us collect in $M - X_0$ all summands of the highest "weight" and denote them by M^*. If now we succeed in reducing $X_0 + M^*$ to the normal form (up to small summands) then we suggest taking its principal part for the new leading operator. Thus we assume, as in the case of a nilpotent operator, that only operators of the highest weight might affect the new leading operator.

Remark 2.5.10 Trivial example: in the algebraic case of a diagonal X_0 (Chapter 1) we have $M' = 0$; i.e., the height μ_j of M_j is 1. Therefore the "weight" of $\varepsilon^j M_j$ is

$$\delta = (\varepsilon^j)^{2/1+\mu_j} = \varepsilon^j$$

and εM_1 is a highest weight operator. The new leading operator is $X_0 + \varepsilon M_1$.

In conclusion note that we were speaking of reconstruction (choosing of the new leading operator) in the comparatively simple case where X_0 is not qualified as the leading operator in the sense that a solution of the perturbation theory ($M^{(\mu)} = 0$) does not produce, for some reasons, sufficient progress in the study of X. Still more difficult to handle are cases when there exist in D singular manifolds (cf. Remark 2.4.2) which preclude us from constructing S on the whole of D. They will be considered separately in Chapter 5, where reconstruction is understood in a broader sense than here.

2.6 A Connection with N. N. Bogolyubov's Ideas

Let us rewrite (2.1.4) and (2.1.8) once more in the following form: from $F(x'(t,x)) = e^{tX}F(x)$ it follows that

$$F(e^{-S}(e^{S}(x'(t, e^{-S}x)))) = e^{-S}e^{tX}e^{S}F(e^{-S}x) \tag{2.6.1}$$

or, assuming

$$e^{-S}Xe^{S} = M, \quad F(e^{-S}x) = G(x), \quad x^*(t,x) = e^{S}x'(t, e^{-S}x), \tag{2.6.2}$$

where in the last formula e^S is applied to x' (i.e., x' is considered as independent), it follows that

$$G(x^*(t,x)) = e^{tM}G(x). \tag{2.6.3}$$

The last formula means that if $x'(t,x)$ satisfies a system of differential equations defined by X, then $x^*(t,x)$ satisfies the system of differential equations defined by M.

Now let us consider, for simplicity's sake, the case when there is a basic system of invariants and eigenfunctions

$$X_0 e_i = 0, \qquad X_0\varphi_k = \lambda_k(e)\varphi_k,$$

where $i = 1,\ldots,p$, $k = 1,\ldots,q$, and M is of the form

$$\begin{aligned} Me_i &= \varepsilon m_{i0}(e) + \varepsilon^2 m_{i1}(e) + \cdots = m_i(e), \\ M\varphi_k &= (\lambda_k(e) + \varepsilon\lambda_{k,1}(e) + \varepsilon^2\lambda_{k,2}(e) + \cdots)\varphi_k = \bar{\lambda}_k(e)\varphi_k. \end{aligned}$$

Take e and φ as independent variables. Then we may write equations for e and φ corresponding to X in the form

$$\begin{aligned} \varepsilon^\alpha \frac{d\varphi_k'}{dt} &= \lambda_k(e')\varphi_k' + \varepsilon A_{k1}(e',\varphi') + \varepsilon^2 A_{k2}(e',\varphi') + \cdots, \\ \varepsilon^\alpha \frac{de_i'}{dt} &= eB_{i1}(e',\varphi') + \varepsilon^2 B_{i2}(e',\varphi') + \cdots. \end{aligned} \tag{2.6.4}$$

Besides, (2.6.2) implies that

$$\begin{aligned} \varphi_k' &= \varphi_k^* + \varepsilon P_{k1}(e^*,\varphi^*) + \varepsilon^2 P_{k2}(e^*,\varphi^*) + \cdots, \\ e_i' &= e_i^* + \varepsilon Q_{i1}(e^*,\varphi^*) + \varepsilon^2 Q_{i2}(e^*,\varphi^*) + \cdots \end{aligned} \tag{2.6.5}$$

and

$$\begin{aligned} \varepsilon^\alpha \frac{d\varphi_k^*}{dt} &= (\lambda_k(e^*) + \varepsilon\lambda_{k,1}(e^*) + \varepsilon^2\lambda_{k,2}(e^*) + \cdots)\varphi_k^*, \\ \varepsilon^\alpha \frac{de_i^*}{dt} &= \varepsilon m_{i,0}(e^*) + \varepsilon^2 m_{i,1}(e^*) + \cdots. \end{aligned} \tag{2.6.6}$$

Now we may pose the following problem: find functions $P_{k\nu}$, $Q_{i\mu}$, $\lambda_{k,\nu}$, $m_{i,\mu}$ so that (2.6.4) and (2.6.5) will imply that φ^* and e^* satisfy (2.6.6). This is just Bogolyubov's formulation of the problem. Note that *if we do not know beforehand invariants and eigenfunctions of X_0 or lack the corresponding algebraic formulation, then it is difficult to say in which form the desired canonical equations (2.6.6) should appear.*

Remark 2.6.1 There exists an ambiguity in the choice of λ, M, P, and Q. It completely matches the already encountered ambiguity of S_i and additions to eigenvalues $\lambda_{k,1}, \lambda_{k,2}, \ldots$. Note also that, as in the search for S_i, both the time interval and the domain are defined by the problem considered and play an important role. The change of variable performed in one domain D may turn out to be inapplicable in a wider domain D' and since for a sufficiently large time a trajectory may leave D and come to D', the ambiguity depends also on the time interval.

2.7 The Motion Near the Stationary Manifold. Reduction of Information

Consider the system

$$\varepsilon \frac{dx}{dt} = a_0(x) + \varepsilon a_1(x) + \cdots, \tag{2.7.1}$$

where X is a vector and a a vector function, and let the solution of $a_0(x) = 0$ be $\bar{x}_0(c)$, depending on p arbitrary constants $c_1, \ldots, c_p$. Suppose that the p-dimensional manifold $\mathcal{D}_p = \bar{x}_0(c)$ which arises when $c = (c_1, \ldots, c_p)$ runs over a p-dimensional domain D' is such that it is possible to seek a solution in the form

$$x = \bar{x}_0(c) + \varepsilon \bar{x}_1(t) + \cdots \tag{2.7.2}$$

when initial values of (2.7.1) belong to an n−dimensional domain $\mathcal{D}_n$ sufficiently close to $\mathcal{D}_p$. Here we suppose that $c = c(t)$ is a vector function in t, vectors $\bar{x}_i(t)$ are bounded as $t \to \infty$, and the series (2.7.2) is an asymptotic one in ε. This section is devoted to the exposition of two principally different methods of defining $c(t)$ and $\bar{x}_i(t)$ under the following condition:

Condition. The matrix

$$\left.\frac{\partial a_0}{\partial x}\right|_{x=\bar{x}_0(c)} = A(c)$$

for $c \in \mathcal{D}'$ has exactly p linearly independent vectors $\psi_i(c)$, where $i = 1, \ldots, p$, such that $A\psi_i = 0$ [and $n - p$ linearly independent vectors $\psi_j(c)$, where $j = p+1, \ldots, n$, such that $A\psi_j = \lambda_j(c)\psi_j$] and $\operatorname{Re}\lambda_j(c) < \alpha < 0$.

Remark 2.7.1 If $\mathcal{D}_p$ were one point, $\bar{x}_0$, then we would have the problem of motion in a neighborhood of the stationary point. The requirement of negativeness of real parts of eigenvalues of A would turn out to be the Lyapunov stability condition. In our case ($p \neq 0$) the condition on eigenvalues should guarantee formally that the point will not leave the vicinity of $\mathcal{D}_p$.

Method A. First notice that since $a_0(\bar{x}_0(c)) = 0$, then

$$\frac{\partial a_0(\bar{x}_0)}{\partial \bar{x}_0} \frac{\partial \bar{x}_0}{\partial c_s} = 0, \qquad \text{where } s = 1, \ldots, p. \tag{2.7.3}$$

Since vectors $\partial\bar{x}_0/\partial c_s$ should be linearly independent, then the condition implies that in what follows we may assume

$$\psi_s = \frac{\partial \bar{x}_0}{\partial c_s}, \tag{2.7.4}$$

where $s = 1, \ldots, p$. Now let us substitute (2.7.2) into (2.7.1), expand everything with respect to powers of ε and compare the expressions obtained. We will have

$$\begin{aligned}\frac{d\bar{x}_0}{dt} &= \sum_{1\le s\le p} \psi_s \frac{dc_s}{dt} = \frac{\partial a_0}{\partial \bar{x}_0}\bar{x}_1 + \bar{b}_1,\\ \frac{d\bar{x}_1}{dt} &= \frac{\partial a_0}{\partial \bar{x}_0}\bar{x}_2 + \bar{b}_2,\\ &\vdots\end{aligned} \tag{2.7.5}$$

where $\bar{b}_1 = a_1(\bar{x}_0)$ is known if $\bar{x}_0$ is known, b_2 is expressed in terms of $\bar{x}_0$ and $\bar{x}_1$, etc. Since, by assumption, c_s is a function in t, the first equation gives

$$\sum_{1\le s\le p} \psi_s \frac{dc_s}{dt} - \bar{b}_1(\bar{x}_0) = A\bar{x}_1. \tag{2.7.6}$$

Let us denote by $\tilde{\psi}_s$ such p vectors $s = 1, \ldots, p$, that $A^*\tilde{\psi}_s = 0$, where A^* is the transpose of A, and assume for the sake of definiteness that $(\tilde{\psi}_\alpha, \psi_\beta) = \delta_{\alpha\beta}$. Then (2.7.6) yields

$$\frac{dc_s}{dt} = (\bar{b}_1(\bar{x}_0(c)), \tilde{\psi}_s). \tag{2.7.7}$$

If the c_s's are defined from (2.7.7), then the left-hand side of (2.7.6) is known and orthogonal to all $\tilde{\psi}_s$ for $s = 1, \ldots, p$. Therefore, (2.7.6) makes it possible to define the vector $\bar{x}_1(t)$ in the form

$$\bar{x}_1(t) = \bar{y}_1 + \sum_{1\le s\le p} r_s\psi_s, \tag{2.7.8}$$

where $\bar{y}_1$ is uniquely defined by the condition $(\bar{y}_1, \tilde{\psi}_s) = 0$ for $s = 1, \ldots, p$ and r_s are unknown functions in t. Substituting now $\bar{x}_1(t)$ in the second equation in (2.7.5) we get

$$\frac{d\bar{y}_1}{dt} + \sum_{1\le s\le p} \frac{dr_s}{dt}\psi_s + r\frac{d\psi_s}{dt} = \frac{\partial a_0}{\partial \bar{x}_0}\bar{x}_2 + \bar{b}_2. \tag{2.7.9}$$

That b_2 is a known function in r_s and t makes it possible to repeat the process which gave us (2.7.7) and to get differential equations which define r_s. Finding r_s we define $\bar{x}_2$ from (2.7.9) in a form similar to that of x_1 from

(2.7.8). Iterating the process, we get the formal asymptotic development of a solution in the form

$$x(t) = x_0(c(t)) + \varepsilon\bar{y}_1 + \varepsilon^2\bar{y}_2 + \cdots + \sum_{1\le s\le p} (\varepsilon r_s^{(1)} + \varepsilon^2 r_s^{(2)} + \cdots)\psi_s, \quad (2.7.10)$$

where $y = \varepsilon\bar{y}_1 + \varepsilon^2\bar{y}_2 + \cdots$ is orthogonal to $\tilde{\psi}_1, \dots, \tilde{\psi}_p$.

Now notice that the real parts of nonzero eigenvalues of A are negative and y belongs to the subspace spanned by the corresponding eigenvectors. Therefore we should expect that after a sufficiently large period of time (which in the scale that is of interest to the observer who solves the problem might be quite small) the vector y will turn infinitesimally small. In this connection it is desirable to investigate the problem in order to better study the role of the nonzero eigenvalues of A. Besides, as we have already seen, the solution in the form of the formal decomposition with respect to powers of ε usually exists in a very narrow domain of parameters. All these remarks justify the application of our standard method to this problem.

Method B. Let us seek X in the form

$$x = \bar{x}_0(c) + \varepsilon \sum_{p+1\le s\le n} r_s\psi_s, \quad (2.7.11)$$

where $c_1, \dots, c_p$ and r_s are desired functions in t and ε. [We may always assume that X is normed so that in the representation

$$x = \bar{x}_0(c) + \varepsilon \sum_{1\le s\le n} r_s\psi_s$$

we may "kill" the first p summands $\sum_{1\le s\le p} r_s\psi_s$, replacing c_i by $c_i + \varepsilon d_i$.] Substitute (2.7.11) into (2.7.1). Let us develop $a_j(x)$ with respect to powers of ε and express the results as linear combinations of ψ_μ. Comparison of coefficients of linearly independent vectors ψ_μ, where $\mu = 1, \dots, n$, gives equations for dc_i/dt and dr_s/dt in the form of power series in ε. We get the following system of equations (from the very beginning we may seek the right-hand sides of dc_i/dt and dr_s/dt in the suggested form)

$$\varepsilon\frac{dc_i}{dt} = \varepsilon b_i(c) + \varepsilon^2 d_i(c,r) + \cdots \qquad \text{where } i = 1, \dots, p, \quad (2.7.12)$$

$$\varepsilon\frac{dr_s}{dt} = \lambda_s r_s + b_s(c) + \varepsilon d_s(c,r) + \cdots, \qquad \text{where } s = p+1, \dots, n, \quad (2.7.12')$$

where right-hand sides are known and are polynomials in r. To the system (2.7.12) corresponds the operator $X = X_0 + \varepsilon X_1 + \cdots$, where

$$X_0 = \sum_s (\lambda_s r_s + b_s(c))\frac{\partial}{\partial r_s}.$$

Invariants and eigenfunctions of X_0 are c_i, where $i = 1, \ldots, p$, and $\varphi_s = r_s + b_s(c)/\lambda_s(c)$, where $s = p+1, \ldots, n$, respectively. Therefore, following the Bogolyubov scheme, we may search for a solution of the form

$$c_i = c_i^* + \varepsilon q_{i,1}(\varphi^*, c^*) + \varepsilon^2 q + i, 2(\varphi^*, c^*) + \cdots, \tag{2.7.13}$$

$$\varphi_s = \varphi_s^* + \varepsilon p_{s,1}(\varphi^*, c^*) + \varepsilon^2 p_{s,2}(\varphi^*, c^*) + \cdots, \tag{2.7.13$'$}$$

where c^*, φ^* is a solution of the system

$$\varepsilon \frac{d\varphi_s^*}{dt} = (\lambda_s(c^*) + \varepsilon a_{s,1}(c^*) + \cdots)\varphi_s^*, \tag{2.7.14}$$

$$\frac{dc_i^*}{dt} = b_i(c^*) + \varepsilon m_{i,1}(c^*) + \cdots. \tag{2.7.14$'$}$$

Now it is clear that for $t \gg \varepsilon$ the functions φ_s^* are exponentially small since $Re\lambda_s < \alpha < 0$ and they can be dropped; after that only Equation (2.7.14$'$) for c^* and (2.7.13) will remain. But (2.7.13) gives $\varphi_s = \varepsilon p_{s,1}(0, c^*) + \varepsilon^2 p_{s,2}(0, c^*) + \cdots$ if we drop φ^*. Inverting (2.7.13) we obtain, on the other hand, equations of the form

$$c_i^* = c_i + \varepsilon q_{i,1}^*(c^*) + \varepsilon^2 q_{i,2}^*(c^*) + \cdots;$$

hence r_s is a function in C. Consequently, we see that for $t \gg \varepsilon$ all quantities are expressed only in terms of p parameters c which satisfy a system of differential equations. This produces a method of Reduction of Information. For $t \gg \varepsilon$ it is possible to seek a solution of (2.7.1) in the form $x = \bar{x}_0(c) + \varepsilon \bar{x}_1(c) + \varepsilon^2 \bar{x}_2(c) + \cdots$, where c_i satisfies

$$\frac{dc_i}{dt} = b_i(c) + \varepsilon m_{i,1}(c) + \cdots \quad \text{and} \quad (\bar{x}_j, \tilde{\psi}_i) = 0 \qquad \text{for } i = 1, 2, \ldots, p.$$

This method was developed in a series of papers by N. N. Bogolyubov and other authors. We illustrated it for the case of ordinary differential equations only to avoid formal complications. The reader acquainted, for instance, with the deduction of hydrodynamical Euler and Navier–Stokes equations from the Boltzmann equation will easily recognize in our Method A the Hilbert method and in the reduction of information method that of Chapman–Enskog.

Let us look once more at the computation procedure. Substituting $x = \bar{x}(c) + \varepsilon \bar{x}(c) + \cdots$ into (2.7.1) we get

$$\varepsilon \frac{\partial \bar{x}_0}{\partial c_i} \frac{dc_i}{dt} + \varepsilon^2 \frac{\partial \bar{x}_1}{\partial c_i} \frac{dc_i}{dt} + \cdots$$
$$= \varepsilon \left(\frac{\partial a_0}{\partial \bar{x}_0} \bar{x}_1 + b_1 \right) + \varepsilon \left(\frac{\partial a_0}{\partial \bar{x}_0} \bar{x}_2 + b_2 \right) + \cdots,$$

where b_1 is known if $\bar{x}_0$ is known, b_2 is known if $\bar{x}_0$ and $\bar{x}_1$ are known, etc. Taking the desired equations for c in the form

$$\frac{dc_i}{dt} = m_i^{(0)} + \varepsilon m_i^{(1)} + \cdots$$

and substituting them into the above equality we get

$$\frac{\partial \bar{x}_0}{\partial c_i} m_i^{(0)} = \frac{\partial a_0}{\partial \bar{x}_0}\bar{x}_1 + b_1, \qquad \frac{\partial \bar{x}_0}{\partial c_i} m_i^{(1)} = \left(\frac{\partial \bar{x}_1}{\partial c_i} m_i^{(0)} + b_2\right) + \frac{\partial a_0}{\partial \bar{x}_0}\bar{x}_2, \qquad \ldots.$$

Recall that $\partial \bar{x}_0/\partial c_i = \psi_i$, then $m_i^{(0)} = (b_1, \tilde{\psi}_i)$, which uniquely defines $\bar{x}_1$. Now we can analogously define $m_i^{(1)}$ and $\bar{x}_2, \ldots$, etc. Since this problem is important, we will formulate it in a more general form, not restricting ourselves to ordinary differential equations.

Let us consider the equation

$$\varepsilon\left(\frac{\partial u}{\partial t} + A\frac{\partial u}{\partial x}\right) = T_0(u) + \varepsilon T_1(u). \tag{2.7.15}$$

Here $u(t,x,y)$ is a vector function which depends on t, on the vector of variables X, and the vector of parameters y, and T_0 and T_1 are nonlinear operators. The matrix A depends on X and y. Conditions on T_0 are formulated below. First suppose that the general solution of

$$T_0(u) = 0 \tag{2.7.16}$$

is of the form $\bar{u} = \bar{u}(x, y; c_1, \ldots, c_p)$, where c_i are arbitrary constants. Let us see how to extend the reduction of information method to this case. Let us seek u in the form

$$u = \bar{u}(x, y; c) + \varepsilon u_1, \tag{2.7.17}$$

where c_i is a function in X and t. Substituting in (2.7.15) we have

$$\varepsilon\left(\frac{\partial \bar{u}}{\partial c_s}\frac{\partial c_s}{\partial t} + A\left(\frac{\partial \bar{u}}{\partial x} + \frac{\partial \bar{u}}{\partial c_s}\frac{\partial c_s}{\partial x}\right)\right) + \cdots = \varepsilon\left(\frac{\delta T_0(\bar{u})}{\delta \bar{u}}u_1 + T_1(\bar{u})\right) + \cdots.$$

The first equation (to which we will restrict ourselves) is

$$\frac{\partial \bar{u}}{\partial c_s}\frac{\partial c_s}{\partial t} + A\left(\frac{\partial \bar{u}}{\partial x} + \frac{\partial \bar{u}}{\partial c_s}\frac{\partial c_s}{\partial x}\right) = \frac{\delta T_0(\bar{u})}{\delta \bar{u}}u_1 + T_1(\bar{u}). \tag{2.7.18}$$

Note that

$$\frac{\delta T_0(\bar{u})}{\delta \bar{u}}u_1 = \frac{d}{d\varepsilon}T_0(\bar{u} + \varepsilon u_1)\bigg|_{\varepsilon=0} = 0.$$

Now suppose that

(a) T_0 is a nonlinear operator which depends on X as on a parameter and acts on the function $f(x,y)$ as on the function in y only [e.g., $T_0(u) = \int k(x,y,z)u^2(x,z)\,dz$].

(b) There are exactly p linearly independent functions $\psi_i(x,y,c)$ such that $B\psi_i = 0$ for $i = 1,\ldots,p$, where $B = \delta T_0(\bar{u})/\delta\bar{u}$ (of course, B acts in the space of functions that only depend on y).

(c) There exist only p functions $\tilde{\psi}_i(x,y,c)$ such that $B^*\tilde{\psi}_i = 0$ and the equation $Bf = g$ is solvable if and only if $(g,\tilde{\psi}_i) = 0$ for $i = 1,\ldots,p$. (We do not define here the scalar product, assuming that in a concrete problem it naturally arises, but it has to be independent of y.)

It is clear now that when these conditions are satisfied, we may repeat what has been said above. Instead of ψ_s we may take $\partial\bar{u}/\partial c_s$, where $s = 1,\ldots,p$. It follows from (2.7.18) that

$$\frac{\partial c_s}{\partial t} + \left(\tilde{\psi}_s, A\frac{\partial \bar{u}}{\partial x}\right) + \sum_i \left(\tilde{\psi}_s, A\frac{\partial \bar{u}}{\partial c_i}\frac{\partial c_i}{\partial x}\right) = (\tilde{\psi}_s, T_1(\bar{u})) \qquad \text{for } s = 1,\ldots,p. \tag{2.7.19}$$

[We suppose that $(\tilde{\psi}_i, \psi_j) = \delta_{ij}$.] We have obtained a differential equation for c_s. When we know c_s, we may determine $\bar{u}_1$ uniquely from (2.7.18) if $(\tilde{\psi}_s, \bar{u}_1) = 0$ for $s = 1,\ldots,p$. The process can be evidently iterated ad infinitum. As above, it is easy to modify this method and seek a solution of the form

$$u = \bar{u}(c) + \varepsilon\bar{u}_1(c) + \cdots, \tag{2.7.20}$$

where c satisfies

$$\frac{\partial c_s}{\partial t} + M_{0,s}(c) + \varepsilon M_{1,s}(c) + \cdots = 0.$$

Here $M_{i,s}(c)$ are operators over c depending on X and $(\tilde{\psi}_s, \bar{u}_i) = 0$. This is the *reduction of information method.*

Remark 2.7.2 The case of ordinary differential equations considered above shows that it is necessary to require additionally that all points of the spectrum of $\delta T_0(\bar{u})/\delta(\bar{u})$ except zero would belong to the region $Re\lambda < \alpha < 0$. If only Re $\lambda < 0$ is satisfied the usefulness of the resulting equations is doubtful and it is impossible, generally speaking, to seek a solution of the form (2.7.20). In the sequel several problems of the kind will be discussed.

2.8 Hamiltonian Systems

If the considered system is Hamiltonian, i.e., of the form

$$\varepsilon^{\nu}\frac{d\psi}{dt} = \frac{\partial H}{\partial \pi}, \qquad \varepsilon^{\nu}\frac{d\pi}{dt} = \frac{\partial H}{\partial \psi}, \tag{2.8.1}$$

where ψ and π are canonically conjugate and the Hamiltonian is $H = h_0 + \varepsilon h_1 + \cdots$, then the corresponding operator is $X = X_0 + \varepsilon X_1 + \cdots$, where

$$X_i f = \sum_k \left(\frac{\partial h_i}{\partial \pi_k} \frac{\partial f}{\partial \psi_k} - \frac{\partial h_i}{\partial \psi_k} \frac{\partial f}{\partial \pi_k} \right),$$

or, in the notations used in mechanics, $X_i f = \{h_i, f\}$, where

$$\{f, g\} = \left(\frac{\partial f}{\partial \pi_k} \frac{\partial g}{\partial \psi_k} - \frac{\partial f}{\partial \psi_k} \frac{\partial g}{\partial \pi_k} \right)$$

is the Poisson bracket of f and g.

In the reduction of $e^{-\tilde{S}} X e^{\tilde{S}} = X'$ to normal form it is natural to remain in the class of Hamiltonian systems, i.e., apply only canonical transformations. This technique is well known. Let us write the Jacobi identity in the form

$$\{H, \{G, F\}\} - \{G, \{H, F\}\} = \{\{H, G\}, F\}. \tag{2.8.2}$$

The operator of the Hamiltonian system with Hamiltonian H, which we will denote by H_p, is given by

$$H_p F = \{H, F\}. \tag{2.8.3}$$

Due to (2.8.3), the Jacobi identity (2.8.2) can be rewritten in the form

$$H_p(G_p F) - G_p(H_p F) = \{H, G\}_p F$$

or

$$[H_p, G_p] = \{H, G\}_p; \tag{2.8.4}$$

i.e., the commutator of two Hamiltonian operators is a Hamiltonian operator. But then if we assume that $\tilde{S}$ is a Hamiltonian operator, i.e., $\tilde{S} = S_p$, we get, due to (2.8.3) and (2.8.4),

$$\begin{aligned} e^{-S_p} H_p e^{S_p} &= H_p + [H_p, S_p] + \frac{1}{2!}[[H_p, S_p], S_p] + \cdots \\ &= H_p + \{H, S\}_p + \frac{1}{2!}\{\{H, S\}, S\}_p + \cdots \\ &= \{e^{-S_p} H\}_p. \end{aligned} \tag{2.8.5}$$

Thus the system remains Hamiltonian and the new Hamiltonian is

$$H' = \{e^{-S_p} H\}. \tag{2.8.6}$$

Now consider the problem of reducing $(h_0 + \varepsilon h_1)_p$ to normal form. We will deal with the simplest case when $e_1, \ldots, e_p; \varphi_1, \ldots, \varphi_q$, where $p + q = n$, constitute a basic system of invariants and eigenfunctions with respect to $(h_0)_p$ respectively:

$$(h_0)_p e_i = \{h_0, e_i\} = 0, \quad (h_0)_p \varphi_j = \{h_0, \varphi_j\} = \lambda_j(e) \varphi_j; \quad \lambda_j(e) \neq \lambda_k(e)$$

for $j \neq k$. The Jacobi identity implies $(h_0)_p\{e_\alpha, \varphi_\beta\} = \{(h_0)_p e_\alpha, \varphi_\beta\} + \{e_\alpha, (h_0)_p \varphi_\beta\}$ or

$$(h_0)_p\{e_\alpha, \varphi_\beta\} = \lambda_\beta\{e_\alpha, \varphi_\beta\}. \tag{2.8.7}$$

Now let us show that to reduce $(h_0 + \varepsilon h_1)_p$ to normal form it suffices to find S so that

$$H' = e^{-S_p}(h_0 + \varepsilon h_1) = h_0 + \varepsilon M(\varepsilon) + \cdots. \tag{2.8.8}$$

We must show that $\{H', e_\alpha\}$ is invariant and $\{H', \varphi_\beta\}$ corresponds to λ_β. But

$$\begin{aligned}\{H', e_\alpha\} &= \{h_0, e_\alpha\} + \varepsilon\{M(e), e_\alpha\} + \cdots \\ &= \varepsilon \sum_s \frac{\partial M}{\partial e_s}\{e_s, e_\alpha\} + \cdots, \\ \{H', \varphi_\beta\} &= \{h_0, \varphi_\beta\} + \varepsilon\{M(e), \varphi_\beta\} + \cdots \\ &= \lambda_\beta \varphi_\beta + \varepsilon \sum_s \frac{\partial M}{\partial e_s}\{e_s, \varphi_\beta\} + \cdots.\end{aligned}$$

These equations show the validity of our assertion. Thus it remains only to verify the possibility of defining S in order to satisfy (2.8.8). As always, we will consider only the equation $\{h_0, S\} + h_1 = M(e)$. But if $h_1 = \gamma(e) + \sum \tau_s(e)\varphi_s$, then for

$$S = -\sum_s \frac{\tau_s(e)}{h_s(e)}\varphi_s$$

we will have $\{h_0, S\} + h_1 = \gamma(e)$.

Remark 2.8.1 Variables e and φ are not, generally speaking, canonically conjugate.

Remark 2.8.2 As P. Dirac has shown, the system $dx_i/dt = a_i(x)$ may be reduced to a Hamiltonian one if we double the number of variables. In fact, put

$$H = \sum_{1 \le i \le n} a_i(x) y_i$$

and let y_i be the canonical conjugate of x_i. Then the Hamiltonian system is

$$\frac{dx_i}{dt} = a_i(x) = \frac{\partial H}{\partial y_i}, \qquad \frac{dy_i}{dt} = a_i(x) = -\frac{\partial H}{\partial x_i}.$$

Such a reduction enables one to make computations in a uniform way. Consider, as before, the simple case of an operator

$$X_0 + \varepsilon X_1 = \sum_{1 \le j \le q} \lambda_j(e)\varphi_j \frac{\partial F}{\partial \varphi_j}$$

$$+\varepsilon\left(\sum_{1\le i\le p} a_i(e,\varphi)\frac{\partial}{\partial e_i}+\sum_{1\le j\le q} b_j(e,\varphi)\frac{\partial}{\partial \varphi_i}\right),$$

where e_i, for $i = 1, \ldots, p$, are invariants of X_0. Let us assign to invariants e_i the canonically conjugate variables y_i and to eigenfunctions φ_j the variables z_j. Then

$$H = H_0 + \varepsilon H_1 = \sum_{1\le j\le q} \lambda_j(e)\varphi_j z_j + \varepsilon\left(\sum_{1\le i\le p} a_i y_i + \sum_{1\le k\le q} b_j z_j\right). \quad (2.8.9)$$

The operator $(H_0)_p F$ is

$$(H_0)_p F = \{H_0, F\} = \sum_{1\le j\le q} \lambda_j(e)\varphi_j \frac{\partial}{\partial \varphi_j}$$
$$- \sum_{1\le s\le p}\sum_{q\le j\le q} \frac{\partial \lambda_j(e)}{\partial e_s}\varphi_j z_j \frac{\partial F}{\partial y_s} - \sum_{1\le j\le q} \lambda_j z_j \frac{\partial F}{\partial z_j}. \quad (2.8.10)$$

It is clear from (2.8.10) that

$$\begin{aligned}
(H_0)_p e_\alpha &= 0,\\
(H_0)_p \varphi_\beta &= \lambda_\beta \varphi_\beta,\\
(H_0)_p z_\beta &= -\lambda_\beta z_\beta, \qquad (2.8.11)\\
(H_0)_p y_\alpha &= -\frac{\partial}{\partial e_\alpha}\sum_{1\le j\le q}\lambda_j\varphi_j z_j,
\end{aligned}$$

i.e., invariants of $(H_0)_p$ are e_α, eigenfunctions are φ_β, and z_β and y_α's correspond to zero since the $\varphi_j z_j$'s do. It is not difficult to see that the procedure of constructing $H' = H_0 + \varepsilon H_1' + \cdots = e^{-S_p} H$ produces H_1' in the form

$$H_1' = \sum_{1\le i\le p} u_i(e,\varphi) y_i + \sum_{1\le j\le q} r_j(e,\varphi) z_j, \quad (2.8.12)$$

where the $u_i(e,\varphi)$'s correspond to zero and $r_j(e,\varphi_j)$ corresponds to λ_j. It is necessary to have in mind, though, that the Dirac theorem does not eliminate the problem of finding eigenfunctions and invariants of X_0.

3

Examples

We will begin with the now classical example of motion of the pendulum of variable length. This examples illustrates the simplest and at the same time the most essential methods of computation.

3.1 Example: The Pendulum of Variable Length

We consider a mathematical pendulum whose length slowly varies under a sufficiently smooth law $\ell = \ell(\varepsilon t)$, where ε is a small parameter. It is intuitively clear that for a period of time of order 1 we will observe with accuracy of order 1 harmonic oscillations with frequency $\omega = \sqrt{g/\ell}$ and total energy

$$E = \tfrac{1}{2}m\ell^2(\dot{\alpha}^2 + \omega^2\alpha^2)$$

(here $\sin\alpha \approx \alpha$) which defines the amplitude of the motion. The dependence of E on ω, which for large periods of time ($\sim 1/\varepsilon$) changes by a finite value, is of the main interest. Let us write the equation of motion. The kinetic energy of the pendulum is

$$T = \tfrac{1}{2}mv^2 = \tfrac{1}{2}m(\ell^2\alpha^2 + \ell^2).$$

The work of the external force over a virtual displacement permitted by constraints is $\delta A = mg\delta(\ell\cos\alpha) = -mg\ell\sin\alpha\delta d$. The generalized force is $Q_\alpha = -mg\ell\sin\alpha \approx -mg\ell\alpha$. Substituting T, Q_α into the Lagrange equations

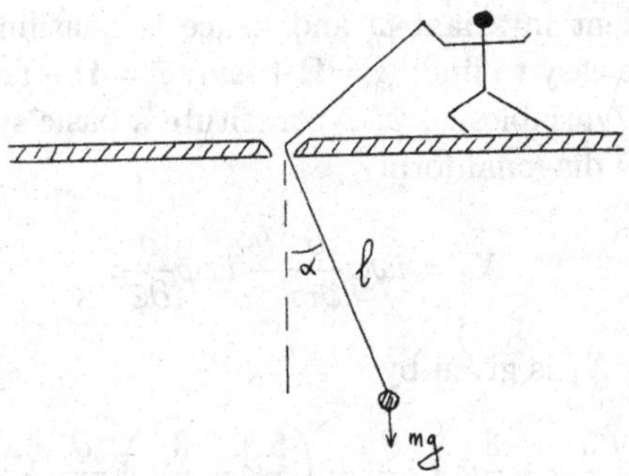

FIGURE 3.1.

$$\frac{d}{dt}\left(\frac{\partial T}{\partial \dot{\alpha}}\right) - \frac{\partial T}{\partial \alpha} = Q_\alpha$$

and dividing by $m\ell^2$ we get

$$\ddot{\alpha} + 2\frac{\dot{\ell}}{\ell}\dot{\alpha} + \frac{g}{\ell}\alpha = 0. \tag{3.1.1}$$

Introduce ω instead of ℓ and write the law of dependence of ℓ on t in the form

$$\frac{d\omega}{dt} = \varepsilon\omega h(\omega), \tag{3.1.2}$$

where $h(\omega)$ is a sufficiently smooth function. (The convenience of such a representation will soon be clear.) Then (3.1.1) will take the form $\ddot{\alpha} - 4\varepsilon h(\omega)\alpha + \omega^2\alpha = 0$. Putting $\dot{\alpha} = \Omega$, we get

$$\frac{d\alpha}{dt} = \Omega, \quad \frac{d\Omega}{dt} = -\omega^2\alpha + 4\varepsilon h(\omega)\Omega, \quad \frac{d\omega}{dt} = \varepsilon\omega h(\omega). \tag{3.1.3}$$

Remark 3.1.1 The adding of (3.1.2) should not be considered as a specific trick; since ω depends on t [recall that $\omega = \omega(\varepsilon t)$], the system of the first two equations is nonautonomous. We could pass to an autonomous system adding $d\tau/dt = \varepsilon$ so that $\tau = \varepsilon t$, $\omega = \omega(\tau)$. Here ω is more convenient than τ.

Thus we have obtained the system (3.1.3) in the standard form, for which

$$X = X_0 + \varepsilon X_1 = \Omega\frac{\partial}{\partial\alpha} - \omega^2\alpha\frac{\partial}{\partial\Omega} + \varepsilon h(\omega)\left\{4\Omega\frac{\partial}{\partial\Omega} + \omega\frac{\partial}{\partial\omega}\right\}.$$

In this example we will restrict ourselves to the first approximation of perturbation theory, computing M_1 and S_1 and showing three methods of solution; each of them in its own way illustrates the method proposed in this book.

(1) The operator

$$X_0 = \Omega\frac{\partial}{\partial\alpha} - \omega^2\alpha\frac{\partial}{\partial\Omega}$$

possesses an evident invariant ω and hence is quasilinear. Therefore its eigenfunctions are easy to find: $\varphi = \Omega + i\omega\alpha$, $\bar{\varphi} = \Omega - i\omega\alpha$, where $i^2 = -1$, $\lambda = i\omega$, $\bar{\lambda} = -i\omega$. Variables ω, φ, $\bar{\varphi}$ constitute a basic system in which X_0 is of the canonical diagonal form

$$X_0 = i\omega\varphi\frac{\partial}{\partial\varpi} - i\omega\bar{\varphi}\frac{\partial}{\partial\bar{\varphi}}. \tag{3.1.4}$$

In these variables X_1 is given by

$$X_1 = h(\omega)\left\{\left(\frac{5}{2}\varphi + \frac{3}{2}\bar{\varphi}\right)\frac{\partial}{\partial\varphi} + \left(\frac{5}{2}\bar{\varphi} + \frac{3}{2}\varphi\right)\frac{\partial}{\partial\bar{\varphi}} + \omega\frac{\partial}{\partial\omega}\right\}. \tag{3.1.5}$$

Its coefficients are already expanded into a "series" of eigenfunctions and we may pass to the construction of M_1 and S_1. Let us find $M_1\omega$ and $S_1\omega$. We have

$$M_1\omega = [X_0S_1]\omega + X_1\omega = X_0(S_1\omega) + \omega h(\omega).$$

The function $M_1\omega$ should correspond to zero. Since $\omega h(\omega)$ corresponds to zero, we may put $S_1\omega = 0$, hence

$$M_1\omega = \omega h(\omega). \tag{3.1.6}$$

Remark 3.1.2 The choice $S_1\omega = 0$ is convenient for two reasons: (1) in the following calculations the eigenvalue behaves like a constant; (2) the variable ω, in which we are interested in this problem, remains unchanged. By (3.1.4)–(3.1.6),

$$X_1\varphi = [X_0S_1]\varphi + X_1\varphi = X_0(S_1\varphi) - i\omega S_1\varphi + \frac{5}{2}h(\omega)\varphi + \frac{3}{2}h(\omega)\bar{\varphi}.$$

The function $M_1\varphi$ should correspond to $i\omega$. Setting $S_1\varphi = A(\omega)\bar{\varphi}$, we get

$$M_1\varphi = \left(-2i\omega A(\omega) + \frac{3}{2}h(\omega)\right)\bar{\varphi} + \frac{5}{2}h(\omega)\varphi$$

and choose $A(\omega)$ so that the coefficient of $\bar{\varphi}$ vanishes:

$$A(\omega) = -\frac{3i}{4}\frac{h(\omega)}{\omega}.$$

Thus

$$S_1\varphi = -\frac{3i}{4}\frac{h(\omega)}{\omega}\bar{\varphi}, \quad M_1\varphi = \frac{5}{2}h(\omega)\varphi.$$

Replacing i by $-i$, we get

$$S_1\bar{\varphi} = \frac{3i}{4}\frac{h(\omega)}{\omega}\varphi, \quad M_1\bar{\varphi} = \frac{5}{2}h(\omega)\bar{\varphi}.$$

Thus

$$M_1 = h(\omega)\left(\frac{5}{2}\phi\frac{\partial}{\partial\varphi} + \frac{5}{2}\bar{\varphi}\frac{\partial}{\partial\bar{\varphi}} + \omega\frac{\partial}{\partial\omega}\right), \quad S_1 = \frac{3ih(\omega)}{4\omega}\left(-\bar{\varphi}\frac{\partial}{\partial\varphi} + \varphi\frac{\partial}{\partial\bar{\varphi}}\right).$$

The system of equations of the first approximation, i.e., the ordinary system corresponding to $X_0 + \varepsilon M_1$, is

$$\frac{d\varphi}{dt} = \left(i\omega + \frac{5}{2}\varepsilon h(\omega)\right)\varphi, \quad \frac{d\bar{\varphi}}{dt} = \left(-i\omega + \frac{5}{2}\varepsilon h(\omega)\right)\bar{\varphi}, \quad \frac{d\omega}{dt} = \varepsilon\omega h(\omega). \tag{3.1.7}$$

It describes the motion of the pendulum up to a quantity of order ε during the time period $0 \leq t \lesssim 1/\varepsilon$ and is easy to integrate. The right-hand sides of this system in the first two equations differ from exact values of φ and $\bar{\varphi}$ by values of order ε^2, and the third equation is exact. Starting from these facts it is easy to prove the validity of the above estimate.

Remark 3.1.3 Recall that old and new variables are denoted by the same signs. In this example we restrict ourselves to the accuracy $0(\varepsilon)$; i.e., we do not take into account summands of order ε, and therefore there is no need to distinguish between old and new variables.

The system (3.1.7) implies in particular that

$$\frac{d}{d\omega}\varphi\bar{\varphi} = \frac{5}{\omega}\varphi\bar{\varphi};$$

i.e., $\varphi\bar{\varphi}/\omega^5 = \text{const}$. Since $\varphi\bar{\varphi} = \Omega^2 + \omega^2\alpha^2 = \dot{\alpha}^2 + \omega^2\alpha^2$, then $\varphi\bar{\varphi}/\omega^4$ is proportional to E at the time t. We thus have the well-known fact that energy is proportional to the frequency: $E/\omega = \text{constant}$ (this constant is called an *adiabatic invariant*).

(2) We might have constructed M_1 and S_1 by making use of the commutation relation since all conditions of Theorem 2.4.7 are satisfied. Namely, computing subsequently $X_1' = [X_0X_1], X_1'', \ldots$, we have

$$X_1' = ,h(\omega)\left\{-4\Omega\frac{\partial}{\partial\alpha} - 2\omega^2\alpha\frac{\partial}{\partial\Omega}\right\},$$

$$X_1'' = 6\omega^2h(\omega)\left\{\alpha\frac{\partial}{\partial\alpha} - \Omega\frac{\partial}{\partial\Omega}\right\},$$

$$X_1''' = 12\omega^2h(\omega)\left\{\Omega\frac{\partial}{\partial\alpha} + \omega^2\alpha\frac{\partial}{\partial\Omega}\right\},$$

$$X_1^{\text{iv}} = -24\omega^4h(\omega)\left\{\alpha\frac{\partial}{\partial\alpha} - \Omega\frac{\partial}{\partial\Omega}\right\}.$$

The commutation relation is

$$4\omega^2X_1'' + X_1^{\text{iv}} = 0. \tag{3.1.8}$$

Now set $S_1 = (1/4\omega^2)X_1'$ so that $M_1 = X_1 + (1/4\omega^2)X_1''$. Then $M_1'' = 0$ by (3.1.8). It is not difficult to verify that the result for M_1 coincides with the above, and that for S_1 differs by $-(h(\omega)/4\omega^2)X_0$, which is not surprising since S is defined up to a summand permutable with X_0.

(3) Sometimes the following trick might be useful. Consider again the system

$$\varepsilon^\alpha\frac{dx}{dt} = f_0 + \varepsilon f_1 + \cdots$$

for which X_0 is diagonal. (The generalization to the Jordan case is trivial.) If we introduce the additional variable τ via the equation $\varepsilon^\alpha(d\tau/dt) = 1$, then the leading operator X_0 will turn into $\tilde{X}_0 = X_0 + (\partial/\partial\tau)$ and the other operators will remain unchanged. Then $\tilde{X}_0$, which now depends on $n+1$ variables, has n invariants of the form $g = \varphi e^{-\lambda\tau}$. If we pass to variables g and τ, then $\tilde{X}_0$ takes the form $\tilde{X}_0 = \partial/\partial\tau$. This operator possesses eigenfunctions of the form $\exp\omega(g)\tau$, functions corresponding to zero of the form τ^m, and, more generally, functions corresponding to eigenvalues of the form $P(\tau)\exp\omega(g)\tau$, where $P(\tau)$ is a polynomial whose coefficients are functions of invariants g. If the coefficients of operators X_i, where $i = 1, 2, \ldots$, are developed into series with respect to functions of the mentioned form, then we choose S to annihilate all summands of coefficients of M that do not correspond to zero (it should go without saying that the requirement of absence of resonances is sustained). In our example, invariants of $\tilde{X}_0$ are ω, $g = \varphi e^{-i\omega\tau}$, and $\bar{g} = \bar{\varphi} e^{i\omega\tau}$. Operators $\tilde{X}_0$ and X_1 in variables ω, g, and $\bar{g}$ are of the form

$$\tilde{X}_0 = \frac{\partial}{\partial\tau},$$

$$X_1 = h(\omega)\left\{\omega\frac{\partial}{\partial\omega} + \left(\frac{5}{2}g + \frac{3}{2}\bar{g}e^{-2i\omega\tau} - i\omega g\tau\right)\frac{\partial}{\partial g} + \left(\frac{5}{2}\bar{g} + \frac{3}{2}ge^{2i\omega\tau} + i\omega\bar{g}\tau\right)\frac{\partial}{\partial\bar{g}}\right\}.$$

Setting

$$S_1 = h(\omega)\left\{\frac{3\bar{g}e^{-2i\omega\tau}}{4i\omega}\frac{\partial}{\partial g} - \frac{3ge^{2i\omega\tau}}{4i\omega}\frac{\partial}{\partial\bar{g}}\right\}$$

we get

$$M_1 = h(\omega)\left\{\omega\frac{\partial}{\partial\omega} + \left(\frac{5}{2}g - i\omega g\tau\right)\frac{\partial}{\partial g} + \left(\frac{5}{2}\bar{g} + i\omega\bar{g}\tau\right)\frac{\partial}{\partial\bar{g}}\right\}.$$

These operators are the same that were found in (1). The reader acquainted with the Bogolyubov–Krylov averaging method will easily recognize both this and the first methods as its direct generalization.

The following well-known problem differs greatly from the previous one both in formulation and in properties of its solution.

3.2 Example: A Second Order Linear Equation

Let us consider the equation

$$\varepsilon y'' + a(x)y' + b(x)y = 0, \quad y(0) = \alpha, \quad y(1) = \beta, \tag{3.2.1}$$

where $a(x)$ and $b(x)$ are smooth functions defined in $[0, 1]$. Suppose that $a(x)$ does not vanish and that $\varepsilon > 0$. Put $\varepsilon y' = z$ and adding $\varepsilon(dx/dt) = z$ let us pass from (3.2.1) to the autonomous system

$$\varepsilon\frac{dx}{dt} = \varepsilon, \qquad \varepsilon\frac{dy}{dt} = z, \quad \varepsilon\frac{dz}{dt} = -a(x)z - \varepsilon b(x)y.$$

The operator X for this system is

$$X = X_0 + \varepsilon X_1 = z\frac{\partial}{\partial y} - a(x)z\frac{\partial}{\partial z} + \varepsilon\left(\frac{\partial}{\partial x} - b(x)y\frac{\partial}{\partial z}\right).$$

Invariants of X_0 are x and $\omega = z + a(x)y$; its eigenfunction corresponding to $\lambda = -a(x)$ is z. In the basic system x, ω, z, the operator X is of the form

$$X = X_0 + \varepsilon X_1 = -az\frac{\partial}{\partial z} + \varepsilon\left\{\frac{\partial}{\partial x} + \frac{a'-b}{a}(\omega - z)\frac{\partial}{\partial \omega} + \frac{b}{a}(z-\omega)\frac{\partial}{\partial z}\right\}, \tag{3.2.2}$$

where for brevity's sake $a = a(x)$, $b = b(x)$, $a' = da/dx$. Operators of the first approximation are easy to find:

$$\begin{aligned} M_1 &= \frac{\partial}{\partial x} + \frac{a'-b}{a}\omega\frac{\partial}{\partial \omega} + \frac{b}{a}z\frac{\partial}{\partial z}, \\ S_1 &= -\frac{a'-b}{a^2}z\frac{\partial}{\partial \omega} + \frac{b}{a^2}\omega\frac{\partial}{\partial z}. \end{aligned} \tag{3.2.3}$$

Computations of the following approximations are similar in principle. We will find M_2 to give the reader an idea of the volume and the technique of computation. It is convenient to express M_2 in the form $M_2 = [X_0, S_2] + \frac{1}{2}[(M_1 + X_1), S_1]$ [see (1.2.3′)]. Now, due to (3.2.2) and (3.2.3),

$$\begin{aligned} \frac{1}{2}(M_1 + X_1) &= \frac{\partial}{\partial x} + \frac{a'-b}{a}\left(\omega - \frac{z}{2}\right)\frac{\partial}{\partial \omega} + \frac{b}{a}\left(z - \frac{\omega}{2}\right)\frac{\partial}{\partial z} \\ \frac{1}{2}((M_1 + X_1)S_1) &= \left(\frac{(a'-b)b}{a^3}\omega + fz\right)\frac{\partial}{\partial \omega} + \left(g\omega - \frac{(a'-b)b}{a^3}z\right)\frac{\partial}{\partial z} \end{aligned}$$

where f and g are expressed in terms of a, b, a', b', and a''. Then

$$\begin{aligned} M_2 &= \frac{(a'-b)b}{a^3}\left(\omega\frac{\partial}{\partial \omega} - z\frac{\partial}{\partial z}\right), \\ S_2 &= \frac{f}{z}z\frac{\partial}{\partial \omega} - \frac{g}{a}\omega\frac{\partial}{\partial z}. \end{aligned}$$

The system corresponding to $M = X_0 + \varepsilon M_1 + \varepsilon^2 M_2 + 0(\varepsilon^3)$ is

$$\begin{aligned} \varepsilon\frac{dx}{dt} &= \varepsilon, \\ \varepsilon\frac{d\omega}{dt} &= \varepsilon\frac{a'-b}{a}\omega + \varepsilon^2\frac{(a'-b)b}{a^3}\omega + 0(\varepsilon^3), \\ \varepsilon\frac{dz}{dt} &= -az + \varepsilon\frac{b}{a}z + \varepsilon^2\frac{(b-a')b}{a^3}z + 0(\varepsilon^3) \end{aligned}$$

or

$$\begin{aligned} \frac{d\omega}{dx} &= \left(\frac{a'-b}{a} + \varepsilon\frac{(a'-b)b}{a^3}\right)\omega + 0(\epsilon^2), \\ \frac{dz}{dx} &= \left(-\frac{a}{\varepsilon} + \frac{b}{a} - \varepsilon\frac{(a'-b)b}{a^3}\right)z + 0(\epsilon^2), \end{aligned} \tag{3.2.4}$$

For definiteness' sake we will assume that $a(x) > 0$ on $[0, 1]$. Set

$$h(x) = \int_0^x a(\zeta)d\zeta,$$

$$\mathcal{T}_0(x) = \int_0^x \frac{b(\zeta)}{a(\zeta)}d\zeta,$$

$$\mathcal{T}_1(x) = \int_0^x \frac{(a'(\zeta) - b(\zeta))b(\zeta)}{a^3(\zeta)}\,d\zeta$$

and express the solution of (3.2.4) in the form

$$\begin{aligned} \omega^* &= Aa(x)e^{-\mathcal{T}_0(x)}(1 + \varepsilon\mathcal{T}_1(x)) + 0(\varepsilon^2), \\ z^* &= Be^{-h(x)/\varepsilon}e^{\mathcal{T}_0(x)}(1 - \varepsilon\mathcal{T}_1(x)) + 0(\varepsilon^2), \end{aligned} \tag{3.2.5}$$

where A and B are constants. Note that due to the presence of a small parameter ε in coefficients of derivatives of the initial system we lose accuracy; finding two approximations we get an error $\sim \varepsilon^2$, not $\sim \varepsilon^3$. Therefore in the operator e^{-S} which returns us to the old variables the operator S_2 may be ignored, putting $e^{-S} \approx E - \varepsilon S_1$.

We are interested only in y. Since $y = \omega - z/a$, then by (3.2.3)

$$S_1 y = -\frac{a' - b}{a^3}z - \frac{b}{a^3}\omega$$

and the solution of the initial equation is

$$y = \frac{\omega^* - z^*}{a} + \varepsilon\left(\frac{a' - b}{a^3}z^* + \frac{b}{a^3}\omega^*\right) + 0(\varepsilon^2)$$

[see (3.2.5)].

Inserting $\omega^* z^*$, and putting $A = A_0 + \varepsilon A_1$, $B = B_0 + \varepsilon B_1$, we get

$$\begin{aligned} y = {} & A_0 e^{-\mathcal{T}_0(x)} - \frac{B_0(x)}{a(x)}e^{-h(x)/\varepsilon}e^{\mathcal{T}_0(x)} \\ & + \varepsilon\Bigg[A_1 e^{-\mathcal{T}_0(x)} - \frac{B_1(x)}{a(x)}e^{-h(x)/\varepsilon}e^{\mathcal{T}_0(x)} \\ & + A_0 e^{-\mathcal{T}_0(x)}\left(\mathcal{T}_1(x) + \frac{b(x)}{a^2(x)}\right) \\ & + \frac{B_0(x)}{a(x)}e^{-h(x)/\varepsilon}e^{\mathcal{T}_0(x)}\left(\mathcal{T}_1(x) + \frac{a'(x) - b(x)}{a^2(x)}\right)\Bigg] \\ & + 0(\varepsilon^2). \end{aligned}$$

Since $e^{-(1/\varepsilon)h(1)} \ll \varepsilon^n$ for $n > 0$ we will satisfy the boundary conditions with accuracy $0(\varepsilon^2)$ putting

$$A_0 - \frac{B_0}{a(0)} = \alpha,$$

$$A_1 - \frac{B_1}{a(0)} + A_0 \frac{b(0)}{a^2(0)} + \frac{B_0}{a(0)} \frac{a'(0) - b(0)}{a^2(0)} = 0,$$
$$A_0 e^{-\mathcal{T}_0(1)} = \beta,$$
$$A_1 e^{-\mathcal{T}_0(1)} + A_0 e^{-\mathcal{T}_0(1)} \left(\mathcal{T}_1(1) + \frac{b(1)}{a^2(1)} \right) = 0.$$

In case $a(x) < 0$ let us replace the lower limit in $h(x)$ [and, for symmetry, in $\mathcal{T}_0(x), \mathcal{T}_1(x)$] by 1 so that $h(x) > 0$, as earlier. The equations for A and B will change accordingly.

Note that we have found a uniform asymptotic without applying the usual techniques of matching the so-called interior and exterior developments.

The procedure of returning to the old variables via e^{-S} is trivial in principle and is only connected with accuracy of computations (nontrivial effects are connected with the new system corresponding to M). In subsequent examples we will not, as a rule, return to the old variables.

In the considered problems all expressions are linear combinations of a finite number of eigenfunctions. When this is not so, the development into an infinite series can often be avoided. The next example demonstrates this. Simultaneously it shows once more the connection with the Bogolyubov–Krylov method.

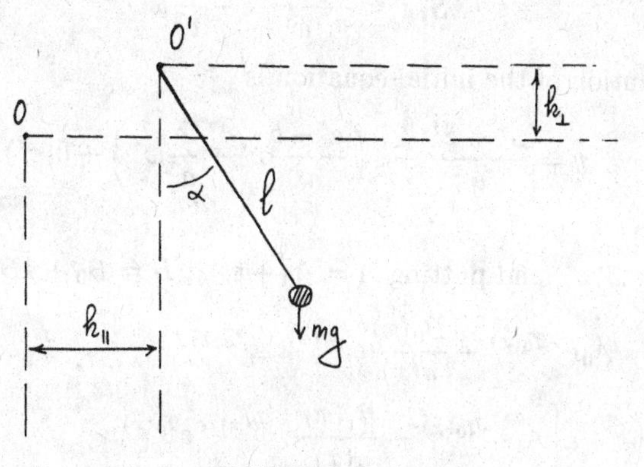

FIGURE 3.2.

3.3 Example: P. L. Kapitsa's Problem: A Pendulum Suspended from an Oscillating Point

Let us consider the mathematical pendulum whose suspension point 0′ oscillates with a small amplitude and high frequency (as compared with the length of the pendulum and the frequency of its oscillations). Let us write the deviations of 0′ from a fixed point 0 in vertical and horizontal directions in the form $h_\perp = \varepsilon \ell f(t/\varepsilon)$, $h_\| = \varepsilon \ell g(t/\varepsilon)$, where ℓ is the length of the pendulum, $\varepsilon > 0$ a small parameter, and $f(\theta)$ and $g(\theta)$ ($\theta = t/\varepsilon$) are smooth periodic periodic functions in θ with a common period which we will take to be equal to 1. (All quantities are assumed to be dimensionless.) It is possible to consider the case of an arbitrary ratio of periods and, generally, restrictions on $f(\theta)$ and $g(\theta)$ can be weakened, as we will see later.

Turning the frame of reference connected with 0′ into an inertial one by adding the inertia force $\{-mh_\perp, -mh_\|\}$, we get an equation of motion

$$\ddot{\alpha} + \left(\omega^2 + \frac{1}{\varepsilon} f''(\theta)\right) \sin\alpha + \frac{1}{\varepsilon} g''(\theta)\cos\alpha = 0, \tag{3.3.1}$$

where $\omega = \sqrt{g/\ell}$ is the frequency of small oscillations of the usual pendulum, $f''(\theta) = d^2 f(\theta)/d\theta^2$, $g''(\theta) = d^2 g(\theta)/d\theta^2$. An autonomous system equivalent to (3.3.1) is

$$\begin{aligned}
\varepsilon \frac{d\theta}{dt} &= 1, \\
\varepsilon \frac{d\alpha}{dt} &= \varepsilon\Omega, \\
\varepsilon \frac{d\Omega}{dt} &= -f''(\theta)\sin\alpha - g''(\theta)\cos\alpha - \varepsilon\omega^2 \sin\alpha.
\end{aligned}$$

The operator X is given by

$$\begin{aligned}
X &= X_0 + \varepsilon X_1 \\
&= \frac{\partial}{\partial\theta} - [f''(\theta)\sin\alpha + g''(\theta)\cos\alpha]\frac{\partial}{\partial\Omega} + \varepsilon\left(\Omega\frac{\partial}{\partial\alpha} - \omega^2 \sin\alpha \frac{\partial}{\partial\Omega}\right).
\end{aligned}$$

Invariants of the leading operator X_0 are α and $e = \Omega + f'(\theta)\sin\alpha + g'(\theta)\cos\alpha$, and θ is the function corresponding to 0. In the basic system of variables θ, α, e, operators X_0 and X_1 are of the form

$$\begin{aligned}
X_0 &= \frac{\partial}{\partial\theta}, \\
X_1 &= [e - f''(\theta)\sin\alpha - g'(\theta)\cos\alpha]\frac{\partial}{\partial\alpha} \\
&\quad + \left\{ ef'(\theta)\cos\alpha - [eg'(\theta) + \omega^2]\sin\alpha \right. \\
&\quad \left. - \frac{f'^2(\theta) - g'^2(\theta)}{2}\sin 2\alpha - f'(\theta)g'(\theta)\cos 2\alpha \right\}\frac{\partial}{\partial\ell}.
\end{aligned} \tag{3.3.2}$$

Coefficients of $\partial/\partial\alpha$ and $\partial/\partial\ell$ in X_1 are representable by Fourier series of the form $\sum_n k_n(\alpha, e)\exp(2\pi i n\theta)$, where $k_n(\alpha, e)$ are invariants of X_0 and

$\exp(2\pi i n\theta)$ is the eigenfunction of X_0 corresponding to the eigenvalue $2\pi i n$. To reduce M_1 to the canonical form it is not necessary to actually calculate these series. It is only necessary to find free terms (those corresponding to $n = 0$). Then S_1 is found by integration of the remainder along θ. Set

$$a = \frac{1}{2}\int_0^1 [f'^2(\theta) - g'^2(\theta)]\, d\theta, \qquad b = \int_0^1 f'(\theta)g'(\theta)\, d\theta$$

(free terms in decompositions $\frac{1}{2}[f'^2(\theta) - g'^2(\theta)]$ and $f'(\theta)g'(\theta)$) and

$$\begin{aligned} S_1 = {} & [f(\theta)\sin\alpha + g(\theta)\cos\alpha]\frac{\partial}{\partial\alpha} \\ & + \Bigg\{ - ef(\theta)\cos\alpha + eg(\theta)\sin\alpha \\ & + \left(\int_0^\theta \left(\frac{f'^2(\zeta) - g'^2(\zeta)}{2} - a\right) d\zeta\right) \sin 2\alpha \\ & + \left(\int_0^\theta [f'(\zeta)g'(\zeta) - b]d\zeta\right) \cos 2\alpha \Bigg\} \frac{\partial}{\partial e}. \end{aligned}$$

Then, due to (3.3.2),

$$M_1 = e\frac{\partial}{\partial\alpha} - (\omega^2 \sin\alpha + a\sin 2\alpha + b\cos 2\alpha)\frac{\partial}{\partial e}$$

and $M_1' = 0$. Computations of other operators are similar.

Remark 3.3.1 It is clear that trying to achieve $M' = 0$ by the choice of S we solve each time one and the same problem: how to represent a function $F(\theta)$ in the form $F(\theta) = c + dG(\theta)/d\theta$, where $c = \text{const}$ and $G(\theta)$ is a bounded function such that

$$c = \lim_{\theta\to\infty} \frac{1}{\theta} \int_0^\theta F(\zeta)d\zeta.$$

The possibility of such a representation for all $F(\theta)$ that arise is a weaker condition on $f(\theta)$, $g(\theta)$ than the periodicity condition.

The system of equations corresponding to $X_0 + \varepsilon M_1$, i.e.,

$$\begin{aligned} \varepsilon\frac{d\theta}{dt} &= 1, \\ \varepsilon\frac{d\alpha}{dt} &= \varepsilon e, \\ \varepsilon\frac{de}{dt} &= -\varepsilon(\omega^2\sin\alpha + a\sin 2\alpha + b\cos 2\alpha), \end{aligned} \tag{3.3.3}$$

describes the motion of the pendulum with accuracy $0(\varepsilon)$ during the time $t \lesssim 1$. Rejecting the first (trivial) equation and eliminating e we may write instead of (3.3.3) this second order equation for the angle α:

$$\ddot{\alpha} + \omega^2 \sin\alpha + a \sin 2\alpha + b \cos 2\alpha = 0,$$

where constants a and b defined above are, generally speaking, arbitrary (if, however, $g = 0$ so that only vertical oscillations take place, then $a > 0$ and $b = 0$). Let us mention several interesting corollaries.

(1) Unlike the usual pendulum, the pendulum with the oscillating point of suspension may have any given *a priori* steady equilibrium state $\alpha = \alpha_0$. This takes place if a and b satisfy

$$\omega_0^2 \sin\alpha_0 + a \sin 2\alpha_0 + b \cos 2\alpha_0 = 0,$$

$$\omega_0^2 \cos\alpha_0 + 2a \cos 2\alpha_0 - 2b \sin 2\alpha_0 > 0.$$

Small oscillations around this position are harmonic with frequency

$$(\omega^2 \cos\alpha_0 + 2a \cos 2\alpha_0 - 2b \sin 2\alpha_0)^{1/2}.$$

(2) In the case of vertical oscillations of the suspension point ($b = 0$, $a > 0$) for $a < \omega^2/2$ the pendulum has only one steady point of equilibrium, like the usual one, but the frequency of small oscillations is higher and is equal to $(\omega^2 + 2a)^{1/2}$.

(3) For $a > \omega^2/2$ the vertical oscillating of the suspension point stabilizes the upper ($\alpha_0 = \pi$) equilibrium. Practically, in this case (as in the general case of a sufficiently large a) a large dynamical load arises in the rod and in the oscillating support. P. L. Kapitsa showed that it is quite easy to realize stabilization of the pendulum in the upper or inclined position.

In previous problems it was possible to get a clear understanding of the structure of M_j and S_j, finding only the first order approximation (M_1, S_1). This often helps in obtaining estimates as, for instance, in the following problem.

3.4 Example: Van der Pol Oscillator with Small Damping

$$\frac{d^2u}{dt^2} + u = \varepsilon(1 - u^2)\frac{du}{dt} \qquad (\varepsilon > 0).$$

The system is $du/dt = v$, $dv/dt = -u + \varepsilon(1 - u^2)v$ and the corresponding operator is

$$X = X_0 + \varepsilon X_1 = v\frac{\partial}{\partial u} - u\frac{\partial}{\partial v} + \varepsilon(1 - u^2)v\frac{\partial}{\partial v}. \qquad (3.4.1)$$

Eigenfunctions of X_0 are $\varphi = v + iu$, $\bar{\varphi} = v - iu$, and $\lambda = i$, $\bar{\lambda} = -i$. In the basic system, φ and $\bar{\varphi}$ operators (3.4.1) are of the form

$$X_0 = i\varphi\frac{\partial}{\partial\varphi} - i\bar{\varphi}\frac{\partial}{\partial\bar{\varphi}},$$

$$X_1 = (\psi_i + \psi_{-i} + \psi_{3i} + \psi_{-3i})\left(\frac{\partial}{\partial\varphi} + \frac{\partial}{\partial\bar{\varphi}}\right),$$

where

$$\psi_i = \left(\frac{1}{2} - \frac{1}{8}\varphi\bar{\varphi}\right)\varphi, \qquad \psi_{-i} = \bar{\psi}_i, \qquad \psi_{3i} = \frac{1}{8}\varphi^3, \qquad \psi_{-3i} = \bar{\psi}_{3i},$$

ψ_μ corresponds to μ, and a bar stands for complex conjugation. It is easy to deduce, making use of the known technique, that

$$M_1 = \psi_i\frac{\partial}{\partial\varphi} + \psi_{-i}\frac{\partial}{\partial\bar{\varphi}} = \left(\frac{1}{2} - \frac{1}{8}\varphi\bar{\varphi}\right)\left(\varphi\frac{\partial}{\partial\varphi} + \bar{\varphi}\frac{\partial}{\partial\bar{\varphi}}\right),$$

$$S_1 = \left(-\frac{i}{2}\psi_{-i} + \frac{i}{2}\psi_{3i} - \frac{i}{4}\psi_{-3i}\right)\frac{\partial}{\partial\varphi} + \left(\frac{i}{2}\psi_i - \frac{i}{2}\psi_{-3i} + \frac{i}{4}\psi_{3i}\right)\frac{\partial}{\partial\bar{\varphi}}. \tag{3.4.2}$$

We will not compute further approximations but will mention their evident properties.

The arbitrariness of S makes it possible to choose S so that coefficients of all operators of $\partial/\partial\varphi$ and $\partial/\partial\bar{\varphi}$ are conjugate to each other. Each time these coefficients are linear combinations of eigenfunctions of the form $(\varphi\bar{\varphi})^m\varphi$ and $(\varphi\bar{\varphi})^m\bar{\varphi}$, where $\varphi\bar{\varphi}$ is an invariant of X_0. These polynomials in $\varphi\bar{\varphi}$ have real coefficients in M_{2k+1} and S_{2k} and imaginary coefficients in M_{2k} and S_{2k+1}. Coefficients of M are of the form $\Lambda(\varepsilon, \varphi\bar{\varphi})\varphi$, $\bar{\Lambda}(\varepsilon, \varphi\bar{\varphi})\bar{\varphi}$, where

$$i\Lambda(\varepsilon, \varphi\bar{\varphi}) = \sum_{k\geq 0}(i\varepsilon)^k P_k(\varphi\bar{\varphi})$$

and $P_k(\varphi\bar{\varphi})$ are polynomials. In particular, since $\bar{\varphi}M_2\varphi + \varphi M_2\bar{\varphi} = 0$, the energy $E = \frac{1}{2}\varphi\bar{\varphi} = \frac{1}{2}(u^2 + v^2)$ is defined by the equation $dE/dt = \varepsilon E(1 - \frac{1}{2}E)$ which follows from the system of the first order approximation,

$$\frac{d\varphi}{dt} = \left(i + \varepsilon\varphi\left(\frac{1}{2} - \frac{\varphi\bar{\varphi}}{8}\right)\right), \qquad \frac{d\bar{\varphi}}{dt} = \left(-i + \varepsilon\varphi\left(\frac{1}{2} - \frac{\varphi\bar{\varphi}}{8}\right)\right)\bar{\varphi}, \tag{3.4.3}$$

and the energy equation holds with accuracy $0(\varepsilon^3)$, but not $0(\varepsilon^2)$. The formula

$$E(t) = \frac{2E_0e^{\varepsilon t}}{2 - E_0 + E_0e^{\varepsilon t}},$$

where $E_0 = E(0)$, which is derived from (3.4.3), holds with accuracy $0(\varepsilon)$ for $t \lesssim 1/\varepsilon^2$. Inserting $\varphi\bar{\varphi} = 2E(t)$ in the right-hand sides of (3.4.3) we get the principal term of the asymptotics for $t \gg 1/\varepsilon$ (but not, strictly speaking, for $t \ll 1/\varepsilon^2$)

$$\varphi = \sqrt{2/E_0}\varphi_0 e^{it}, \qquad \bar{\varphi} = \sqrt{2/E_0}\bar{\varphi}_0 e^{-it}, \tag{3.4.4}$$

where $\varphi_0 = \varphi(0)$, $\bar{\varphi}_0 = \bar{\varphi}_0(0)$, $\varphi_0\bar{\varphi}_0 = 2E$ since

$$\exp \varepsilon \int_0^t \left(\frac{1}{2} - \frac{1}{8}\varphi\bar{\varphi}\right) dt = \exp \varepsilon \int_0^t \left(\frac{1}{2} - \frac{1}{8}E(t)\right) dt \approx \sqrt{2/E_0}$$

for $t \gg 1/\varepsilon$. Thus after the time $t \gg 1/\varepsilon$ the amplitude of oscillations becomes ≈ 2. [It is known that for any fixed ε there is a limit cycle. The trajectory defined by (3.4.4) is the limit position of the cycles as $\varepsilon \to 0$; i.e., the limit of limit cycles.]

Sometimes studies of M and S similar to those carried out in the above problem make it possible to present a rather detailed structure of the desired solution and seek it in a corresponding form.

3.5 Example: Duffing Oscillator

$$\frac{d^2}{dt^2} + u + \varepsilon u^3 = 0. \tag{3.5.1}$$

As in Example 3.4, let us pass to the system

$$\frac{du}{dt} = v,$$

$$\frac{dv}{dt} = -u - \varepsilon u^3,$$

$$X = X_0 + \varepsilon X_1 = v\frac{\partial}{\partial u} - u\frac{\partial}{\partial v} - \varepsilon u^3 \frac{\partial}{\partial v}.$$

In the basic system of eigenfunctions $\varphi = v + iu$, $\bar{\varphi} = v - iu$ of X_0 we have

$$X_0 = i\left(\varphi\frac{\partial}{\partial\varphi} - \bar{\varphi}\frac{\partial}{\partial\bar{\varphi}}\right),$$

$$X_1 = \frac{i}{8}(-\varphi^3 + 3(\varphi\bar{\varphi})\bar{\varphi} - 3(\varphi\bar{\varphi})\varphi + \bar{\varphi}^3)\left(\frac{\partial}{\partial\varphi} + \frac{\partial}{\partial\bar{\varphi}}\right).$$

The structure of the operators M and S is almost the same as in Example 3.4. The difference is that polynomials contained in M have purely imaginary coefficients and those contained in S have real ones (in all approximations). Since we have in this basic system that

$$M = i\omega(\varepsilon, \varphi\bar{\varphi})\left(\varphi\frac{\partial}{\partial\varphi} - \bar{\varphi}\frac{\partial}{\partial\bar{\varphi}}\right),$$

where $\omega(\varepsilon, \varphi\bar{\varphi})$ is representable by a series in ε, then in new variables the asymptotics are of the form $u^* = A\cos\omega^* t + B\sin\omega^* t$, where $\omega^* = 1 + \varepsilon\omega_1^* + \varepsilon^2\omega_2^* + \cdots$ and A, B, and ω^* only depend on initial conditions ($\omega^* = \text{const}$ since in any approximation we have an integral $\varphi\bar{\varphi} = \text{const}$).

Furthermore, it is not difficult to deduce, considering the form of X_1 and formulas (1.2.3′), that coefficients of S_k, where $k = 1, 2, \ldots$, are homogeneous polynomials in φ and $\bar{\varphi}$ of degree $2k + 1$. Since the solution of the system corresponding to M is $\varphi = \varphi_0 e^{i\omega t}$ and $\bar{\varphi} = \bar{\varphi}_0 e^{-i\omega t}$, then the operator giving the old variables, e^{-S}, yields the formal Fourier series in frequencies $(2k + 1)\omega$. Therefore we conclude that the asymptotics are representable in the form

$$u = a\cos(\omega t + \psi) + \sum_{k \geq 1} c_k \cos[(2k + 1)(\omega t + \psi)], \qquad (3.5.2)$$

where

$$\omega = 1 + \sum_{m \geq 1} \varepsilon^m \omega_m, \qquad c_k = 0(\varepsilon^k),$$

and a and ψ are arbitrary constants (ω and c_k depend on a). Inserting (3.5.2) into (3.5.1) it is easy to compute, for instance, that

$$u = a\cos(\omega t + \psi) + \frac{\varepsilon a^3}{2^5}\left(1 - \varepsilon\frac{21a^2}{2^5} + \varepsilon^2\frac{417a^4}{2^{10}}\right)\cos 3(\omega t + \psi)$$
$$+ \frac{\varepsilon^2 a^5}{2^{10}}\left(1 - \varepsilon\frac{43a^2}{2^5}\right)\cos 5(\omega t + \psi) + \frac{\varepsilon^3 a^7}{2^{15}}\cos 7(\omega t + \psi) + 0(\varepsilon^4),$$
$$\omega = 1 + \varepsilon\frac{3a^2}{2^3} - \varepsilon^2\frac{15a^4}{2^8} + \varepsilon^3\frac{123a^6}{2^{13}} + 0(\varepsilon^4).$$

Let us give an example of a comparatively cumbersome problem where in the solution of the perturbation theory problem eigenoperators will be used instead of eigenfunctions.

3.6 Example: Drift of a Charged Particle in an Electromagnetic Field

We will discuss very briefly the physical meaning of the problem. Details may be found, e.g., in Ref. [37].

The equations of motion of a charged particle in an electromagnetic field are of the form

$$\frac{d(mv)}{dt} = eE + \frac{e}{c}[V \times B], \qquad \frac{dr}{dt} = v. \qquad (3.6.1)$$

Here r, v, $B = B(r, t)$, and $E = E(r, t)$ are vectors that define the position of the particle, its velocity, and magnetic and electric fields, respectively;

e is the charge of the particle and m is its mass; $m = m_0(1 - v^2/c^2)^{-1/2}$, where m_0 is the rest mass and c is the speed of light.

In the simplest case $E = 0$ and $B = \text{const}$ ($m = \text{constant}$) the system (3.6.1) is linear and is easily integrated. Evident integrals $v^2 = \text{const}$, $v_\parallel \equiv (1/|B|)(B, V) = \text{const}$, make it clear that the particle moves along a spiral line with constant speed $V_\parallel$ along force lines of the magnetic field and with constant angular speed of rotation around a fixed force line,

$$|\omega| = \frac{|e||B|}{mc} \tag{3.6.2}$$

(ω is referred to as the *cyclotron frequency*). The radius of rotation (the so-called *Larmor radius*) accordingly equals $V_\perp/|\omega|$, where $V^2 = v_\parallel^2 + v_\perp^2 = \text{const}$. In an arbitrary field the notion of a *local* cyclotron frequency ω and a *local* Larmor radius $V_\perp/|\omega|$ for $|B| \neq 0$ is introduced. If the magnetic field is "strong" (they say that the Larmor radius is small as compared with the characteristic value of nonhomogeneity of the field), then the motion of the particle may be considered as a swift rotation along a Larmor circle of a small (variable) radius which slowly drifts in space. The so-called drift equations describe the motion of the circle; more exactly, they describe the variation of r, $v_\perp$ and $v_\parallel$ averaged with respect to the "period" of rotation.

The idea of drift description of motion belongs to H. Alfwen [1]. The complete derivation of drift equations by the Bogolyubov–Krylov method was given by Bogolyubov and Mitropolski [10] and Braginski [11]. A detailed presentation is in [37].

Let us rewrite (3.6.1) in a more convenient form. We introduce the unit vector tangent to the force line of the magnetic field and the momentum vector instead of v:

$$\tau = \frac{B}{|B|}, \qquad p = \frac{c}{e} mv. \tag{3.6.3}$$

Due to (3.6.2) and (3.6.3), Equations (3.6.1) can be rewritten as

$$\frac{1}{\omega}\frac{dp}{dt} = \frac{e}{\omega}E + [\tau \times p], \qquad \frac{1}{\omega}\frac{dr}{dt} = -\frac{1}{|B|}p, \qquad \text{where } \omega = \frac{e|B|}{mc}.$$

The small parameter ε may be introduced by the formal transformation $B \mapsto \frac{1}{\varepsilon}B$ ($\omega \mapsto \frac{1}{\varepsilon}\omega$). We get

$$\varepsilon\frac{dp}{d\tilde{t}} = [\tau \times p] + \varepsilon\frac{e}{\omega}E, \qquad \varepsilon\frac{dr}{d\tilde{t}} = -\frac{\varepsilon}{|B|}p, \qquad \varepsilon\frac{dt}{d\tilde{t}} = \frac{\varepsilon}{\omega}, \tag{3.6.4}$$

where the last equation is added in order to consider, as usual, an autonomous system. To the system (3.6.4) the operator $X = X_0 + \varepsilon X_1$ corresponds, where

$$X_0 = ([\tau \times p], \nabla_p), \qquad X_1 = \frac{c}{\omega}(E, \nabla_p) - \frac{1}{|B|}(p, \nabla) + \frac{1}{\omega}\frac{\partial}{\partial t} \tag{3.6.5}$$

[see (2.1.2); ∇_p is the gradient with respect to momenta, ∇ is the gradient with respect to spatial variables]. The leading operator X_0 possesses the full set of six invariants[1] (the total number of variables is seven):

$$t,\ r,\ p^2,\ p_{\|} = (\tau, p) \tag{3.6.6}$$

(τ, E, $|B|$, and ω are also invariants of X_0). Below we will present a method of constructing the change of variables operator S such that $M = e^{-S} X e^{S}$ commutes with X_0; i.e.,

$$M' = 0. \tag{3.6.7}$$

Meanwhile, note that if g is any of the invariants in (3.6.6), then by (3.6.7) Mg is an invariant of X_0; i.e., Mg is a function in the variables (3.6.6). Therefore in new variables $dg/dt = Mg/Mt = f(t, r, p^2, p_{\|})$.

Thus in new variables we obtain the closed system of equations:

$$\frac{dr}{dt} = \frac{Mr}{Mt}, \qquad \frac{d(p^2)}{dt} = \frac{M(p^2)}{Mt}, \qquad \frac{dp_{\|}}{dt} = \frac{Mp_{\|}}{Mt}. \tag{3.6.8}$$

The system (3.6.8) is the desired *system of drift equations.*

Remark 3.6.1 Since the system of (five) equations (3.6.8) is written in new variables, then p and r no longer define the momentum and position of the particle. The variables r, p^2, and $p_{\|}$ define the averaged position of the particle, averaged energy and averaged parallel momentum. Considering (3.6.8), we "forget" about the change of variables, putting the particle into the "mean" position. Now the speed equals Mr/Mt (it is the drift speed), and p^2 and $p_{\|}$ play the role of parameters.

Computing a functional of real trajectory of the particle (e.g., of the flow of particles) it is impossible to use (3.6.8) only, ignoring the change of variables, though it is close to the identity. We should consider the full set of (six) equations corresponding to M and take the parameter transformation into account.

For example, the known (see [29]) effect that the flow of particles does not coincide with the flow of "leading centers," i.e., the flow of particles put onto the drift trajectories, is formally the result of identification of the old and new variables. Details are in [5]

The Construction of S and M

Let us find the commutation relation for X_1. Making use of the formulas

$$\begin{aligned} \tau^2 &= 1, \\ X_0 p &= [\tau \times p], \\ X_0^2 p &= [\tau \times X_0 p] = p_{11}\tau - p, \\ X_0^3 p &= -X_0 p, \end{aligned} \tag{3.6.9}$$

---

[1] Each component of the vector r is an invariant.

and ordinary rules of vector analysis to compute derivatives of X_1 with respect to X_0 [see also (2.1.9)], we get

$$
\begin{aligned}
X_1' &= \left(-\frac{c}{\omega}[\tau \times E] + \frac{1}{|B|}[(p,\nabla)\tau \times p] - \frac{1}{\omega}\left(\frac{\partial \tau}{\partial t} \times p\right), \nabla_p\right) - \frac{1}{|B|}(X_0 p, \nabla),\\
X_1'' &= \left(\frac{c}{\omega}[\tau \times [\tau \times E]] + \frac{2}{|B|}[(X_0 p,\nabla)\tau \times p] + \frac{1}{|B|}[[(p,\nabla)\tau \times \tau] \times p]\right.\\
&\quad \left. - \frac{1}{\omega}\left(\left(\frac{\partial \tau}{\partial t} \times \tau\right) \times p\right), \nabla p\right)\\
&\quad - \frac{1}{|B|}(X_0 p^2, \nabla p),\\
X_1''' &= -X_1' + \frac{3}{|B|}([(X_0^2 p,\nabla)\tau \times p] + [[(X_0 p,\nabla)\tau \times \tau] \times p], \nabla p)\\
X_1^{IV} &= -X_1'' + \frac{6}{|B|}(-[(X_0 p,\nabla)\tau \times p] + [[(X_0^2 p,\nabla)\tau \times \tau] \times p], \nabla p),\\
X_1^{V} &= -X_1''' - \frac{12}{|B|}([(X_0^2 p,\nabla)\tau \times p] + [[(X_0 p,\nabla)\tau \times \tau] \times p], \nabla p).
\end{aligned}
\tag{3.6.10}
$$

From here we deduce the commutation relation

$$4X_1' + 5X_1''' + X_1^V = 0. \tag{3.6.11}$$

Making use of (3.6.11) we get eigenoperators H_λ defined by the condition $H_\lambda' = \lambda H_\lambda$. We get the following formulas to be used especially in the *second* order approximation [the answer for M_1 is evident from (3.6.11)]:

$$
\begin{aligned}
H_0 &= 4X_1 + 5X_1'' + X_1^{IV},\\
H_{\pm i} &= 4X_1'' + X_1^{IV} \pm i(4X_1' + X_1'''),\\
H_{\pm 2i} &= X_1'' + X_1^{IV} \pm 2i(X_1' + X_1''').
\end{aligned}
$$

The development of X_1 with respect to eigenoperators is

$$X_1 = \frac{1}{4}H_0 - \frac{1}{6}(H_i + H_{-i}) + \frac{1}{24}(H_{2i} + H_{-2i}).$$

We have

$$M_1 = \frac{1}{4}H_0, \qquad S_1 = -\frac{i}{6}(H_i - H_{-i}) + \frac{i}{48}(H_{2i} - H_{-2i}) + \bar{S}_1, \tag{3.6.12}$$

where $\bar{S}_1$ is any operator which commutes with X_0. We will see later that it is convenient to take this arbitrariness into account. Making use of (3.6.12), the fact that $[H_\lambda, H_\mu]$ corresponds to $\lambda + \mu$, and using the Jacobi identity, we find

$$M_2 = -\frac{i}{36}[H_i, H_{-i}] - \frac{i}{1152}[H_{2i}, H_{-2i}] + [M_1, \bar{S}_1]$$

(we do not write out S_2), etc. Let us restrict ourselves to the approximation $M \approx X_0 + \varepsilon M_1 + \varepsilon^2 M_2$. Substitution for H_0, $H_{\pm i}$, and $H_{\pm 2i}$ gives

$$\begin{aligned} M_1 &= X_1 + \frac{5}{4} X_1^{II} + \frac{1}{4} X_1^{IV}, \\ M_2 &= \frac{1}{18}[4X_1^I + X_1^{II}, 4X_1^{II} + X_1^{IV}] \\ &\quad + \frac{1}{288}[X_1^I + X_1^{III}, X_1^{II} + X_1^{IV}] + [M_1, \bar{S}_1]. \end{aligned} \tag{3.6.13}$$

To obtain drift equations (3.6.8) with the assumed accuracy it remains to choose $\bar{S}_1$ and to compute M on the invariant in (3.6.6). We will omit these computations and only sketch how we choose $\bar{S}_1$. Clearly, $\bar{S}_1$ should be chosen so that M_2 would be "simplest." Here we are speaking about iterating the perturbation theory procedure with the new leading operator M_1 [see (1.1.3)]. In order to perform this we have to know M_1 effectively (e.g., to construct a basic system of its eigenfunctions and invariants). For arbitrary B and E it is impossible to study M_1 effectively and it remains to take $\bar{S}_1$ so that (3.6.8) are of the most convenient form for physical interpretation. (From calculations' point of view the choice of $\bar{S}_1$ is of no consequence since any two ε-near drift trajectories are equally "good.") If we set

$$\bar{S}_1 t = \bar{S}_1 p^2 = 0, \qquad \bar{S}_1 r = 0, \qquad \bar{S}_1 p_\| = \frac{p_\perp^2}{2|B|}(\tau, rot\tau), \tag{3.6.14}$$

where $p^2 = p_\|^2 + p_\perp^2$, then such a choice of $\bar{S}_1$ would correspond to the Braginski corrections [11]. Computations of M (rather prolonged) on functions (3.6.6) give, with (3.6.14) taken into account:

$$\begin{aligned} Mt &= \frac{\varepsilon}{\omega}, \\ M_2 &= -\frac{\varepsilon p_\|}{|B|}\tau \\ &\quad + \varepsilon^2 \left\{ \frac{c}{\omega|B|}[E \times \tau] - \frac{p_\|^2}{|B|^2}[\tau \times (\tau, \nabla)\tau] - \frac{p_\perp}{2|B|^3}[\tau \times \nabla|B|] \right\}, \\ M(p^2) &= -\frac{2c|B|}{\omega}(E, Mr) + \frac{p_\perp^2 \varepsilon}{\omega|B|}\frac{\partial|B|}{\partial t} \\ Mp_\| &= \varepsilon \left\{ \frac{c}{\omega}(\tau, E) + \frac{p_\perp^2}{2|B|^2}(\tau, \nabla|B|) \right\} \\ &\quad + \varepsilon^2 \left(\frac{cp_\|}{\omega|B|}E + \frac{p_\| p_\perp^2}{2|B|^3}\nabla|B|, [\tau \times (\tau, \nabla)\tau] \right). \end{aligned} \tag{3.6.15}$$

In the computation of (3.6.15) the Maxwell equations

$$\operatorname{div} B = 0, \qquad \operatorname{curl} E = -\frac{1}{c}\frac{1}{\varepsilon}\frac{\partial B}{\partial t}$$

are used (they imply, in particular, that $\partial B/\partial t \sim \varepsilon$, making it possible to ignore some terms in M_2).

Equations (3.6.15), after simple transformations, may be written in the form (here the parameter ε introduced artificially is set to be equal to 1):

$$\begin{aligned}
\frac{dr}{dt} &= \frac{B}{|B|}V_{\parallel} + \frac{c}{B^2}[E \times B] \\
&\quad + \frac{mcV_{\parallel}^2}{e|B|^4}[B \times (B\nabla)B] + 2\frac{mcV_{\perp}^2}{e|B|^3}[B \times \nabla|B|], \\
\frac{dE}{dt} &= e\left(E, \frac{dr}{dt}\right) + \frac{mcV_{\perp}^2}{2|B|}\frac{\partial|B|}{\partial B}, \\
&\qquad \frac{d}{dt}\left(\frac{mV_{\perp}^2}{|B|}\right) = 0
\end{aligned} \tag{3.6.16}$$

where $\varepsilon = mc^2$, and $V^2 = V_{\parallel}^2 + V_{\perp}^2$ is the mean kinetic energy.

Summands of the drift velocity dr/dt are usually interpreted as follows: the first is the *longitudinal motion*, the second is the *electric drift*, the third is the *centrifugal drift*, and the fourth is the "*gradient*" *drift*. The dimension of the system (3.6.16) is actually equal to 4, since $m^2V_{\perp}^2/|B| = \text{const}$. This constant is called the *transverse adiabatic invariant*.

The following very simple example will be given to illustrate the idea of an extended operator. Simultaneously we note one important circumstance not mentioned yet: the possible loss of accuracy caused by variability of eigenvalues.

3.7 Example: Nonlinear System: Example of an Extension of an Operator

Consider the system

$$\varepsilon\frac{dx}{dt} = -(x^2 + y^2)y - \frac{2}{3}\varepsilon x^3, \qquad \varepsilon\frac{dy}{dt} = (x^2 + y^2)x - \frac{2}{3}\varepsilon y^3.$$

The corresponding operator is

$$\begin{aligned}
X &= X_0 + \varepsilon X_1 \\
&= -(x^2 + y^2)\left(y\frac{\partial}{\partial x} - x\frac{\partial}{\partial y}\right) - \frac{2}{3}\varepsilon\left(x^3\frac{\partial}{\partial x} + y^3\frac{\partial}{\partial y}\right).
\end{aligned}$$

Putting $\varphi = x + iy$ and passing to φ and $\bar{\varphi}$ we get

$$\begin{aligned}
X_0 &= i\lambda\left(\varphi\frac{\partial}{\partial\varphi} - \bar{\varphi}\frac{\partial}{\partial\varphi}\right), \\
X_1 &= -\left(\frac{\bar{\varphi}^3}{6} + \frac{\lambda\varphi}{2}\right)\frac{\partial}{\partial\varphi} - \left(\frac{\varphi^3}{6} + \frac{\lambda\bar{\varphi}}{2}\right)\frac{\partial}{\partial\bar{\varphi}},
\end{aligned}$$

where $\lambda = \varphi\bar{\varphi}$ is an invariant of X_0. Variables φ and $\bar{\varphi}$ constitute the basic system of eigenfunctions which belong to eigenvalues $\pm i\lambda$ depending on φ and $\bar{\varphi}$. Let us introduce a supplementary variable λ:

$$\varepsilon\frac{d\lambda}{dt} = \varepsilon\frac{d(\varphi\bar{\varphi})}{dt} = X(\varphi\bar{\varphi}) = -\varepsilon\left(\lambda^2 + \frac{\varphi^4}{6} + \frac{\bar{\varphi}^4}{6}\right).$$

Extended operators are

$$X_0^* = i\lambda\left(\varphi\frac{\partial}{\partial\varphi} - \bar{\varphi}\frac{\partial}{\partial\bar{\varphi}}\right),$$

$$X_1^* = -\left(\frac{\bar{\varphi}^3}{6} + \frac{\lambda\varphi}{2}\right)\frac{\partial}{\partial\varphi} - \left(\frac{\varphi^3}{6} + \frac{\lambda\bar{\varphi}}{2}\right)\frac{\partial}{\partial\bar{\varphi}} - \left(\lambda^2 + \frac{\varphi^4}{6} + \frac{\bar{\varphi}^4}{6}\right)\frac{\partial}{\partial\lambda},$$

where λ is considered as independent.

Remark 3.7.1 Coefficients of X_1^* as functions in λ, φ, $\bar{\varphi}$ are not defined unambiguously; e.g., we could write

$$X_1^*\lambda = -\left(\varphi^2\bar{\varphi}^2 + \frac{\varphi^4}{6} + \frac{\bar{\varphi}^4}{6}\right)$$

instead of

$$X_1^*\lambda = -\left(\lambda^2 + \frac{\varphi^4}{6} + \frac{\bar{\varphi}^4}{6}\right).$$

In the sequel only the possibility of expanding $X_1^*\varphi$, $X_1^*\bar{\varphi}$, and $X_1^*\lambda$ in eigenfunctions of X_0 is important. Further, we restrict ourselves, for simplicity's sake, to computation of the main term of the asymptotics. It is easy to verify that we may put

$$M^* = -\frac{\lambda}{2}\left(\varphi\frac{\partial}{\partial\varphi} + \bar{\varphi}\frac{\partial}{\partial\bar{\varphi}}\right) - \lambda^2\frac{\partial}{\partial\lambda},$$

$$S^* = \frac{i}{24\lambda}\left\{\left(\bar{\varphi}^3 - \frac{\varphi\bar{\varphi}^4}{4\lambda} - \frac{\varphi^5}{4\lambda}\right)\frac{\partial}{\partial\varphi} + \left(-\varphi^3 + \frac{\bar{\varphi}\varphi^4}{4\lambda} + \frac{\bar{\varphi}^5}{4\lambda}\right)\frac{\partial}{\partial\bar{\varphi}} + (\bar{\varphi}^4 - \varphi^4)\frac{\partial}{\partial\lambda}\right\}.$$

These formulas are obtained by the standard computation described in Theorem 2.4.3.

We may note that M_1^* and S_1^* are extended operators in the same sense that X_0^* and X_1^* are. In fact,

$$M_1^*(\lambda - \varphi\bar{\varphi})|_{\lambda=\varphi\bar{\varphi}} = 0, \qquad S_1^*(\lambda - \varphi\bar{\varphi})|_{\lambda=\varphi\bar{\varphi}} = 0.$$

Operators M_1 and S_1 whose extensions are M_1^* and S_1^* are obtained from M_1^* and S_1^* by ignoring terms with $\partial/\partial\lambda$ and substituting $\lambda = \varphi\bar{\varphi}$:

$$M_1 = -\frac{\lambda}{2}\left(\varphi\frac{\partial}{\partial\varphi} + \bar{\varphi}\frac{\partial}{\partial\bar{\varphi}}\right),$$

$$S_1 = \frac{i}{24\lambda}\left(\left(\frac{3}{4}\bar{\varphi}^3 - \frac{\varphi^5}{4\lambda}\right)\frac{\partial}{\partial\varphi} + \left(-\frac{3}{4}\varphi^3 + \frac{\bar{\varphi}^5}{4\lambda}\right)\frac{\partial}{\partial\bar{\varphi}}\right).$$

Therefore in this approximation and similarly in the following ones we may return to the unextended system. This is not necessary since we may (as had been already said in Section 2.3) match initial values for the extended system and not worry about the conditions $M(\lambda - \varphi\bar{\varphi})|_{\lambda=\varphi\bar{\varphi}} = 0$, $S(\lambda - \varphi\bar{\varphi})|_{\lambda=\varphi\bar{\varphi}} = 0$.

By the way, note that during the computation of S_1^*, M_1^* we have ignored certain summands which can be arbitrary (see the proof of Theorem 2.4.3). It is clear that to S_1^* we could have added

$$\omega_1\frac{\partial}{\partial\lambda} + \omega_2\varphi\frac{\partial}{\partial\varphi} + \omega_3\bar{\varphi}\frac{\partial}{\partial\bar{\varphi}},$$

where ω_1, ω_2, ω_3 are arbitrary invariants of X_0, which adds to M_1^* the summand

$$-i\omega_1\left(\varphi\frac{\partial}{\partial\varphi} - \bar{\varphi}\frac{\partial}{\partial\bar{\varphi}}\right).$$

The operator M_1^* will be, as earlier, an extended one, but S_1^* will be an extended one only if $\omega_1 - \varphi\bar{\varphi}(\omega_2 + \omega_3) = 0$ for $\lambda = \varphi\bar{\varphi}$.

Note another fact (not connected with the extension of the system). As was mentioned in Chapter 2, for variable eigenvalues we have $M_1^{II} = 0$ (in the case of diagonal X_0) but not $M_1^I = 0$, and M_1 is not uniquely defined. If, as in this case, we are interested only in the main term of the asymptotics, then the change of variables may be ignored and the nonuniqueness of M_1 shows that the first approximation is insufficient to find the main term of the asymptotics. The reason is clear: since $\varphi = \exp(i/\varepsilon)\int\lambda\,dt$, then error in computing λ leads to loss of accuracy in computing φ. Thus *equations for eigenvalues should be computed with higher accuracy*. In the preceding examples we always had $S\lambda = 0$ and equations for λ were exact; hence there was no error.

Since $M_2^* = [X_0^*, S_2^*] + \frac{1}{2}[M_1^* + X_1^*, S_1^*]$, then

$$M_2^*\lambda = [X_0^*, S_2^*]\lambda + \frac{i}{24}(\bar{\varphi}^4 - \varphi^4) + \frac{i}{144\lambda^2}(\bar{\varphi}^8 - \varphi^8).$$

Therefore it is possible to choose S_2^* so that $M_2^*\lambda = 0$. In the principal order the equation for λ is $d\lambda/dt = (M_1^* + \varepsilon M_2^*)\lambda = -\lambda^2$. Hence $\lambda = \lambda_0/(1+\lambda_0 t)$, $\lambda_0 = \lambda|_{t=0} = \varphi_0\bar{\varphi}_0$. Equations for φ and $\bar{\varphi}$ are

$$\varepsilon\frac{d\varphi}{dt} = \left(i\lambda - \frac{\lambda}{2}\varepsilon\right)\varphi, \qquad \varepsilon\frac{d\bar{\varphi}}{dt} = \left(-i\lambda - \frac{\lambda}{2}\varepsilon\right)\bar{\varphi}.$$

Hence

$$\varphi = \frac{\varphi_0}{\sqrt{1+\lambda_0 t}} e^{i\ln(1+\lambda_0 t)/\varepsilon}, \qquad \bar{\varphi} = \frac{\bar{\varphi}_0}{\sqrt{1+\lambda_0 t}} e^{-i\ln(1+\lambda_0 t)/\varepsilon},$$

where $\lambda_0 = \varphi_0\bar{\varphi}_0$.

Remark 3.7.2 Since the right-hand sides of the initial system are homogeneous in x and y, the fact that $\lambda \to 0$ as $t \to \infty$ does not mean that the obtained asymptotics are of restricted usage. It is valid for all t.

The following simple example shows once more that the knowledge of expansions in series of eigenfunctions is not necessary in order to calculate M and S: the purpose of the choice of M consists of the possibility of solving the equation $[X_0, S] + X = M$ regularly in D.

3.8 Example: Nonlinear Oscillator with Small Mass and Damping

This is described by the equation

$$\varepsilon\frac{d^2u}{dt^2} + \frac{du}{dt} + \sin u = 0.$$

The corresponding system is

$$\varepsilon\frac{du}{dt} = v, \qquad \varepsilon\frac{dv}{dt} = -v - \varepsilon\sin u,$$

and the operator is

$$X = X_0 + \varepsilon X_1 = v\frac{\partial}{\partial u} - v\frac{\partial}{\partial v}\varepsilon\sin u\frac{\partial}{\partial v}.$$

The invariant of X_0 is $g = u + v$ and the eigenfunction is v. In variables g and v we have

$$X_0 = -v\frac{\partial}{\partial v}, \qquad X_1 = -\sin(g-v)\left(\frac{\partial}{\partial g} + \frac{\partial}{\partial v}\right).$$

For $M_1 g$ and $M_1 v$ we have

$$M_1 g = -v\frac{\partial(S_1 g)}{\partial v} - \sin(g-v),$$
$$M_1 v = -v\frac{\partial(S_1 v)}{\partial v} + S_1 v - \sin(g-v). \tag{3.8.1}$$

The function $\sin(g - v)$ may be expanded into series in powers of v, i.e., with respect to eigenfunctions v^n corresponding to eigenvalues $-n$, and it is possible to search for $S_1 g$ and $S_1 v$ in the same form. Since the free term and the term corresponding to -1 in the development of $\sin(g-v)$ are equal to $-\sin g$ and $v\cos g$ respectively, we have $M_1 g = -\sin g$ and $M_1 v = v\cos g$. Note that Equations (3.8.1) rewritten in the form

$$-v\frac{\partial r}{\partial v} = \sin(g-v) + \psi_1(g), \qquad M_1 g = \psi_1(g), \tag{3.8.1'}$$

$$-v\frac{\partial R}{\partial v} + R = \sin(g-v) + v\psi_2(g), \qquad M_1 v = v\psi_2(g), \tag{3.8.1''}$$

are solvable for "any" functions ψ_i in the domain $|v| > \alpha > 0$. But a neighborhood of 0 is important in this problem since trajectories of X_0 quickly reach 0. Therefore it is necessary to choose $\psi_1(g)$ and $\psi_2(g)$ so that the system (3.8.1′) has a regular solution in the domain $|v| < \alpha$. Clearly, it is regular if $[\sin(g-v) + \psi_1(g)]_{v=0} = 0$, and if

$$\left[\frac{\partial}{\partial v}(\sin(g-v) + v\psi_2(g))\right]_{v=0} = 0,$$

implying $M_1 g = -\sin g$, $M_1 v = v\cos g$.

It is worth mentioning that here we encounter the "averaging principle" in a different situation;

$$S_1 g = \int_0^v \frac{\sin g - \sin(g-\zeta)}{\zeta}\,d\zeta,$$

$$S_1 v = \sin g + v\int_0^v \frac{\sin g - \sin(g-\zeta) - \zeta\cos g}{\zeta^2}\,d\zeta.$$

If we restrict ourselves to the principal term of the asymptotics, then we may ignore S_1, and the system of equations corresponding to $X_0 + \varepsilon M_1$ is

$$\varepsilon\frac{dg}{dt} = -\varepsilon\sin g, \qquad \varepsilon\frac{dv}{dt} = -v + \varepsilon v\cos g.$$

Hence

$$tg\frac{g}{2} = c_1 e^{-t}, \qquad v = c_2 e^{-t}(c_1^2 + e^{2t})e^{-t/\varepsilon},$$

where c_1 and c_2 are constants. As in Example 3.3, "main events" take place in the narrow boundary layer $0 \le t \le t_0 \sim \varepsilon$. Later, v becomes exponentially small, $g \sim u$, and u exponentially decreases; damping quenches oscillation.

Until now we have considered problems with a linear leading part (in the initial formulation) and the question of how to introduce eigenfunctions had a trivial answer. The following example may be considered as essentially nonlinear.

3.9 Example: A Nonlinear Equation; Boundary-Layer-Type Problem

The equation is

$$\varepsilon \frac{d^2v}{dt^2} + v\frac{dv}{dt} - v = 0. \tag{3.9.1}$$

The corresponding system is

$$\varepsilon \frac{dv}{dt} = u, \qquad \varepsilon \frac{du}{dt} = -vu + \varepsilon v,$$

and the operator is

$$X = X_0 + \varepsilon X_1 = u\frac{\partial}{\partial v} - vu\frac{\partial}{\partial u} + \varepsilon v \frac{\partial}{\partial u}.$$

Here it is convenient to pass first to variables v and g, where $g = v^2 + 2u$ is an invariant of X_0;

$$X_0 = \frac{1}{2}(g - v^2)\frac{\partial}{\partial v}, \qquad X_1 = 2v\frac{\partial}{\partial g}. \tag{3.9.2}$$

We may restrict ourselves to the domain $g > 0$, $v + \sqrt{g} > 0$, where $g \sim 1$, $v + \sqrt{g} \sim 1$. (In fact, the asymptotics obtained will also serve for $g \gg \varepsilon$, $v + \sqrt{g} \gg \varepsilon$.) In this domain the phase trajectory of the point $P(g, v)$ lies near the curve consisting of a segment of the straight line $g =$ const and a branch of the parabola $\sqrt{g} = v$: starting from the point P_0 (see Fig. 3) the point P soon reaches a small neighborhood of the branch $\sqrt{g} = v$ and then with speed ~ 1 moves upward in the vicinity of this branch (this follows immediately from the equations of motion

$$\frac{dv}{dt} = \frac{1}{2\varepsilon}(g - v^2), \qquad \frac{dg}{dt} = 2v.$$

Domains $g < 0$ or $v < -\sqrt{g}$ are not interesting since here P will quickly go to infinity.

Remark 3.9.1 We may also be interested in the case when at the initial moment P is ε-near the boundary of the above domains which consists of the branch $v = -\sqrt{g}$ and the ray $v > 0$, $g = 0$. It is clear that the situation near the branch $v = -\sqrt{g}$ is unstable. If at the initial moment we have $v \gtrsim 1$, $g \sim \varepsilon$, then after a small ($\sim \sqrt{\varepsilon}$) period of time the point P reaches a position when $v \sim \sqrt{\varepsilon}$ and $g \sim \varepsilon$; i.e., $v \sim \sqrt{\varepsilon}$ and $\dot{v} \sim 1$. Substituting $v = \sqrt{\varepsilon}\tilde{v}$, $t = \sqrt{\varepsilon}\tilde{t}$ in (3.9.1), we get

$$\ddot{\tilde{v}} + \tilde{v}\dot{\tilde{v}} - \tilde{v} = 0$$

which does not contain a small parameter, and in this case the problem is reduced to an exact integration.

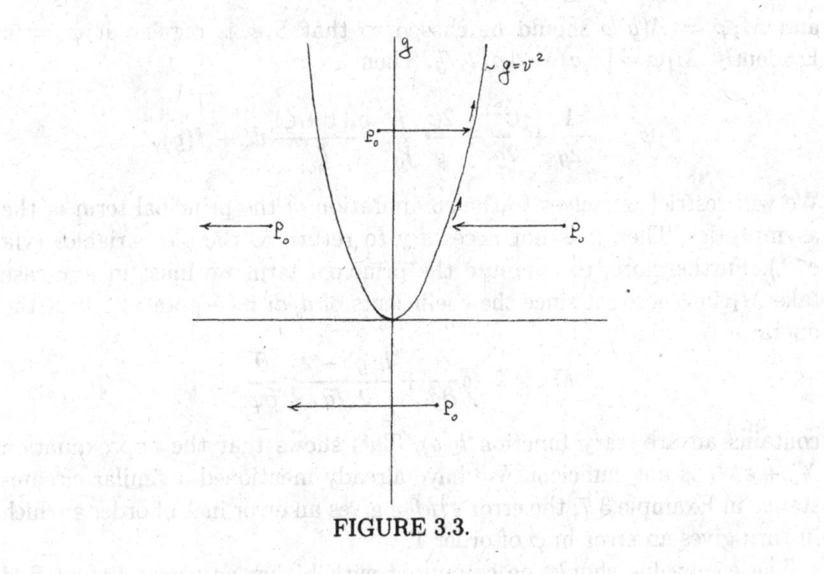

FIGURE 3.3.

Let us seek an eigenfunction of X_0 from the equation

$$\frac{1}{2}(g - v^2)\frac{\partial \varphi}{\partial v} = \lambda(g)\varphi.$$

We have

$$\varphi = [(\sqrt{g} - v)(\sqrt{g} + v)]^{-\lambda(g)/\sqrt{g}}.$$

In order to have a regular transformation at $v = \sqrt{g}$ we should get

$$\lambda(g) = -\sqrt{g}, \qquad \varphi = \frac{\sqrt{g} - v}{\sqrt{g} + v}. \tag{3.9.3}$$

Operators (3.9.2) in variables g and φ are

$$X_0 = -\sqrt{g}\varphi\frac{\partial}{\partial \varphi}, \qquad X_1 = 2\sqrt{g}\frac{1-\varphi}{1+\varphi}\frac{\partial}{\partial g} + \frac{(1-\varphi)^2}{2\sqrt{g}}\frac{\partial}{\partial \varphi}.$$

Let us find $S_1 g$. Clearly $M_1 g = 2\sqrt{g}$ since $X_1 g|_{\varphi=0} = 2\sqrt{g}$ is the constant term in the expansion of $X_1 g$ with respect to eigenfunctions (powers of φ). The equation $M_1 g = [X_0, S_1]g + X_1 g$ is of the form

$$-\sqrt{g}\varphi\frac{\partial(S_1 g)}{\partial \varphi} - \frac{4\sqrt{g}\varphi}{1+\varphi} = 0;$$

hence $S_1 g = -4\ln(1+\varphi) + h(g)$ [we might put $h(g) = 0$ but in what follows it will be more convenient to choose $h(g)$ otherwise]. Now

$$M_1\varphi = -\sqrt{g}\varphi\frac{\partial(S_1\varphi)}{\partial \varphi} + \sqrt{g}S_1\varphi + \frac{h(g) - 4\ln(1+\varphi)}{2\sqrt{g}}\varphi + \frac{(1-\varphi)^2}{2\sqrt{g}}$$

and $M_1\varphi = A(g)\varphi$ should be chosen so that $S_1\varphi$ is regular at $\varphi = 0$. Evidently, $M_1\varphi = [h(g) - 2]\varphi/2\sqrt{g}$. Then

$$S_1\varphi = -\frac{1}{2g} + \frac{\varphi^2}{2g} - \frac{2\varphi}{g}\int_0^{\varphi} \frac{\ln(1+\zeta)}{\zeta}\, d\zeta + f(g)\varphi.$$

We will restrict ourselves to the computation of the principal term of the asymptotics. Then it is not necessary to return to the old variables (via e^{-S}). Furthermore, to compute the principal term we must in any case take M_1 into account since the coefficients of d/dt incorporates ε. But the operator

$$M_1 = 2\sqrt{g}\frac{\partial}{\partial g} + \frac{h(g)-2}{2\sqrt{g}}\varphi\frac{\partial}{\partial \varphi}$$

contains an arbitrary function $h(g)$. This shows that the approximation $X_0 + \varepsilon M_1$ is not sufficient. We have already mentioned a similar circumstance in Example 3.7; the error $\varepsilon^2 M_2 g$ gives an error in λ of order ε which in turn gives an error in φ of order 1.

The eigenvalue should be computed with higher accuracy. Let us find $M_2 g$. Since $M_2 = [X_0, S_2] + \frac{1}{2}[M_1 + X_1, S_1]$, we have

$$M_2 g = \frac{1}{2}([M_1 + X_1, S_1])|_{\varphi=0} = -\frac{h(g)+2}{\sqrt{g}} + 2\sqrt{g}\frac{dh(g)}{dg}.$$

The principal term of the asymptotics is defined by equations

$$\varepsilon\frac{d\varphi}{dt} = (X_0 + \varepsilon M_1)\varphi, \qquad \varepsilon\frac{dg}{dt} = \varepsilon M_1 g + \varepsilon^2 M_2 g.$$

If we compute g with accuracy $0(\varepsilon)$ inclusively, then we will find φ with accuracy $0(1)$ inclusively. Then the answer for g may be taken with accuracy $0(1)$. The result $g + 0(\varepsilon)$, $\varphi + 0(\varepsilon)$ does not depend on $h(g)$. It is natural to choose $h(g)$ so that$M_2 g = 0$; that will simplify equations for the principal term of the asymptotics. Evidently, we may put $h(g) = -2$. Then

$$\varepsilon\frac{d\varphi}{dt} = \left(-\sqrt{g} - \frac{2\varepsilon}{\sqrt{g}}\right)\varphi, \qquad \frac{dg}{dt} = 2\sqrt{g},$$

implying

$$g = (c_1 + t)^2, \qquad \varphi = \frac{c_2}{(c_1+t)^2}\exp\left(-\frac{1}{2\varepsilon}[(c_1+t)^2 - c_1^2]\right),$$

where c_1 and c_2 are constants. Since $v = \sqrt{g}(1-\varphi)/(1+\varphi)$ [see (3.9.3)], the principal term v_0 of the asymptotics for v is

$$v_0 = \omega\frac{1-(c_2/\omega^2)\zeta}{1+(c_2/\omega^2)}, \qquad \omega = c_1+t, \qquad \zeta = e^{-\theta/\varepsilon}, \qquad \theta = \tfrac{1}{2}[(c_1+t)^2 - c_1^2]. \tag{3.9.4}$$

In problems with a boundary layer ($\lambda < 0$) like this one, it is often easy to guess the general form of the asymptotics and seek a solution in this form. We intend now to illustrate with this example a sufficiently general trick for making computations of this kind. Later on it will be applied to a considerably more difficult problem.

The above makes it clear that the asymptotics are of the form

$$v = v_0(t,\zeta) + \varepsilon v_1(t,\zeta) + \varepsilon^2 v_2(t,\zeta) + \cdots, \tag{3.9.5}$$

where $\zeta = e^{-\theta/\varepsilon}$, $\theta = \theta(t), \theta(0) = 0$, $\theta(t) > 0$ for $t > 0$, and the functions $v_j(t,\zeta)$ are regular as $\zeta \to 0$:

$$v_j(t,\zeta) = \bar{v}_j(t) + \hat{v}_j(t)\zeta + 0(\zeta^2) \quad \text{and} \quad \hat{v}_0(t) \neq 0. \tag{3.9.6}$$

Here $-\theta = \int \lambda\, dt$ is a "boundary layer variable" $(0 \leq \theta \lesssim \varepsilon)$ and ζ is an eigenfunction (in the first order approximation). The regularity condition is an analogue of that of decomposability with respect to eigenfunctions.

As we will see a little later, regularity conditions (3.9.6) make it possible (and it is a sufficiently general situation) to find θ and v_j; i.e., to construct uniform asymptotics (3.9.5).

The general scheme of computation is as follows. For the sake of brevity set $\dot{\theta} = \omega$ so that $\dot{\zeta} = -(\omega/\varepsilon)\zeta$. The (complete) derivative with respect to t of the function F in t and ζ is of the form

$$\dot{F} = -\frac{\omega}{\varepsilon}\zeta\frac{\partial F}{\partial \zeta} + \frac{\partial F}{\partial t}. \tag{3.9.7}$$

Substituting (3.9.5), making use of (3.9.7), into (3.9.1) and considering ζ and t as independent variables, let us collect terms with the same powers of ε (i.e., $\varepsilon^{-1}, \varepsilon^0, \varepsilon^1, \ldots$) and equate them to zero so that (3.9.5) will become a solution of (3.9.1). We will evidently obtain a chain of partial differential equations for the functions $v_j(t,\zeta)$. It is clear due to (3.9.7) that the j-th equation contains only derivatives of v_j with respect to ζ. Hence it is a chain of (second order) ordinary differential equations in ζ. Functions in the parameter t which enter (3.9.6), namely, $\bar{v}_j(t)$ and $\hat{v}_j(t)$, are initial values at $\zeta = 0$ for the j-th equation (for v_j). Regularity conditions (3.9.6) allow one to find equations for $\theta(t)$, $\bar{v}_j(t)$, and $\hat{v}_j(t)$ since $\zeta = 0$ is the singular point for each equation of the chain [and therefore conditions (3.9.6) are sufficiently restrictive]. Namely, the 0-th equation (for v_0) gives a constraint on ω and $\bar{v}_0$; the first equation (for $\bar{v}_0$) gives a constraint on $\hat{v}_0$, $\dot{\hat{v}}_0$, and $\bar{v}_1$; the second equation (for $\bar{v}_1$) gives a constraint on $\dot{\hat{v}}_1$, $\hat{v}_1$, $\hat{v}_0$, $\dot{\hat{v}}_0$, $\ddot{\hat{v}}_0$, $\bar{v}_1$, $\bar{v}_2$, etc. Let us demonstrate the details via our example. Equations of the chain are of the form (for convenience's sake each equation is divided by

$\omega\zeta$):

$$
\begin{aligned}
(\varepsilon^{-1}) \qquad & \frac{\partial}{\partial\zeta}\left(\omega\zeta\frac{\partial v_0}{\partial\xi} - \frac{\partial v_0^2}{\partial 2}\right) = 0, \\
(\varepsilon^{0}) \qquad & \frac{\partial}{\partial\zeta}\left(\omega\zeta\frac{\partial v_1}{\partial\xi} - v_0 v_1\right) = \frac{\partial}{\partial\zeta}\left(2\frac{\partial v_0}{\partial t} + \frac{\dot\omega}{\omega}v_0\right) \\
& \qquad - \frac{1}{\omega\zeta}\left(\frac{\partial}{\partial t}\left(\frac{v_0^2}{2}\right) - v_0\right) \\
(\varepsilon^{1}) \qquad & \frac{\partial}{\partial\zeta}\left(\omega\zeta\frac{\partial v_2}{\partial\xi} - v_0 v_2\right) = \frac{\partial}{\partial\zeta}\left(2\frac{\partial v_1}{\partial t} + \frac{\dot\omega}{\omega}v_1 + \frac{v_1^2}{2}\right) \\
& \qquad - \frac{1}{\omega\zeta}\left(\frac{\partial}{\partial t}(v_0 v_1) - v_1 + \frac{\partial^2 v_0}{\partial t^2}\right).
\end{aligned}
\tag{3.9.8}
$$

Let us substitute (3.9.6) into (3.9.8). For $\bar v_j$ we get the chain of equations [since in the right-hand side of (3.9.8) no singularity must be present as $\zeta \to 0$]:

$$
\frac{d}{dt}\left(\frac{\bar v_0^2}{2}\right) - \bar v_0 = 0, \qquad \frac{d}{dt}(\bar v_0, \bar v_1) - \bar v_1 = -\frac{d^2 \bar v_0}{dt^2}. \tag{3.9.9}
$$

We would have obtained the same equations by substituting $\bar v = \bar v_0 + \varepsilon\bar v_1 + \varepsilon^2\bar v_2$ into (3.9.1) with the requirement that $\bar v$ was a solution. This solution is "far" ($\theta \gg \varepsilon$) from the boundary layer. It is called an *exterior* solution. Besides (3.9.9) we will also get

$$
\begin{aligned}
\omega &= \bar v_0, \\
-\hat v_0 \bar v_1 &= \frac{d\hat v_0}{dt} + \frac{\hat v_0}{\omega}, \\
-\hat v_0 \bar v_2 &= \frac{d\hat v_1}{dt} + \frac{\hat v_1}{\omega} + \bar v_1 \hat v_1 + \bar v_1 \hat v_1 + \frac{\dot{\hat v}_0}{\omega^2} - \frac{\hat v_0 \dot\omega}{\omega^3}, \\
&\ \ \vdots
\end{aligned}
\tag{3.9.10}
$$

These equations can be simplified somewhat: the second one with the use of the first one, the third one with the use of the second one, etc. Chains (3.9.9) and (3.9.10) are all corollaries of (3.9.6) and (3.9.8).

Since $\omega = \theta$, then the first equation from (3.9.10) is an equation for θ since $\hat v_0$ is defined from (3.9.9). It is possible to call this equation an *eikonal* equation. Other equations in (3.9.10) are linear equations for $\hat v_j$ [only the first equations in (3.9.8) and (3.9.9) are nonlinear and, generally speaking, in other problems the first equation in (3.9.10) may be nonlinear].

Let us give an example of computation of two terms of asymptotics. From (3.9.9) we get $\bar v_0 = c_1 + t$, $\bar v_1 = \bar v_2 = \cdots = 0$. Such a simple answer is typical of this problem; $c_1 + t$ is an exact particular solution. It is possible to retain a constant only in $\bar v_0$ and for $\bar v_j$, where $j \geq 1$, take particular solutions; the same is true for the other equations. The first (eikonal) equation in (3.9.10) yields

$$\theta = \frac{1}{2}[(c_1 + t)^2 - c_1^2], \qquad \text{i.e., } \theta(0) = 0,$$

and further, due to the next equations in (3.9.10), taking the previously found $\bar{v}_j$ into account, we have

$$\hat{v}_0 = -\frac{2c_2}{\omega}, \qquad \hat{v}_1 = \frac{2c_2}{\omega^3}, \qquad \text{where } c_2 = \text{const.}$$

The first equation in (3.9.8), due to $v_0|_{\zeta=0} = \bar{v}_0 = \omega$, takes the form

$$\omega\zeta\frac{\partial v_0}{\partial \zeta} = \frac{1}{2}(v_0^2 - \omega^2),$$

implying

$$v_0 = \omega\frac{1 - \frac{c_2}{\omega^2}\zeta}{1 + \frac{c_2}{\omega^2}\zeta}, \qquad \omega = \bar{v}_0 = c_1 + t, \tag{3.9.11}$$

due to $\partial v_0/\partial\zeta|_{\zeta=0} = \hat{v}_0 = -2c_2/\omega$, cf. (3.9.4). Further, using $\dot{\omega} = 1$, (3.9.11) and

$$-\frac{1}{\omega\zeta} = \frac{2}{\omega^2 - v_0^2}\frac{\partial v_0}{\partial \zeta},$$

it is possible to write the term $-(1/\omega\zeta)\{\partial[\partial(v_0/2)/\partial t] - v_0^2\}$ in (3.9.8) in the form

$$\frac{\partial}{\partial\zeta}\left(\frac{1}{\omega^2}(\omega - v_0)^2 + 2\ln(\omega + v_0)\right)$$

and express $2(\partial v_0/\partial t) + (\dot{\omega}/\omega)v_0$ in terms of v_0 and ω. Integrating the second equation in (3.9.8) with respect to ζ we have

$$\omega\zeta\frac{\partial v_1}{\partial\zeta} - v_0 v_1 = \frac{v_0}{\omega} - \frac{v_0^2}{\omega^2} + 2\ln\left(\frac{\omega + v_0}{2\omega}\right)$$

due to the fact that $v_0|_{\zeta=0} = \omega$, $\bar{v}_1|_{\zeta=0}$. Then, considering ζ_1 as a function in v_0, where

$$\omega\zeta\frac{\partial v_1}{\partial\zeta} = \frac{v_0^2 - \omega^2}{2}\frac{\partial v_1}{\partial v_0}$$

and recalculating accordingly the initial value $\hat{v}_1 = 2c_2/\omega^3$, we get

$$v_1 = \frac{v_0^2 - \omega^2}{2\omega^3}\left(\ln^2\left(\frac{v_0 + \omega}{2\omega}\right) - 2\int_0^{v_0 - \omega/2\omega}\frac{\ln(1+\eta)}{\eta}\,d\eta\right) - \frac{2v_0}{\omega^2}\ln\left(\frac{\tau_0 + \omega}{2\omega}\right).$$

3.10 Example: Resonances. Particular Solutions

Various problems, either directly or after simple transformations, give rise to systems with a linear leading part and polynomial perturbations. Coefficients of the system in such a problem often depend on one or several parameters. The qualitative behavior of solutions may abruptly alter at certain values of parameters. Below we will consider several examples. The passage of a parameter through a critical (resonance) value which causes the change of the nature of the solution yields, as we will see, spasmodic alteration of the normal form of the system. Here we will discuss several aspects of reduction to the normal form of the systems mentioned, restricting ourselves, for simplicity's sake, to systems depending on a single parameter. Consider the system

$$\begin{gathered}\varepsilon^\alpha \frac{dy}{dt} = A(\mu)y + \varepsilon P_1(y,\mu) + \varepsilon^2 P_2(y,\mu) + \cdots, \\ y = \{y_1,\ldots,y_n\}\end{gathered} \tag{3.10.1}$$

where $A(\mu)$ is a constant matrix, components $p_{jk}(y,\mu)$ of vector functions $P_j(y,\mu)$ are polynomials in y, $p_{jk}(0,\mu) = 0$, where $k = 1,\ldots,n$, μ is a parameter, and $\alpha = \text{const}$.

To the system (3.10.1) the operator $X = X_0 + \varepsilon X_1 + \varepsilon^2 X_2 + \cdots$ corresponds, where $X_j = (P_j(y,\mu), \nabla)$,

$$\nabla = \left\{\frac{\partial}{\partial y_1},\ldots,\frac{\partial}{\partial y_n}\right\}, \qquad j = 1,2,\ldots, \quad X_0 = (A(\mu)y, \nabla)$$

so that $\varepsilon^\alpha(dy/dt) = Xy$. Denote by $\lambda_1(\mu),\ldots,\lambda_n(\lambda)$ the roots of the characteristic polynomial $\det\|A(\mu) - \lambda E\|$, where E is the identity matrix.

In the sequel we will perform various changes of variables which preserve the form of the system (3.10.1) as a system with linear leading part and polynomial perturbations. For brevity's sake we will not introduce new notations, indicating only corresponding alterations in $A(\mu)$ and in $p_{jk}(y,\mu)$.

(1) The Normal Form for a Fixed μ

Here we actually repeat the known arguments and apply them to the concrete system (3.10.1). Fix μ (i.e., consider μ as independent of ε). Let us perform the change of variables $y \mapsto c(\mu)y$, choosing the nondegenerate matrix $c(\mu)$ in such a way that in new variables the matrix $A(\mu)$ of (3.10.1) is in Jordan form. [Note an ambiguity in the choice of $c(\mu)$.] Now y_k, where $k = 1,\ldots,n$, is the function corresponding to an eigenvalue of X_0, i.e., to one of $\lambda_1(\mu),\ldots,\lambda_n(\mu)$. Choose the indexing of roots so that y_k corresponds to $\lambda_k(\mu)$, where a root of multiplicity m is counted m times. Then the monomial

$$h(\mu)y_1^{m_1}y_2^{m_2}\cdots y_n^{m_n}, \tag{3.10.2}$$

where $m_k \geq 0$ are integers such that $m_1 + \cdots + m_n \geq 1$, is the eigenfunction corresponding to the eigenvalue $m_1\lambda_1(\mu) + \cdots + m_n\lambda_n(\mu)$.

Now make a new transformation $y \mapsto e^{-S}y$, where $S = \varepsilon S_1 + \varepsilon^2 S_2 + \cdots$ and $S_j y_k = q_{jk}(y, \mu)$ are polynomials in y, $q_{jk}(0, \mu) = 0$, and the S_j's are first order operators. We get the system $\varepsilon^\alpha(dg/dt) = My$, where $M = e^{-S}Xe^S = X_0 + \varepsilon M_1 + \varepsilon^2 M_2 + \cdots$, and $M_j = [X_0, S_j] + Y_j$, where $Y_1 = X_i$, and Y_j for $j \geq 2$ are computed via the now known formulas (1.2.3). Evidently, $Y_j y_k$ are polynomials in y so that we again have a system of the form (3.10.1).

Consider the operator $M_j = [X_0, S_j] + Y_j$, where Y_j is known. Since $Y_j y_k$'s are polynomials in y; i.e., sums of functions which are monomials of the form (3.10.2) corresponding to constant eigenvalues of X_0, we see that the conditions of Theorem 2.4.3 are satisfied. Hence there exists S_j which solves the perturbation theory problem (2.4.1) for M_j. The method of constructing S_j given in the proof of Theorem 2.4.3 shows that it is possible to seek $S_j y_k$ in the form of a polynomial (sum of monomials); i.e., in the form proposed from the very beginning. [Here, uniqueness is again lacking: in $S_j y_k$ any monomials can be added that satisfy (2.4.1) for M_j.]

To solve (2.4.1) for M_j means that $M_j y_k$'s are sums of monomials of the form (3.10.2) which correspond to $\lambda_k(\mu)$. The double transformation $y \mapsto c(\mu)y \mapsto c(\mu)e^{-S}y$ that solves the problem (2.4.1)' for M will be denoted by $W(\mu)$ [recall once more that $W(\mu)$ is not unique] and the system (3.10.1) after this transformation will be said to be in *the normal form.*

Thus the system in the normal form is such that $A(\mu)$ is a Jordan matrix, y_k corresponds to $\lambda_k(\mu)$, where $k = 1, \ldots, n$, and monomials of the form (3.10.2) which constitute $p_{jk}(y, \mu)$ satisfy

$$m_1\lambda_1(\mu) + m_2\lambda_2(\mu) + \cdots + m_m\lambda_n(\mu) = \lambda_k(\mu). \tag{3.10.3}$$

Remark 3.10.1 It is possible to propose the following, additional, canonization of the system for a fixed μ. Let us reduce (3.10.1) to the normal form (in the just stated sense) and perform the new transformation $y_k \mapsto y_k \exp \lambda_k(\mu)t/\varepsilon^\alpha$, where $k = 1, \ldots, n$. Clearly, due to (3.10.3), after dividing each equation by $\exp \lambda_k(\mu)t/\varepsilon^\alpha$ we will obtain the system of the form (3.10.1) with the canonical nilpotent matrix in the leading part.

In section 2.5 we have given a normal form of an operator with respect to nilpotent X_0. After the corresponding transformation $y \mapsto e^{-S}y$ we may now "finally" normalize the system. However, it is possible to avoid the new transformation $y \mapsto e^{-S}y$ if in the $W(\mu)$-procedure we lessen the arbitrariness of $S_j y_k$ accordingly.

(2) Resonance

Evidently, for $\mu = \mu(\varepsilon)$ the above normal form is generally unapplicable: relation (3.10.3) may hold for some $\mu = \mu_0$ but fail at points arbitrarily close to μ_0. Coefficients of the normal form as functions in μ may be discontinuous. This may be so even in the principal order since the Jordan form of $A(\mu)$ is, generally speaking, discontinuous. (Example: the matrix

$$\begin{pmatrix} \mu & 1 \\ 0 & \mu^2 \end{pmatrix}$$

for $\mu \neq 0$ is equivalent to the diagonal matrix and for $\mu = 0$ is Jordan.)

We will say that a system of the form (3.10.1) is *continuous* in μ at $\mu = \mu_0$ *with accuracy* ε^j, where $j = 1, 2, \ldots$, if all coefficients of polynomials $X_0 y_k, \ldots, X_{j-1} y_k$, where $k = 1, \ldots, n$, are continuous functions in μ at $\mu = \mu_0$. With accuracy ε^0 the system will be considered continuous at any μ.

The initial system (3.10.1) will be supposed to be continuous in μ at any μ with the accuracy required by the concrete problem; i.e., with any given (but finite) accuracy.

Definition 3.10.2 *The point* $\mu = \mu_0$ *is called an* i-point of resonance *(or just* i*-point) if for any* $W(\mu)$ *the system (3.10.1) in the normal form is not continuous at* $\mu = \mu_0$ *with accuracy* ε^{i+1} *and if there is* $W(\mu)$ *such that that (3.10.1) in the normal form is continuous at* $\mu = \mu_0$ *with accuracy* ε^i, *where* $i = 0, 1, 2, \ldots$.

Remark 3.10.3 In concrete problems we are usually completely sure that if there is a discontinuity of the normal form for some $W(\mu)$, then we are dealing with a resonance and there is no necessity to adhere rigidly to the above definition.

Remark 3.10.4 In nonlinear problems with small parameters one has to deal with two essentially different kinds of resonances: one, indicated just now for constant eigenvalues (it would be to the point to call this resonance *parametric* if this term were new), and another one, when eigenvalues which are invariants of X_0 are variables. And though the second one is sometimes reducible to the first one, the difference is of a principle nature. Resonance of this kind is much more cumbersome; see Chapter 4.

Further, suppose that i-points of resonance are isolated for every fixed i (recall that the accuracy is assumed to be bounded) and consider a sufficiently small segment containing only a single 1-point. Suppose for the sake of definiteness that $-\tilde{\mu} < \mu < \tilde{\mu}$ and $\mu = 0$ is the only i-point.

Remark 3.10.5 The assumption on isolatedness of the i-point for any fixed i is not a very restrictive one, though there are important problems when it fails. But an assumption of uniform isolatedness of i-points for all i would serve only rare problems (though such problems do exist; see further).

Now we will show that it is possible to define, in a sense "naturally," a normal form of Equations (3.10.1) which serves on the whole open segment $(-\tilde{\mu}, \tilde{\mu})$ for a finite (sufficiently small) $\tilde{\mu}$ if $\mu = 0$ is not a 0-point and is valid for any i if $\tilde{\mu} = \tilde{\mu}(\varepsilon) \to 0$ as $\varepsilon \to 0$ [$\tilde{\mu}(\varepsilon)$ is supposed to be continuous].

(3) The Normal Form in a Vicinity of a Resonance. (the Uniform Normal Form)

Let $\mu = 0$ be not a 0-point; i.e., $i \geq 1$. This means that a linear, nondegenerate, and continuous with respect to μ change of variables reduces the initial system to a form such that (in new variables) the matrix $A(\mu)$ of the leading linear part is Jordan and the system is continuous in μ at $\mu = 0$ with the needed accuracy.

We will assume from the very beginning that this transformation has been already performed. Let us now make a transformation $y \mapsto e^{-S}y$, where $S = \varepsilon S_1 + \varepsilon^2 S_2 + \cdots$ and S_j depends on μ, not imposing for the time being any restrictions on S except continuous dependence on μ.

To our system in new variables the operator $M = e^{-S}Xe^{S} = X_0 + \varepsilon M_1 + \cdots$ corresponds, where [see (1.2.3)] $M_j = [X_0, S_j] + Y_j$ and Y_j is expressed via $S_1, \ldots, S_{j-1}$ and $X_1, \ldots, X_j$ and $X_0 = (A(\mu)y, \nabla)$, where $A(\mu)$ is Jordan and is continuous in μ at $\mu = 0$.

We have defined above the notion of a normal form of M_j at the fixed μ to be

$$M_j^{(\mu_j)} = [X_0, [X_0, \ldots [X_0, M_j] \ldots]] = 0, \qquad \mu_j \text{ times.}$$

Suppose that $\mu = 0$ is a resonance point and introduce the new notion of the normal form. Set $\bar{X}_0 = (A(0)y, \nabla)$.

Definition 3.10.6 *The operator M possesses a (uniform) normal form with respect to $\bar{X}_0$ if there is ν_j such that*

$$M_j^{(\bar{\nu}_j)} \equiv [\bar{X}_0, [\bar{X}_0, \ldots, [\bar{X}_0, M_j] \ldots]] = 0, \qquad \nu_j \textit{ times}, \tag{3.10.4}$$

and $M_0 = X_0$ satisfies this condition.

Theorem 3.10.7 *There is an operator S which continuously depends on μ and reduces X to the uniform normal form M with respect to X_0.*

We will give two variants of the proof: the first exploits the fact that $y_1, \ldots, y_n$ is the basic system for X_0 and $\bar{X}_0$ simultaneously and gives a simple recipe for construction of S while the second one is less constructive but is (partially) useful for the case of a 0-point.

PROOF 1

The monomial (3.10.2) corresponds to the eigenvalue $m_1\lambda_1(0) + \cdots + m_n\lambda_n(0)$ with respect to $\bar{X}_0$. Let $\overline{Y_j y_k}$ be the sum of all monomials in

$Y_j y_k$'s corresponding to $\lambda_k(0)$ with respect to $\bar{X}_0$, then $\widetilde{Y_j y_k} = Y_j y_k - \overline{Y_j y_k}$ is the sum of monomials corresponding to eigenvalues $\lambda(\mu)$ with respect to X_0 such that $|\lambda(\mu) - \lambda_k(\mu)| > \gamma_j > 0$, where $\gamma_j = \text{const}$ for any sufficiently small μ. Hence it is possible to find $S_j y_k$ such that

$$[X_0 S_j] y_k + Y_j \widetilde{y_k} = 0, \qquad \text{where } k = 1, 2, \ldots, n. \tag{3.10.5}$$

Then $M_j y_k = \overline{Y_j y_k}$, which is equivalent to (3.10.4) by Theorem 2.4.1. Due to the same theorem, X_0 is normal with respect to $\bar{X}_0$.

Proof 2

Let us express X_0 in the form

$$X = \bar{X}_0 + \tilde{X}_0, \qquad \tilde{X}_0 = ((A(\mu) - A(0))y, \nabla). \tag{3.10.6}$$

Then clearly $[\bar{X}_0, X_0] = 0$ since $A(\mu)$ is in Jordan form. Let us express M_j in the form

$$M_j = [X_0, S_j] + Y_j = S_j^{(1)} + [\tilde{X}_0, S_j] + Y_j,$$

and conditions (3.10.4) in the form

$$(S_{j(0)}^{(\nu_j+1)} + [\tilde{X}_0, S_j]_{(0)}^{(\nu_j)}) y_k = -Y_{j(0)}^{(\nu_j)} y_k \qquad \text{for } k = 1, \ldots, n. \tag{3.10.4'}$$

It is a system of linear equations with respect to coefficients of polynomials $S_j y_m$, where $m = 1, \ldots, n$ (let their degree be no higher than the maximal degree of $Y_j y_k$). Let Z be a first order operator such that $Z y_k$ are polynomials of degree no higher than the maximal degree in $Y_j y_m$, where $m = 1, \ldots, n$. Since the linear space of such operators Z is finite dimensional, then by Theorem 2.4.3 there is a ν_j such that the system

$$S_{j(0)}^{(\nu_j+1)} y_k = Z_{(0)}^{(\nu_j)} y_k,$$

where $k = 1, \ldots, n$, is solvable for any Z; i.e., the rank R of its matrix coincides with the rank of the extended matrix for any Z. Comparing this system for a given ν_j with (3.10.4′) we have

(a) the rank of the extended matrix (3.10.4′) is no more than R,

(b) the rank of the matrix (3.10.4′) at $\mu = 0$ equals R,

(c) the rank of (3.10.4′) at $\mu \neq 0$ when μ is sufficiently small is no less than R.

This implies that for any sufficiently small μ ranks of matrices of the system (3.10.4′) and its extended matrix coincide and by Kronecker–Capelli's theorem (3.10.4′) is solvable. The theorem is proved.

Now let $\mu = 0$ be a 0-point of the resonance. We may assume that $A(0)$ is in Jordan form and y_k, where $k = 1,\ldots,n$, corresponds to the eigenvalue $\lambda_k(0)$ of $\bar{X}_0 = (A(0)y, \nabla)$ (if this is not so, we may choose a corresponding linear transformation which does not depend on μ).

Let us express X_0 in the form (3.10.6). As earlier, we could have reduced M_j to the normal form with respect to $\bar{X}_0$ [the matrix $A(\mu)$ is continuous but not in the Jordan form]. But due to the summand $\tilde{X}_0$ which has, generally speaking, no normal form with respect to $\bar{X}_0$, i.e., the relation $[\bar{X}_0, \ldots, [\bar{X}_0, \tilde{X}_0] \ldots] = 0$ fails, the "whole" operator M has no normal form. However, taking into account that all our considerations are only needed when μ tends to zero together with ε (since for a fixed μ we deal with the usual reduction procedure) let us suppose that $\tilde{\mu} = \tilde{\mu}(\varepsilon) \to 0$ as $\varepsilon \to 0$. We will show how to reduce X to the normal form with respect to $\bar{X}_0$ under this assumption.

But let us first make the transformation $y \mapsto e^{-\tilde{S}}y$, where $\tilde{S} = (B(\mu)y, \nabla)$, $B(0) = 0$ and consider the operator

$$\tilde{M} = e^{-\tilde{S}} X e^{\tilde{S}} = e^{-\tilde{S}} X_0 e^{\tilde{S}} + e^{-\tilde{S}} (\varepsilon X_1 + \varepsilon^2 X_2 + \cdots) e^{\tilde{S}}.$$

Let us assume that $\tilde{X}_0$ is an operator of the first order of smallness as compared with $\bar{X}_0$ since $\tilde{X}_0 = ((A(\mu) - A(0))y, \nabla)$ and $A(\mu) - A(0)$ continuously depends on μ as $\mu \to 0$. Set $\tilde{S} = \tilde{S}_1 + \tilde{S}_2 + \cdots$ so that $\tilde{S}_j$ is a j-th order of smallness operator. This formally means that operators $\tilde{M}_j$ in the decomposition

$$e^{-\tilde{S}} X_0 e^{\tilde{S}} = e^{-\tilde{S}} (\bar{X}_0 + \tilde{X}_0) e^{\tilde{S}} = \bar{X}_0 + \tilde{M}_1 + \tilde{M}_2 + \cdots$$

are computed via formulas (1.2.3):

$$\tilde{M}_1 = [\bar{X}_0, \tilde{S}_1] + \tilde{X}_0, \qquad \tilde{M}_2 = [\bar{X}_0, \tilde{S}_2] + [\tilde{X}_0, \tilde{S}_1] + \tfrac{1}{2}[[\bar{X}_0, \tilde{S}_1], \tilde{S}_1], \ldots$$

If $\tilde{\mu}(\varepsilon)$ is given, then we may reduce $\tilde{X}_0 \equiv \tilde{M}_1 + \tilde{M}_2 + \cdots$ to the normal form with respect to $\bar{X}_0$ with the needed accuracy. Knowing $\tilde{S}$, we may compute $\tilde{X}_j = e^{-\tilde{S}} X_j e^{\tilde{S}}$ with the needed accuracy. We will get $\tilde{M} = \bar{X}_0 + \tilde{\tilde{X}}_0 + \varepsilon \tilde{X}_1 + \varepsilon^2 \tilde{X}_2 + \cdots$, where $\tilde{\tilde{X}}_0$ is normal with respect to $\bar{X}_0$. Now the operator X may be reduced to the normal form with respect to $\bar{X}_0$ making use of the above procedure (of solution of the system of linear equations; see Proof 2) for $i \neq 0$.

(4) Particular Solutions

Suppose that the roots $\lambda_1(\mu), \ldots, \lambda_n(\mu)$ of the characteristic polynomial of $A(\mu)$ are divisible into two sets $\lambda_1(\mu), \ldots, \lambda_2(\mu)$ and $\lambda_{r+1}(\mu), \ldots, \lambda_n(\mu)$ so that

$$\lambda_s(0) \neq m_1\lambda_1(0) + \cdots + m_r\lambda_r(0) \qquad \text{for } s = r+1, \ldots, n, \tag{3.10.7}$$

where $m_q \geq 0$ are integers and $m_1 + \cdots + m_r \geq 1$ [a typical and most important example is $\operatorname{Re} \lambda_q(0) = 0$ for $q = 1, \ldots, r$ and $\operatorname{Re} \lambda_s(0) < 0$ for $s = r+1, \ldots, n$].

Consider Equations (3.10.1) in the uniform normal form. In the right-hand sides of equations for y_s, where $s = r + 1, \ldots, n$, there enter only monomials corresponding to eigenvalues $\lambda_s(0)$. Due to (3.10.7), each of them contains at least one of the variables $y_{r+1}, \ldots, y_n$. This means that Equations (3.10.1) in the normal form have a particular solution

$$y_{r+1} = \cdots = y_n = 0. \tag{3.10.8}$$

If for some reasons only such particular solutions are of interest (in the example given a little above, $y_{r+1}, \ldots, y_n$ decrease exponentially), the system simplifies considerably since usually r is rather small.

As above, let μ vary in a sufficiently small segment with 0 as its center and let (3.10.7) hold. The initial system (3.10.1) is continuous in μ. We will not discuss resonances for a while. By continuity, (3.10.7) implies

$$\lambda_s(\mu) \neq m_1\lambda_1(\mu) + \cdots + m_2\lambda_2(\mu) \qquad \text{for } s = r+1, \ldots, n. \tag{3.10.9}$$

From (3.10.9) we deduce that there exists a nondegenerate continuous transformation $y \mapsto c(\mu)y$ such that in new variables the matrix $A(\mu)$ of the system (3.10.1) is of the form

$$A(\mu) = \begin{pmatrix} A^{(r)}(\mu) & 0 \\ 0 & A^{(n-r)}(\mu) \end{pmatrix}, \tag{3.10.10}$$

where the eigenvalues of $A^{(r)}(\mu)$ are $\lambda_1(\mu), \ldots, \lambda_r(\mu)$ and those of $A^{(n-r)}(\mu)$ are $\lambda_{n+1}(\mu), \ldots, \lambda_n(\mu)$.

Theorem 3.10.8 *Suppose the matrix $A(\mu)$ is of the form (3.10.10) and (3.10.7) holds. Then there exists a change of variables close to the identity and obtained by a finite number of arithmetic operations such that the system (3.10.1) in the new variables possesses the particular solution (3.10.8). Actual knowledge of the roots is not necessary in order to perform calculations.*

PROOF

Let us make the change of variables $y \mapsto e^{-S}y$, where $S = \varepsilon S_1 + \cdots$ and $S_j y_k$ are polynomials which depend only on $y_1, \ldots, y_r$, where $k = 1, \ldots, n$. Consider equalities $M_j y_s = [X_0, S_j]y_s + Y_j y_s$, where $s = r+1, \ldots, n$. Let us show that it is possible to "kill" monomials in $M_j y_s$ which only depend on $y_1, \ldots, y_r$. Note that by restricting the choice of $S_j y_k$'s we did not lose the possibility of solving the problem. Indeed, the form of $A(\mu)$ immediately shows that adding to $S_j y_k$ monomials which depend on $y_{r+1}, \ldots, y_n$, we add to $M_j y_s$ monomials that also depend on $y_{r+1}, \ldots, y_n$.

Further, to define coefficients of polynomials $S_j y_k$ we have a system of linear equations whose solvability is obvious due to (3.19.7) for a sufficiently small μ.

The knowledge of $\lambda(\mu)$ is naturally not necessary for solution of this system and for reduction of $A(\mu)$ to the canonical form. The theorem is proved.

Note that for small r computations are not cumbersome since Sy_k only depends on $y_1, \dots, y_r$. Finally, note that the reduction (3.10.8) is performed regardless of the presence of resonances. The reduction (3.10.8) is closely connected with the so-called center manifold theorem and Lyapunov–Schmidt's reduction. The "shortened" system (for $y_1, \dots, y_r$) is now subjected to reduction to the normal form as indicated above. The notion of a resonance is naturally applied only to this system.

Let us give some applications of considerations exposed here. The next very simple example admits an easy and complete investigation of all possible resonances.

3.11 Example: The Mathieu Equation

$$\frac{d^2u}{dt^2} + (\omega^2 + \varepsilon \cos 2t)u = 0. \tag{3.11.1}$$

The corresponding system is $\dot{u} = v$, $\dot{v} = -(\omega^2 + \varepsilon \cos 2t)u$. This system is nonautonomous. Put $\psi = e^{2it}$, $\bar{\psi} = e^{-2it}$ and extend the system. [We will see later that all operators Z which will be encountered satisfy $Z(\psi\bar{\psi})|_{\psi\bar{\psi}=1} = 0$; therefore the matching of initial conditions, $(\psi\bar{\psi})|_{t=0} = 1$, holds automatically.] We get

$$\frac{du}{dt} = v, \qquad \frac{dv}{dt} = -\left(\omega^2 + \varepsilon\frac{\psi + \bar{\psi}}{2}\right)u, \qquad \frac{\partial \psi}{\partial t} = 2i\psi, \qquad \frac{\partial \bar{\psi}}{\partial t} = -2i\bar{\psi}.$$

The corresponding operator is $X = X_0 + \varepsilon X$, where

$$X_0 = v\frac{\partial}{\partial u} - \omega^2 u \frac{\partial}{\partial v} + 2i\psi\frac{\partial}{\partial \psi} - 2i\bar{\psi}\frac{\partial}{\partial \bar{\psi}},$$

$$X_1 = \frac{1}{2}(\psi + \bar{\psi})u\frac{\partial}{\partial v}. \tag{3.11.2}$$

Let $\omega \neq 0$, where ω is "far from 0." Assume that $\omega > 0$. Eigenfunctions of X_0 are ψ, $\bar{\psi}$; $\varphi = v + i\omega u$, $\bar{\varphi} = v - i\omega u$. In the sequel we drop complex conjugate formulas.

We have

$$X_0\varphi = i\omega\varphi, \qquad X_0\psi = 2i\psi, \qquad X_1\varphi = \frac{i}{4\omega}(\psi + \bar{\psi})(\varphi - \bar{\varphi}), \qquad X_1\psi = 0. \tag{3.11.3}$$

The linearity of the initial equation (3.11.1) makes it easy to understand the general shape of the normal form up to a numerical coefficient; its computation will be put aside for the time being.

It is not difficult to notice that $S_j\varphi$ and $M_j\varphi$ may be sought in the form of polynomials linear in φ and $\bar{\varphi}$:

$$S_j\varphi = A_j(\psi,\bar{\psi})\varphi + B_j(\psi,\bar{\psi})\bar{\varphi},$$
$$M_j\varphi = iC_j(\psi,\bar{\psi})\varphi + iD_j(\psi,\bar{\psi})\bar{\varphi},$$

and $S_j\psi = M_j\psi = 0$. Since ψ and $\bar{\psi}$ are invariants of X_1, we have $S_j\psi = S_j\bar{\psi} = M_j\psi = M_j\bar{\psi} = 0$; besides, $X_1\varphi$, $X_1\bar{\varphi}$ are linear in φ, $\bar{\varphi}$. Here A_j, B_j, C_j, D_j are homogeneous polynomials in ψ, $\bar{\psi}$ of degree j with real coefficients. This is clear from the form of X_1 and (1.2.3).

Let us look at the monomials that enter $M_j\varphi$. They are either monomials of the form

$$i\alpha\varphi\psi^{m_1}\bar{\psi}^{m_2}$$

or

$$i\beta\bar{\varphi}\psi^{m_1}\bar{\psi}^{m_2},$$

where $m_1 + m_2 = j$ and α, β are real. Corresponding eigenvalues are $i(\omega + 2m_1 - 2m_2)$ and $i(-\omega + 2m_1 - 2m_2)$. Since in $M_j\varphi$ only monomials corresponding to $i\omega$ can enter, we get two possibilities:

$$i(\omega + 2m_1 - 2m_2) = i\omega \quad \text{or} \quad i(-\omega + 2m_1 - 2m_2) = i\omega. \tag{3.11.4}$$

The first of these equalities holds for any ω and only for $m_1 = m_2 = j/2$ where j is even. Hence for any ω in M_φ a summand of the form $i\{\varepsilon^2\alpha_1(\psi\bar{\psi}) + \varepsilon^4\alpha_2(\psi\bar{\psi})^2 + \cdots\}\varphi$ might be present. But since $\psi\bar{\psi} = 1$, this summand equals $i\varepsilon^2\alpha(\varepsilon^2)\varphi$, where $\alpha(\varepsilon^2) = \alpha_1 + \varepsilon^2\alpha_2 + \cdots$ is real.

Furthermore, the second of equations (3.11.4) holds only for an integer $\omega = n$ and $m_1 - m_2 = n$, where $m_1 > m_2$. If $\omega \neq n$, then monomials $i\beta\bar{\varphi}\psi^{m_1}\psi^{m_2}$ will not enter$M_j\varphi$, where $\beta = 0$, and if $\omega = n$, then monomials $i\beta\bar{\varphi}\psi^{m_1}\psi^{m_2}$, where

$$m_1 = \frac{j+n}{2}, \qquad m_2 = \frac{j-n}{2}$$

(i.e., monomials of the form $i\beta\bar{\varphi}\psi^{j+n/2}\psi^{j-n/2}$), will enter $M_j\varphi$. Since $m_2 \geq 0$, then $j \geq n$ and for the first time such a monomial might occur in $M_n\varphi$. In $M\varphi$ we will get the summand $i(\varepsilon^n\beta_1\psi^n + \varepsilon^{n+2}\beta_2\psi^{n+1}\bar{\psi} + \cdots)\bar{\varphi}$ or $i\varepsilon^n\beta(\varepsilon^2)\psi^n\bar{\varphi}$, where $\beta(\varepsilon^2) = \beta_1 + \varepsilon^2\beta_2 + \cdots, \beta(\varepsilon^2)$ is real after substituting $\psi\bar{\psi} = 1$.

Thus for $\omega = n$ the normal form is discontinuous so that $\omega = n$ is an n-point of the resonance (see Remark 3.10.2). Now let $n - 1 < \omega < n + 1$, where ω is "far from $n - 1$ and $n + 1$." To obtain the uniform normal form applicable for any ω (and for ω which is ε-near to n, also) it suffices only to change slightly the procedure of constructing S and M. We must not try to "kill" monomials of the form $i\beta\bar{\varphi}\psi^{m_1}\bar{\psi}^{m_2}$ in $M_j\varphi$ if $m_1 - m_2 = n$ for any ω [see (3) in Section 3.10]. We will get

$$M\varphi|_{\varphi\bar{\varphi}=1} = i(\omega + \varepsilon^2\alpha(\varepsilon^2))\varphi + i\varepsilon^n\beta(\varepsilon^2)\psi^n\bar{\varphi}, \qquad M\psi = X_0\psi = 2i\psi.$$

Remark 3.11.1 Of course $\alpha(\varepsilon^2)$ and $\beta(\varepsilon^2)$ depend on ω. We may substitute $\psi = e^{2it}$ into the equation $d\varphi/dt = M\varphi|_{\psi\bar{\psi}=1}$. Thus the equation for φ in new variables takes the form

$$\frac{d\varphi}{dt} = i(\omega + \varepsilon^2\alpha(\varepsilon^2))\varphi + i\varepsilon^n\beta(\varepsilon^2)e^{2int}\bar{\varphi}.$$

Putting now $\varphi = \chi e^{int}$, we get the linear system with constant coefficients

$$\frac{d\chi}{dt} = i(\omega - n + \varepsilon^2\alpha(\varepsilon^2))\chi + i\varepsilon^n\beta(\varepsilon^2)\bar{\chi},$$
$$\frac{d\bar{\chi}}{dt} = -i(\omega - n + \varepsilon^2\alpha(\varepsilon^2))\bar{\chi} - i\varepsilon^n\beta(\varepsilon^2)\chi. \tag{3.11.5}$$

From here we derive exhaustive information on the nature of solutions of Mathieu's equation (3.11.1) for $n - 1 < \omega < n + 1$. In particular, from the equation for eigenvalues of the matrix of the system (3.11.5),

$$\lambda^2 + (\omega - n + \varepsilon^2\alpha(\varepsilon^2))^2 - \varepsilon^{2n}\beta^2(\varepsilon^2) = 0,$$

we derive in the plane $\omega = \omega(\varepsilon)$ the so-called *transitional curves*, defined by the condition $\lambda = 0$, separating stability and instability domains:

$$\omega = n - \varepsilon^2\alpha(\varepsilon^2) \pm \varepsilon^n\beta(\varepsilon^2). \tag{3.11.6}$$

Remark 3.11.2 From (3.11.6) we may see that principal resonances are $n = 1$ and $n = 2$; for $n \geq 3$, the principal term in $\omega - n$ is defined by the nonresonance summand $\alpha(0)$.

Let us give an example of computations up to ε^2 inclusively. Making use of (3.11.3), we find

$$n = 1: \qquad S_1\varphi = \frac{1}{8\omega}(\bar{\psi} - \psi)\varphi - \frac{1}{8\omega(\omega+1)}\bar{\psi}\bar{\varphi},$$
$$S_1\psi = 0,$$
$$M_1\varphi = -\frac{i}{4\omega}\psi\bar{\varphi};$$
$$M_1\psi = 0; \tag{3.11.7}$$

$$n > 1: \qquad S_1\varphi = \frac{1}{8\omega}(\bar{\psi} - \psi)\varphi - \frac{1}{8\omega}\left(\frac{\bar{\psi}}{\omega+1} + \frac{\psi}{\omega-1}\right)\bar{\varphi},$$
$$S_1\psi = 0,$$
$$M_1\varphi = 0,$$
$$M_1\psi = 0. \tag{3.11.8}$$

Further, with M_2 in the form $M_2 = [X_0, S_2] + \frac{1}{2}[M_1 + \chi_1, S_1]$, we find

$$\begin{aligned} n = 1: \quad & \frac{1}{2}[M_1 + X_1, S_1]\varphi \\ & = \frac{i}{32\omega(\omega+1)} \\ & \quad \times \{-(\psi^2 + \bar{\psi} + \psi\bar{\psi})\varphi + [2(\omega+1)\psi^2 - \omega\bar{\psi}^2 - \omega\psi\bar{\psi}]\bar{\varphi}\}; \\ n > 1: \quad & \frac{1}{2}[M_1 + X_1, S_1]\varphi \\ & = \frac{i}{32\omega(\omega^2-1)} \\ & \quad \times \{-(\psi^2 + \bar{\psi}^2 + 2\psi\bar{\psi})\varphi + [(\omega+1)\psi^2 - (\omega-1)\bar{\psi}^2 + 2\psi\bar{\psi}]\bar{\varphi}\}. \end{aligned}$$

From here we deduce

$$\begin{aligned} n = 1: \quad & M_2\varphi = -\frac{i\psi\bar{\psi}\varphi}{32\omega(\omega^2+1)}; \\ n = 2: \quad & M_2\varphi = -\frac{i\psi\bar{\psi}\varphi}{16\omega(\omega^2-1)}; \\ n > 2: \quad & M_2\varphi = -\frac{i\psi\bar{\psi}\varphi}{16\omega(\omega^2-1)}. \end{aligned}$$

This and (3.11.7), (3.11.8) imply

$$\begin{aligned} \alpha(\varepsilon^2) &= -\frac{1}{32\omega(\omega^2+1)} + 0(\varepsilon^2) \qquad \text{for } n = 1; \\ \alpha(\varepsilon^2) &= -\frac{1}{16\omega(\omega^2-1)} + 0(\varepsilon^2) \qquad \text{for } n > 1; \\ \beta(\varepsilon^2) &= -\frac{1}{4\omega} + 0(\varepsilon^2) \qquad \text{for } n = 1; \\ \beta(\varepsilon^2) &= \frac{1}{32\omega(\omega-1)} + 0(\varepsilon^2) \qquad \text{for } n = 2. \end{aligned}$$

Transitional curves (3.11.6) are

$$\omega = 1 \mp \frac{\varepsilon}{4} - \frac{3\varepsilon^2}{64} + 0(\varepsilon^3), \qquad \omega = 2 + \frac{2 \mp 3}{192}\varepsilon^2 + 0(\varepsilon^4),$$

$$\omega = n + \frac{\varepsilon^2}{16n(n^2-1)} + 0(\varepsilon^3) \qquad \text{for } n \geq 3.$$

Now let us consider the resonance $\omega = 0$. From (3.11.2) we see that $\omega = 0$ is a 0-point of the resonance. We have $X_0 = \bar{X}_0 + \tilde{X}_0$, where

$$\bar{X}_0 = v\frac{\partial}{\partial u} - 2i\psi\frac{\partial}{\partial \psi} - 2i\bar{\psi}\frac{\partial}{\partial \bar{\psi}}, \qquad \tilde{X}_0 = -\omega^2 u\frac{\partial}{\partial v}. \tag{3.11.9}$$

From (3.11.9) it is clear that $\bar{X}_0$ is a Jordan operator. Further, $\tilde{X}_0$ in this case turns out to be normal with respect to $\bar{X}_0$ (clearly it is a coincidence). In fact, v is an invariant of X_0 and u corresponds to zero.

Therefore we can obtain a uniform normal form for $0 \leq \omega^2 < 1$, not only for $\omega \to 0$, as might be expected of the 0-point. Here the general form of the obtained equations is also easy to understand: $S_j u$, $S_j v$, $M_j u$, and $M_j v$ are linear in u and v, and monomials in $M_j u$ and $M_j v$ are of the form $\alpha u \psi^{m_1} \psi^{-m_2}$ and $\beta v \psi^{m_1} \psi^{-m_2}$ and may correspond to 0 only for $m_1 = m_2$. Clearly we get linear equations for u and v similar to (3.11.5).

Let us give an example of computation with accuracy ε^2 included with the appropriate choice of S_1. The operator $M_1 = [X_0, S_1] + X_1$ can be "killed." Computing $S_1 u$ and $S_1 v$ in the form of polynomials linear in u and v with coefficients depending on ψ and $\bar{\psi}$ which are first order polynomials, we get

$$
\begin{aligned}
S_1 &= \frac{1}{8(\omega^2-1)}\left\{(\psi+\bar{\psi})\left(u\frac{\partial}{\partial u} - v\frac{\partial}{\partial v}\right) + i(\psi-\bar{\psi})\left(v\frac{\partial}{\partial u} + u\frac{\partial}{\partial v}\right)\right\} \\
&\quad - \frac{1}{8}(\psi-\bar{\psi})u\frac{\partial}{\partial v}, \\
M_1 &= 0.
\end{aligned}
$$

Further, $M_2 = [X_0, S_2] + \frac{1}{2}[X_1, S_1]$, where

$$
\frac{1}{2}[X_1, S_1] = \frac{1}{32(\omega^2-1)}\left\{2(\psi+\bar{\psi})^2 u\frac{\partial}{\partial v} - i(\psi^2-\bar{\psi}^2)\left(u\frac{\partial}{\partial u} - v\frac{\partial}{\partial v}\right)\right\}.
$$

Hence,

$$
M_2 = \frac{\psi\bar{\psi}}{8(\omega^2-1)} u\frac{\partial}{\partial v}
$$

and $M_3 = 0$. The equations in new variables are

$$
\frac{du}{dt} = v + 0(\varepsilon^4), \qquad \frac{dv}{dt} = \left(-\omega^2 + \frac{\varepsilon^2}{8(\omega^2-1)}\right)u + 0(\varepsilon^4).
$$

The transition curve is $\omega = i\varepsilon/2\sqrt{2} + 0(\varepsilon^3)$ (at $\omega^2 > 0$ there are no transition curves). In conclusion, note that the replacement of $\cos 2t$ in (3.11.1) by any periodic function does not produce (in a sense) anything new.

The following example also illustrates section 3.10 but is much more complicated than the preceding linear example.

3.12 Example: Oscillating Spring

We consider a pendulum with two degrees of freedom: the mass point suspended on a spring oscillates in the vertical plane (see Fig. 4); the spring is massless. Put k for the stiffness factor, ℓ for the length of the spring in the state of equilibrium and m for the mass. Put also $\omega = \sqrt{g/\ell}$ for the frequency of small oscillations of the mathematical pendulum of length ℓ and $\mu = \sqrt{k/mg + 1}$. Introduce a Cartesian coordinate system with the center at the point 0, the equilibrium point of the load, and axes oriented along vertical and horizontal lines (see Fig. 4). Denote the coordinates of the load by ℓx, ℓy. The length of the spring is ℓR, where $R = ((1+x)^2 + y^2)^{1/2}$. The tension of the spring is

$$T = k\frac{\ell R - \ell_0}{\ell_0},$$

where ℓ_0 is the length of unloaded spring. On the other hand, since ℓ is the length in the state of equilibrium, then $k(\ell - \ell_0)/\ell_0 = mg$. Inserting $\ell_0 = k\ell/(k + mg)$ in T we have $T = (k + mg)R - k$. From there it is clear that $[(k + mg)/m\ell]^{1/2}$ is the frequency of oscillations of the load when the spring is in the vertical position and μ is the ratio of this frequency and ω.

Components of the force operating upon the load are

$$F_x = mg - T\frac{1+x}{R}, \qquad F_y = -T\frac{y}{R}.$$

Newtonian equations of motion are $m\ell\ddot{x} = F_x$, $m\ell\ddot{y} = F_y$. Introduce the dimension-free time $\tau = \omega t$. Then equations of motion take the form

$$\frac{d^2x}{d\tau^2} = -\mu^2 x - (\mu^2 - 1)\left(1 - \frac{\partial R}{\partial x}\right),$$
$$\frac{d^2y}{d\tau^2} = -\mu^2 y + (\mu^2 - 1)\frac{\partial R}{\partial y}. \qquad (3.12.1)$$

Remark 3.12.1 For small x and y we have $\partial R/\partial x \approx 1$, $\partial R/\partial y \approx y$ so that equations for small oscillations are $x'' + \mu^2 x = 0$, $y'' + y = 0$.

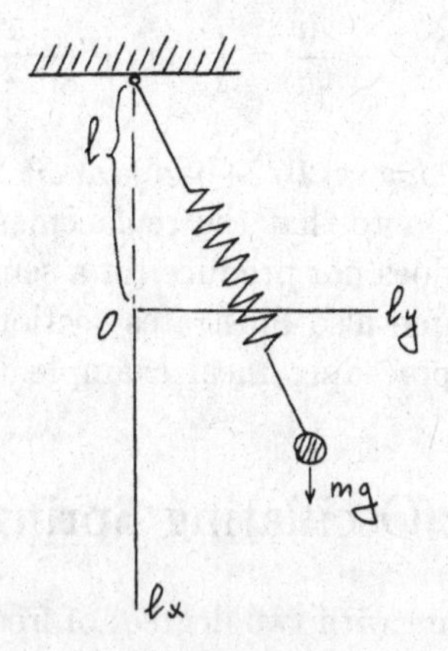

FIGURE 3.4.

Remark 3.12.2 For $\mu = 1$ oscillations are harmonic. The trajectory of the load is an ellipse. Physically, this motion is performed only approximately for a weak spring ($k \ll mg$).

Putting $x' = u$, $y' = v$, we pass to the system of first order equations, which turns out to be Hamiltonian,

$$\frac{dx}{d\tau} = \frac{\partial H}{\partial u}, \qquad \frac{du}{d\tau} = -\frac{\partial H}{\partial x}, \qquad \frac{dy}{d\tau} = \frac{\partial H}{\partial v}, \qquad \frac{dv}{d\tau} = -\frac{\partial H}{\partial y}, \tag{3.12.2}$$

with the Hamiltonian given by

$$H = \frac{1}{2}(u^2 + v^2) + \frac{\mu^2}{2}(R^2 - 1) - (\mu^2 - 1)(R - 1) - x,$$

where the constant in H is chosen so that $H(0,0,0,0) = 0$.

We will study the motion near the equilibrium state during large time τ. Let us replace x, y, u, and v by εx, and εy, εu, εv and H by $\varepsilon^2 H$. The system (3.12.2) remains in the same form and

$$H = H_0 + \varepsilon H_1 + \varepsilon^2 H_2 + \varepsilon^3 H_3 + \cdots,$$

where

$$\begin{aligned} H_0 &= \frac{1}{2}(u^2 + v^2 + \mu^2 x^2 + y^2), \\ H_1 &= \frac{\mu^2 - 1}{2} x y^2, \\ H_2 &= \frac{\mu^2 - 1}{2}\left(\frac{y^4}{4} - x^2 y^2\right), \\ H_3 &= \frac{\mu^2 - 1}{2}\left(x^3 y^2 - \frac{3}{4} x y^4\right), \\ &\vdots \end{aligned} \tag{3.12.3}$$

Remark 3.12.3 The Hamiltonian H_0 is positive definite. The equilibrium is stable. We will consider subsequently $\tau \ll 1/\varepsilon^2$, $\tau \ll 1/\varepsilon^3$, ... (for $\tau \ll 1/\varepsilon$ we have two oscillators whose interaction can be disregarded; see Remark 3.12.1).

The technique of computation which makes use of the fact that the system is Hamiltonian will be illustrated with higher approximations. For $\tau \ll 1/\varepsilon^2$ we write explicitly

$$\frac{dx}{d\tau} = u, \qquad \frac{dy}{d\tau} = v, \qquad \frac{du}{d\tau} = -\mu^2 x - \varepsilon \frac{\mu^2 - 1}{2} y^2,$$

$$\frac{dv}{d\tau} = -y - \varepsilon(\mu^2 - 1) x y.$$

The operator is $X = X_0 + \varepsilon X_1$, where

$$X_0 = u\frac{\partial}{\partial x} - \mu^2 x\frac{\partial}{\partial u} + v\frac{\partial}{\partial y} - y\frac{\partial}{\partial v},$$
$$X_1 = -\frac{\mu^2-1}{2}\left(y^2\frac{\partial}{\partial u} + 2xy\frac{\partial}{\partial v}\right). \tag{3.12.4}$$

The eigenfunctions of X_0 are

$$\varphi = u + i\mu x, \qquad \bar{\varphi} = u - i\mu x, \qquad \psi = v + iy, \qquad \bar{\psi} = v - iy.$$

We have $X_0\varphi = i\mu\varphi$, $X_0\psi = i\psi$,

$$X_1\varphi = \frac{\mu^2-1}{8}(\psi - \bar{\psi})^2, \quad X_1\psi = \frac{\mu^2-1}{4\mu}(\varphi - \bar{\varphi})(\psi - \bar{\psi}).$$

Considering monomials which enter $X_1\varphi$, $X_1\psi$, we find only a single resonance, $\mu = 1$, for $\mu > 0$. We get

$$M_1\varphi = \frac{\mu^2-1}{8}\psi^2, \qquad M_1\psi = -\frac{\mu^2-1}{4}\varphi\bar{\psi}.$$

The system in new variables is

$$\begin{aligned} \frac{d\varphi}{d\tau} &= i\mu\varphi + \varepsilon\frac{\mu^2-1}{8}\psi^2, & \frac{d\psi}{d\tau} &= i\psi - \varepsilon\frac{\mu^2-1}{4\mu}\varphi\bar{\psi}, \\ \frac{d\bar{\varphi}}{d\tau} &= -i\mu\bar{\varphi} + \varepsilon\frac{\mu^1-1}{8}\bar{\psi}^2, & \frac{d\bar{\psi}}{d\tau} &= -i\bar{\psi} - \varepsilon\frac{\mu-1}{4\mu}\bar{\varphi}\psi. \end{aligned} \tag{3.12.5}$$

It describes the main asymptotics for $\tau \ll 1/\varepsilon^2$. The system (3.12.5) possesses two first integrals

$$2\varphi\bar{\varphi} + \mu\psi\bar{\psi} = c_1 = \text{const},$$
$$i\varepsilon\frac{\mu^2-1}{4\mu}(\varphi\bar{\psi}^2 - \bar{\varphi}\psi^2) - (\mu-2)\psi\bar{\psi} = \varepsilon c_2 - (\mu-2)\rho_0$$
$$= \text{const}. \tag{3.12.6}$$

Setting $\psi\bar{\psi} = \rho$ we find from (3.12.5) that

$$\rho' = -\varepsilon\frac{\mu^2-1}{4\mu}(\varphi\bar{\psi}^2 + \bar{\varphi}\psi^2).$$

This and the second first integral (3.12.6) make it possible to express $\varepsilon\varphi\bar{\psi}^2$, $\varepsilon\bar{\varphi}\psi^2$ in terms of ρ, ρ'. Furthermore,

$$(\varepsilon\varphi\bar{\psi}^2)(\varepsilon\bar{\varphi}\psi^2) = \varepsilon^2\varphi\bar{\varphi}\rho^2 = \tfrac{1}{2}\varepsilon^2\rho^2(c_1 - \mu\rho)$$

due to the first of the integrals in (3.12.6). We have obtained the equation for ρ:

$$\rho'^2 = \frac{(\mu^2-1)^2}{8\mu^2}\varepsilon^2(c_1 - \mu\rho)\rho^2 - [\varepsilon c_2 + (\mu - 2)(\rho - \rho_0)]^2. \qquad (3.12.7)$$

where $\rho_0 = \rho(0)$. The problem of integration of (3.12.5) is now reduced to that of (3.12.7). In fact,

$$\begin{aligned} \varepsilon\psi^2 &= \left(\tfrac{\varepsilon\bar{\varphi}\psi^2}{\varphi\bar{\varphi}}\right)\varphi, & \varepsilon\bar{\psi}^2 &= \left(\tfrac{\varepsilon\varphi\bar{\psi}^2}{\varphi\bar{\varphi}}\right)\bar{\varphi}, \\ \varepsilon\varphi\bar{\psi} &= \left(\tfrac{\varepsilon\varphi\bar{\psi}^2}{\psi\bar{\psi}}\right)\psi, & \varepsilon\bar{\varphi}\psi &= \left(\tfrac{\varepsilon\bar{\varphi}\psi^2}{\psi\bar{\psi}}\right)\bar{\psi}. \end{aligned} \qquad (3.12.8)$$

The quantities in parentheses can be expressed as functions of ρ and ρ'. Inserting (3.12.8) into (3.12.5) we find

$$\varphi = \exp\int\left(i\mu + \frac{\mu^2-1}{8}\left(\frac{\varepsilon\bar{\varphi}\psi^2}{\varphi\bar{\varphi}}\right)\right)dt, \qquad \text{etc.}$$

It follows immediately from (3.12.7) that for $|\mu - 1| \gg \varepsilon$ we have $\rho \approx \rho_0 =$ const, as should be expected, and for $|\mu - 2| \leq \varepsilon$ we have $\rho = \rho(\varepsilon\tau)$ in the form of an elliptic Jacobi function.

Now let us demonstrate the technique of computing higher approximations. We will now exclude the resonance $\mu = 2$ from consideration. Set

$$\begin{aligned} \varphi &= \tfrac{e^{-i\pi/4}}{\sqrt{2\mu}}(u + i\mu x), & \psi &= \tfrac{e^{-i\pi/4}}{\sqrt{2}}(v + iy), \\ X_0\varphi &= i\mu\varphi, & X_0\psi &= i\psi, \\ \tilde{\varphi} &= \tfrac{e^{-i\pi/4}}{\sqrt{2\mu}}(u - i\mu x), & \tilde{\psi} &= \tfrac{e^{-i\pi/4}}{\sqrt{2}}(v - iy), \\ X_0\tilde{\varphi} &= -i\mu\tilde{\varphi}, & X_0\tilde{\psi} &= -i\tilde{\psi}. \end{aligned} \qquad (3.12.9)$$

Remark 3.12.4 Factors $(1/\sqrt{2\mu})e^{-i\pi/4}$, $(1/\sqrt{2})e^{-i\pi/4}$ are introduced so that the change of variables (3.12.9) is canonical: which means that the following identity must be true for an arbitrary function K.

$$\frac{\partial K}{\partial u}\frac{\partial}{\partial x} - \frac{\partial K}{\partial x}\frac{\partial}{\partial u} + \frac{\partial K}{\partial v}\frac{\partial}{\partial y} - \frac{\partial K}{\partial y}\frac{\partial}{\partial v} = \frac{\partial K}{\partial\tilde{\varphi}}\frac{\partial}{\partial\varphi} - \frac{\partial K}{\partial\varphi}\frac{\partial}{\partial\tilde{\varphi}} + \frac{\partial K}{\partial\tilde{\psi}}\frac{\partial}{\partial\psi} - \frac{\partial K}{\partial\psi}\frac{\partial\tilde{\psi}}{\partial}.$$

We seek S_1 in the form of a Hamiltonian operator with the Hamiltonian G_1 so that $M_1 = [X_0, S_1] + X_1$ is a Hamiltonian operator with the Hamiltonian $X_0G_1 + H_1$ [see section 2.8]. Due to (3.12.3) and (3.12.9) we have

$$H_1 = \frac{\mu^2-1}{4\sqrt{2\mu}}e^{i\pi/4}(\psi^2\tilde{\varphi} - \tilde{\psi}^2\varphi + \tilde{\psi}^2\tilde{\varphi} - \psi^2\varphi + 2\psi\tilde{\psi}\varphi - 2\psi\tilde{\psi}\tilde{\varphi}).$$

Seeking G_1 as a combination of the same monomials which constitute H_1, we get, for $\mu \neq 2$,

$$G_1 = \frac{i(\mu^2-1)}{4\sqrt{2\mu}}e^{i\pi/4}\left(-\frac{\psi^2\tilde{\varphi} + \tilde{\psi}^2\varphi}{\mu - 2} - \frac{\tilde{\psi}^2\tilde{\varphi} + \psi^2\varphi}{\mu + 2} + \frac{2\psi\tilde{\psi}\varphi + 2\psi\tilde{\psi}\tilde{\varphi}}{\mu}\right),$$

$$X_0G_1 + H_1 = 0 \quad (M_1 = 0).$$

Since $M_1 = 0$, we have

$$M_2 = [X_0, S_2] + X_2 + \tfrac{1}{2}[X_1, S_1],$$

where X_2 is the operator with the Hamiltonian

$$H_2 = -\frac{\mu^2 - 1}{32}(\mu(\psi - \tilde{\psi})^4 - 4(\varphi - \tilde{\varphi})^2(\psi - \tilde{\psi})^2)$$

and $\frac{1}{2}[X_1, S_1]$ is the operator with the Hamiltonian $\frac{1}{2}\{H_1, G_1\}$ ($\{,\}$ is the Poisson brackets), which is easy to compute in the variables in (3.12.9) (see Remark 3.12.4).

We get

$$\begin{aligned}\frac{1}{2}\{H_1, G_1\} = \frac{(\mu^2-1)^2}{32\mu^2(\mu^2-4)}\Big[&- 4(\mu - 2)(\tilde{\varphi}^2\tilde{\psi}^2 + \varphi^2\psi^2) \\ &- 4(\mu + 2)(\varphi^2\tilde{\psi}^2 + \tilde{\varphi}^2\psi^2) + \mu^2(\tilde{\psi}^4 + \psi^4) \\ &- 4(\mu^2 - 2)(\psi\tilde{\psi}^3 + \psi^3\tilde{\psi}) + 8\mu\varphi\tilde{\varphi}(\tilde{\psi}^2 + \psi^2) \\ &+ 8\mu\psi\tilde{\psi}(\tilde{\varphi}^2 + \varphi^2) + 2(3\mu^2 - 8)\psi^2\tilde{\psi}^2 - 16\mu\varphi\tilde{\varphi}\psi\Big],\end{aligned}$$

$$\begin{aligned}H_2 + \frac{1}{2}\{H_1, G_1\} = \frac{\mu^2 - 1}{32\mu^2(\mu^2-4)}\Big[&4(2\mu + 1)(\mu - 2)(\tilde{\varphi}^2\tilde{\psi}^2 + \varphi^2\psi^2) \\ &- 4(2\mu - 1)(\mu + 2)(\varphi^2\tilde{\psi}^2 + \tilde{\varphi}^2\psi^2) + 3\mu^2(\tilde{\psi}^4 + \psi^4) \\ &- 4(\mu^2 + 2)(\psi\tilde{\psi}^3 + \psi^3\tilde{\psi}) + 24\mu\varphi\tilde{\varphi}(\tilde{\psi}^2 + \psi^2) \\ &+ 2(\mu^2 + 8)\psi^2\tilde{\psi}^2 - 48\mu\varphi\tilde{\varphi}\psi\tilde{\psi}\Big].\end{aligned}$$

Besides $\psi^2\tilde{\psi}^2$ and $\varphi\tilde{\varphi}\psi\tilde{\psi}$, which certainly correspond to zero, there are monomials $\varphi^2\tilde{\psi}^2$, $\tilde{\varphi}^2$, and ψ^2 corresponding to zero if $\mu = 1$. But $\mu = 1$ is not a resonance point since these monomials have the factor $\mu - 1$. More generally, $\mu = 1$ is not a resonance point in all orders. Thus, there exists no resonance 2-point if $\mu \neq 2$ and μ is "far" from 2. Seeking the operator S_2 with the Hamiltonian G_2 we get

$$X_0G_2 + H_2 + \frac{1}{2}\{H_1, G_1\} = \frac{\mu^2 - 1}{16\mu^2(\mu^2-4)}[(\mu^2 + 8)\psi^2\tilde{\psi}^2 - 24\mu\varphi\tilde{\varphi}\psi\tilde{\psi}],$$

$$\begin{aligned}G_2 = \frac{i}{16\mu^2}\Big[&\frac{(\mu-1)(2\mu+1)}{\mu+2}(\varphi^2\psi^2 - \tilde{\varphi}^2\tilde{\psi}^2) \\ &+ \frac{(\mu+1)(2\mu-1)}{\mu-2}(\tilde{\varphi}^2\psi^2 - \varphi^2\tilde{\psi}^2)\Big] \\ &+ \frac{i(\mu^2-1)}{64\mu^2(\mu^2-4)}\Big[\frac{3}{2}\mu^2(\psi^4 - \tilde{\psi}^4) + 4(\mu^2 + 2)(\psi\tilde{\psi}^3 - \psi^3\tilde{\psi}) \\ &+ 24\mu\varphi\tilde{\varphi}(\psi^2 - \tilde{\psi}^2) + 24\psi\tilde{\psi}(\varphi^2 - \tilde{\varphi}^2)\Big].\end{aligned}$$

The Hamiltonian

$$H_0 + \varepsilon^2 \left(X_0 G_2 + H_2 + \tfrac{1}{2}\{H_1, G_1\} \right)$$

corresponds to $X_0 + \varepsilon M_1 + \varepsilon^2 M_2$ since $M_1 = 0$. It is easy to see that, in the new variables, integrals of the system are $\varphi\tilde{\varphi} = \text{const}$ and $\psi\tilde{\psi} = \text{const}$, so that when there is no resonance $\mu = 2$ we will get again harmonic oscillations at $\tau \ll 1/\varepsilon^3$ but with frequencies depending on initial values as in the Duffing oscillator. It is useful to notice the difference in the behavior of the system in the presence of a resonance and in its absence.

Consider also the case $\tau \ll 1/\varepsilon^4$. Taking into account that $M_1 = 0$ let us write M_3 in the form $M_3 = [X_0 S_3] + Y_3$, where

$$Y_3 = X_3 + \tfrac{1}{2}[X_1, S_2] + \left(\tfrac{1}{2}(M_2 + X_2) + \tfrac{1}{12}[X_1, S_1], S_1 \right).$$

The Hamiltonian corresponding to Y_3 is

$$\begin{aligned} &H_3 + \tfrac{1}{2}\{H_1, G_2\} \\ &\quad + \left\{ \tfrac{1}{2}\left(X_0 G_2 + H_2 + \tfrac{1}{2}\{H_1, G_1\} \right) + \tfrac{1}{2} H_2 + \tfrac{1}{12}\{H_1, G_1\}, G_1 \right\}, \end{aligned}$$

where G_2, $X_0 G_2 + H_2 + \frac{1}{2}\{H_1, G_1\}$, $\{H_1, G_1\}$, and G_1 are computed above. Examining monomials which constitute this Hamiltonian we find two new resonances: $\mu = 2/3$, $\mu = 4$, only the second of them having a physical meaning. Only two monomials, $\tilde{\varphi}\psi^4$ and $\varphi\tilde{\psi}^4$, can correspond to zero at $\mu = 4$. The Hamiltonian corresponding to M_3 is

$$-\frac{i\gamma e^{i\pi/4}}{\sqrt{2\mu}} (\tilde{\varphi}\psi^4 - \varphi\tilde{\psi}^4),$$

where $\gamma = \gamma(\mu)$ may be evaluated at $\mu = 4$; and (rather long) computation gives $\gamma(4) = 15/256$.

While calculating the system corresponding to $X_0 + \varepsilon^2 M_2 + \varepsilon^3 M_3$, let us perform the transformation

$$\varphi \mapsto (e^{-i\pi/4}/\sqrt{2\mu})\varphi, \qquad \tilde{\varphi} \mapsto (e^{-i\pi/4}/\sqrt{2\mu})\tilde{\varphi},$$

$$\psi \mapsto (e^{-i\pi/4}/\sqrt{2})\psi, \qquad \tilde{\psi} \mapsto (e^{-i\pi/4}/\sqrt{2})\tilde{\psi}.$$

Then, returning to notations $\varphi = u + i\mu x$ and $\psi = v + iy$, we get

$$\begin{aligned} \frac{d\varphi}{d\tau} &= i(\mu + \varepsilon^2 \alpha\psi\bar{\psi})\varphi - \varepsilon^3 \frac{\gamma}{4}\psi^4, \\ \frac{d\psi}{d\tau} &= i\left(1 + \varepsilon^2\left(\frac{\alpha}{\mu}\varphi\bar{\varphi} - \beta\psi\bar{\psi}\right)\right)\psi + \varepsilon^3 \frac{\gamma}{\mu}\varphi\bar{\psi}^3, \\ \frac{d\bar{\varphi}}{d\tau} &= -i(\mu + \varepsilon^2 \alpha\psi\bar{\psi})\bar{\varphi} - \varepsilon^3 \frac{\gamma^{\bar{4}}}{4}\bar{\psi}, \\ \frac{d\bar{\psi}}{d\tau} &= -i\left(1 + \varepsilon^2\left(\frac{\alpha}{\mu}\varphi\bar{\varphi} - \beta\psi\bar{\psi}\right)\right)\bar{\psi} + \varepsilon^3 \frac{\gamma}{\mu}\bar{\varphi}\psi^3, \end{aligned}$$

where

$$\alpha = \frac{3(\mu^2-1)}{4\mu(\mu^2-4)}, \qquad \beta = \frac{(\mu^2-1)(\mu^2+8)}{16\mu^2(\mu^2-4)}.$$

This system describes the principal asymptotics for $\tau \ll 1/\varepsilon^4$ in the absence of the resonance $\mu = 2$ and serves for the resonance $\mu = 4$. The system possesses integrals

$$4\varphi\bar{\varphi} + \mu\psi\bar{\psi} = c_1 = \text{const},$$

$$\frac{i\varepsilon^3}{\mu}(\bar{\varphi}\psi^4 - \varphi\bar{\psi}^4) + (4-\mu)\psi\bar{\psi} - \frac{\varepsilon^2}{2}[(\alpha+4\beta)(\psi\bar{\psi})^2 + \frac{16\alpha}{\mu^2}(\varphi\bar{\varphi})^2]$$
$$= \varepsilon^3 c_2 + (4-\mu)\rho_0 + \varepsilon^2\text{const} = \text{const}$$

and is reduced to quadratures exactly as was done above for the resonance $\mu = 2$.

Remark 3.12.5 The general consideration, which we will avoid, shows that there are perhaps only $(2n+1)$-points of resonance, where $n = 0, 1, 2, \ldots$, so that $\mu = 2(n+1)$ are resonances having physical meaning (we are speaking about *real* resonances, not about all mathematically possible ones).

In the next example considerations from Section 3.10 will be used to construct periodic solutions.

3.13 Example: Periodic Solution (Hopf's Theorem)

Consider the system

$$\frac{dx}{dt} = F(x, \mu),$$

where $x = \{x_1, \ldots, x_n\}$, μ is a parameter, and $F(x,\mu)$ is a vector function smooth in x and μ for sufficiently small x, μ. Let $F(0,\mu) = 0$ so that $x = 0$ is the equilibrium point. Putting $x = \varepsilon y$, where $\varepsilon \ll 1$, $y \sim 1$, we will write our system in the form

$$\frac{dy}{dt} = A(\mu)y + \varepsilon P_1(y,\mu) + \varepsilon^2 P_2(y,\mu) + \cdots. \tag{3.13.1}$$

Suppose that the matrix $A(\mu)$ possesses two simple eigenvalues $\lambda(\mu) = \beta(\mu) + i\alpha(\mu)$ and $\bar{\lambda}(\mu) = \beta(\mu) - i\alpha(\mu)$ which degenerate at $\mu = 0$ into a pair of purely imaginary ones, i.e., $\beta(0) = 0$, $\alpha(0) > 0$, and that other eigenvalues $\nu(\mu)$ are not proportional to $\pm i\alpha(0)$ at $\mu = 0$; e.g., $\operatorname{Re}\nu(0) < 0$. We may assume that

$$A(\mu) = \begin{Vmatrix} \lambda(\mu) & 0 & \\ 0 & \bar{\lambda}(\mu) & 0 \\ & 0 & B(\mu) \end{Vmatrix}.$$

Then the system (3.13.1) at $\varepsilon = 0$ has particular solutions

$$y_3 = \cdots = y_n = 0, \qquad \frac{dy_1}{dt} = \lambda(\mu)y_1, \qquad \frac{dy_2}{dt} = \bar{\lambda}_2(\mu)y$$

which are periodic at $\mu = 0$. It turns out that for sufficiently small $\varepsilon \neq 0$ and sufficiently small μ (perhaps only for $\mu = 0$) and under some other conditions [e.g., $\beta'(0) \neq 0$] which will be discussed below, the system (3.13.1) possesses also nontrivial particular periodic solutions which constitute a two-dimensional manifold Γ situated close to $y_3 = \cdots = y_n = 0$. Note that for a fixed μ such that $\beta(\mu) \neq 0$ and for a sufficiently small ε there are no periodic solutions close to $y_3 = \cdots = y_n = 0$. The existence of these solutions is stated by Hopf's theorem [22] whose different aspects are broadly discussed in [8]. Below we will show how to compute the asymptotics of Hopf's solutions and investigate their stability.

Note that the condition that eigenvalues $\nu(\mu)$ of $B(\mu)$ [see the expression for $A(\mu)$ above] at $\mu = 0$ are not proportional to $\pm i\alpha(0)$ implies $\nu(0) \neq m_1\lambda(0) + m_2\bar{\lambda}(0)$, where m_1 and $m_2 \geq 0$ are integers and $m_1 + m_2 \geq 1$ for any $\nu(\mu)$ (when μ is sufficiently small). Since (3.13.1) is a standard system (see Section 3.10) with a linear leading part and polynomial perturbations, then a change of variables close to the identity can reduce the system (3.13.1) to the form

$$\frac{dy}{dt} = A(\mu)y + \varepsilon\tilde{P}_1(y,\mu) + \varepsilon^2\tilde{P}_2(y,\mu) + \cdots,$$

where the components $\tilde{p}_{sk}(y,\mu)$ of the vector functions $\tilde{P}_s(y,\mu)$ for $s = 3,\ldots,n$ do not contain monomials $y_1^{m_1}y_2^{m_2}$. This means that the system (3.13.1) has a two-dimensional manifold of solutions Γ' whose equations in new variables is $y_s = 0$, where $s = 3,\ldots,n$.

In what follows the problem will be considered on Γ'; i.e., will be reduced to a two-dimensional one, and y stands for the vector $\{y_1, y_2\}$ accordingly.

Remark 3.13.1 We have made use of the knowledge [see subsection (4) of Section 3.10] of some properties of roots of $B(\mu)$ but their exact values will not be needed in computations to follow.

Since there are no conditions imposed on the perturbation, then $\mu = 0$ is, generally speaking, a resonance 1-point for $i > 0$.

Let us reduce the system to the normal form with respect to $A(0)$, where $\bar{X}_0 = (A(0)y, \nabla)$. Since y_1 and y_2 correspond to $i\alpha(0)$, $-i\alpha(0)$ respectively, then $y_1^{m_1}y_2^{m_2}$ corresponds to $i\alpha(0)(m_1 - m_2)$. In the normal form the two-dimensional system (3.13.1) is such that the monomials in the right-hand sides of dy_1/dt and dy_2/dt correspond to $i\alpha(0)$, $-i\alpha(0)$, respectively. Thus we obtain that $m_1 - m_2 = 1$ or -1 for the monomial of degree $j+1 = m_1 + m_2$, i.e., $m_1 = j+2/2$, $m_2 = j/2$ or $m_1 = j/2$, $m_2 = j+2/2$. Hence j must be even. The corresponding monomials are of the form $e_{j/2}(\mu)(y_1y_2)^{j/2}y_1$ and $\bar{e}_{j/2}(\mu)(y_1y_2)^{j/2}y_2$.

Putting $y_1 = \varphi$ and $y_2 = \bar{\varphi}$, where the overbar stands for complex conjugation, set $g = \varphi\bar{\varphi}$. Then the system takes the normal form

$$\begin{aligned}\frac{d\varphi}{dt} &= [\beta(\mu) + i\alpha(\mu) + \varepsilon^2 h(g,\mu,\varepsilon^2)]\varphi,\\ \frac{d\bar{\varphi}}{dt} &= [\beta(\mu) - i\alpha(\mu) + \varepsilon^2 \bar{h}(g,\mu,\varepsilon^2)]\bar{\varphi},\end{aligned} \tag{3.13.2}$$

where $h(g,\mu,\varepsilon^2) = c_1(\mu)g + \varepsilon^2 c_2(\mu)g^2 + \cdots$.

From (3.13.2) the equation for g follows:

$$\frac{dg}{dt} = 2[\beta(\mu) + \varepsilon^2 \operatorname{Re} h(g,\mu,\varepsilon^2)]g. \tag{3.13.3}$$

Equation (3.13.3) possesses a particular solution $g = g_0 = \text{const}$ if $\mu = \mu(\varepsilon)$ satisfies

$$\beta(\mu) + \varepsilon^2 \operatorname{Re} h(g_0,\mu,\varepsilon^2) = 0. \tag{3.13.4}$$

This equation (called the *branching equation*) is quite similar to the equation for transitional curves in Example 3.11. Corresponding solutions φ, $\bar{\varphi}$ are periodic since

$$\frac{d\varphi}{dt} = i[\alpha(\mu) + \varepsilon^2 \sin h(g_0,\mu,\varepsilon^2)]\varphi$$

and the period is, naturally, close to $2\pi/\alpha(0)$.

The condition for existence of such periodic solutions is the solvability condition for the branching equation (3.13.4). Under sufficiently general assumptions, (3.13.4) may be solved with the help of the Newton polygon method [45].

Since it is impossible to list all cases, let us indicate how to obtain the master equation which defines the principal term in the asymptotics of $\mu(\varepsilon)$. Let us expand the left-hand side of (3.13.4) into a series in powers of ε and μ. In what follows we will show that only a finite number of terms will be needed. Let

$$\beta(\mu) + \varepsilon^2 \operatorname{Re} h(g_0,\mu,\varepsilon^2) = \beta(\mu) + \varepsilon^{2m} a_1(\mu) + \varepsilon^{2m+2} a_2(\mu) + \cdots, \tag{3.13.5}$$

where $a_i(\mu)$ are power series in μ. Suppose that there is a monomial $c_k\varepsilon^{2k}$ in this decomposition which does not contain a power of μ. (otherwise we may assume that $\mu(\varepsilon) = 0$) and let $b = \min k$. Let us seek the principal term of the asymptotic in the form $\mu(\varepsilon) = x\varepsilon^\gamma + \cdots$. Substituting it in (3.13.5) we see that the lowest term of the decomposition is of the form

$$c_0\varepsilon^{2b} + \sum_{p,q} c_{pq}\varepsilon^{2p+\gamma q}, \tag{3.13.6}$$

where p, q, and γ are chosen from the condition $2p+\gamma q = 2b$ and $2p'+\gamma q' > 2b$ for other p', q' and the chosen γ. It is clear that after the first term of this form, $c_0\varepsilon^{2b}$, is found for the solution of this problem, further decomposition in powers of ε in (3.13.5) is unnecessary. After this, γ is found using the Newton polygon method. Equating the coefficients of ε^{2b} in (3.13.6) to zero we get the equation for x. Standard mathematical arguments using the implicit function theorem show that the solvability of this equation is equivalent to solvability of (3.13.4), i.e., to the existence of the periodic solution (perhaps not unique; the number of real roots is greater than 1). The case $\beta'(0) \neq 0$ is trivial. Then $\beta(\mu) \sim \beta'(0)\mu$ and the equation for x is $\beta'(0)x + c = 0$, where $c = \text{const}$. The condition $\beta'(0) \neq 0$ was indicated by Hopf. If we drop this condition, we evidently get many opportunities which will not be discussed here.

The stability of the obtained periodic solution is subject to a simple investigation. The question is reduced to the investigation of stability of the solution $g = g_0$ of (3.13.3) which is expressed, due to (3.13.4), in the form

$$\frac{dg}{dt} = 2\varepsilon^2[\operatorname{Re} h(g, \mu, \varepsilon^2) - \operatorname{Re} \varepsilon h(g_0, \mu, \varepsilon^2)]g.$$

In particular, if

$$\frac{d}{dg} \operatorname{Re} h(g, \mu, \varepsilon^2)|_{g=g_0} < 0,$$

then the solution is stable. If $c_1(0) \neq 0$ [see the expression for $h(g, \mu, \varepsilon^2)$ in (3.13.2)], the last condition is equivalent to $\operatorname{Re} c_1(0) < 0$. Evidently this is the condition of the stability of equilibrium for $\mu = 0$ since it is clear that at $\mu \neq 0$ the equilibrium is unstable. Note that the computation of $\operatorname{Re} c_i(0)$ is actually the computation of the so-called Lyapunov values [30].

The previous considerations can be generalized to the essentially nonlinear case. Let us consider briefly the two-dimensional situation. Let the system depend on two parameters μ and ε and be of the form

$$\frac{dx_k}{dt} = a_k(x, \mu) + \varepsilon b_k(x, \mu) \qquad (k = 1, 2). \tag{3.13.7}$$

Consider a domain D, where $x = (x_1, x_2)$ ranges, and let $\lambda, \bar{\lambda}$ and ψ, $\bar{\psi}$ be eigenvalues and eigenfunctions of

$$X_0 = a_1 \frac{\partial}{\partial x_1} + a_2 \frac{\partial}{\partial x_2} \qquad \text{for } x \in D$$

of the form $\psi = \psi(x, \mu)$, $\lambda = i\alpha(e, \mu) + \beta(e, \mu)$, where $e = \psi\bar{\psi}$. Further, let the operator X corresponding to (3.13.7) be reducible to the normal form

$$\tilde{X} = X_0 + \varepsilon N_1 + \varepsilon^2 N_2 + \cdots \tag{3.13.8}$$

and $N_k\psi = (i\alpha_k(e, \mu) + \beta_k(\varepsilon, \mu))\psi$. Next, assume that $\alpha(e, 0) \neq 0$ and $\beta(e, 0) = 0$ for sufficiently small $|\mu|$. But then for $\mu = 0$ to each initial value

ψ_0 there corresponds a periodic solution of the system $d\psi/dt = X_0\psi = \lambda\psi$ with period $\omega_0 = 2\pi/\alpha(e_0, 0)$. Under conditions similar to those introduced earlier we may show that for a sufficiently small $|\varepsilon|$, to each initial value ψ_0, $\bar{\psi}_0$ functions $\mu(\varepsilon, \psi_0, \bar{\psi}_0)$ and $\omega(\varepsilon, \psi_0, \bar{\psi}_0)$ correspond such that for $\mu = \mu(\varepsilon; \psi_0, \bar{\psi}_0)$ there is a periodic solution of (3.13.7) with period ω which passes through ψ_0, $\bar{\psi}_0$ and such that $\omega(0, \cdot) = \omega_0$, $\mu(0, \cdot) = 0$.

3.14 Example: Bifurcation

Let us make use of Example 3.13 to discuss the formulation of bifurcational problems. Consider the problem $f(x, \mu) = 0$, where $f(x, \mu)$ is an operator defined on elements x of the space X, which depends on the parameter μ and perhaps on some other parameters as yet implicit. In Hopf's theorem $f(x, \mu) = dx/dt - F(x, \mu)$ and the class of functions is the class of functions $x(t)$ periodic in t with a specified period. The general problem consists of the study of dependence of a solution, x, on the parameter μ. The bifurcational problem is usually the following one. Fix a known solution $x(\mu_0) = x_0$ and assuming that for $\mu = \mu_0$ the solution x_0 is unique, investigate the solution $x(\mu)$ which depends continuously on μ.

The first question which arises here is that of existence of $x(\mu)$ belonging to the given class X for μ close to μ_0 (e.g., the Hopf theorem), the next question being that of the quantity of solutions $x(\mu)$, and, of course whether they are stable. If there is more than one solution then this phenomenon is called a *bifurcation*. In the Hopf problem the stationary periodic solution $x = 0$ may grow into at least two solutions: the periodic one $x(\mu)$ and $x = 0$. In applications the presence of a bifurcation means a steep change in the behavior of the system, which explains the importance of defining the values of parameter for which a bifurcation of a solution arises. Naturally, the bibliography on bifurcational problems is voluminous. Here we intend only to discuss briefly several aspects of these problems connected with perturbation theory.

Suppose that for $\mu = 0$ we know a solution x_0 of the equation $f(x, 0) = 0$. Putting $x = x_0 + y$ we obtain

$$a_0(\mu) + L(\mu)y + \tilde{L}(\mu, y) = 0, \qquad L(\mu)y = \frac{\partial}{\partial\varepsilon} f(x_0 + \varepsilon y, \mu)|_\varepsilon = 0, \quad (3.14.1)$$

where $a_0(0) = 0$ and in the Hopf theorem the linear operator $L(\mu)y$ is $dy/dt - A(\mu)y$; it acts in the space of vector functions $y(t)$. In other problems, e.g., in problems with partial derivatives, it can act in much more sophisticated spaces. Rewrite (3.14.1) in the form

$$L(0)y + (L(\mu) - L(0))y + a_0(\mu) + \tilde{L}(\mu, y) = L(0)y + B(\mu, y) = 0. \quad (3.14.2)$$

First suppose that the linear operator $L(0)y$ is invertible. Then the equation

(3.14.2) is equivalent to the equation

$$y + L^{-1}(0)B(\mu, y) = 0 \tag{3.14.2'}$$

and we may expect that for small $|\mu|$ there is a unique solution of the problem. (In the mathematical literature this assumption is proved under various conditions.) In this case the solution x_0 is extended "by continuity" to the solution $x(\mu)$ and there is no bifurcation. Clearly, it is essential to study the degenerate case when the equation $L(0)y = 0$ possesses nontrivial solutions. [In the Hopf problem the equation $L(0)y = 0$ is $dy/dt - A(0)y = 0$. It has a $2\pi/\alpha(0)$-periodic solution, $\psi_0 e^{i\alpha(0)t}$.]

Now let us make the following assumptions:

1. Solutions of $L(0)y = 0$ constitute a finite-dimensional linear subspace with the basis $\psi_1, \ldots, \psi_k$.

2. The Fredholm alternative holds, i.e.; equations $L(0)y = b$ are solvable if and only if $(b, \psi_i^*) = 0$, where $i = 1, \ldots, k$, and ψ_i^* constitute a basis of linearly independent vectors which generate all the finite-dimensional space of solutions of the dual equation $L^*(0)y = 0$. (We do not define here notions of the scalar product, the norm, etc. In concrete problems these notions arise naturally.)

Now let us return to the initial problem. Denote by T the space of vectors orthogonal to $\psi_1^*, \ldots, \psi_k^*$. The Fredholm alternative means just that $L(0)$, which evidently maps T into T, possesses an inverse operator, $L^{-1}(0)$, on T and the condition $(b, \psi_i^*) = 0$ [of solvability of the equation $L(0)y = b$] means that $b \in T$, i.e., that $y = L^{-1}(0)b$. [Recall that $L^{-1}(0)$ is defined only on T.] Now put

$$y = \bar{y} + \sum_{1 \le p \le k} \alpha_p \psi_p,$$

where $\bar{y} \in T$. Substituting into (3.14.2) gives

$$L(0)\bar{y} + B\left(\mu, \bar{y} + \sum_{1 \le p \le k} \alpha_p \psi_p\right) = 0.$$

This equation is solvable if

$$\left(B\left(\mu, \bar{y} + \sum_{1 \le p \le k} \alpha_p \psi_p\right), \psi_s^*\right) = 0 \qquad \text{for } s = 1, \ldots, k \tag{3.14.3}$$

is satisfied, and then if (3.14.3) holds, we have

$$\bar{y} = -L^{-1}(0)B\left(\mu, \bar{y} + \sum_{1 \le p \le k} \alpha_p \psi_p\right). \tag{3.14.4}$$

For sufficiently small $|\mu|$ Equation (3.14.4) is usually solvable with respect to $\bar{y}$ [again if (3.14.3) holds]. Substituting this solution into (3.14.3) we get k equations to define α, the so-called *branching equation*; the investigation of them constitutes the subject of the theory. However, before we solve them we should write them in a convenient form. The very question of how to do this is illustrated in what follows with the help of some classical examples.

3.15 Example: Problem of a Periodic Solution of an Autonomous System

$$\frac{dx}{dt} = Ax + \varepsilon g(x), \qquad \text{where } A \text{ is a real matrix.} \tag{3.15.1}$$

Suppose that A possesses $p = 2m$ imaginary eigenvalues $\lambda_k = q_k i$, where q_k's are integers, other eigenvalues λ_s, where $s = p+1, \dots, n$, are not pure imaginary integers, and A is diagonal. In what follows we assume that A has no zero eigenvalues and nontrivial Jordan cells. Both these assumptions are made only for simplicity's sake.

Let us seek a periodic solution of (3.15.1) with period $2\pi(1+\varepsilon\omega(\varepsilon))$ which at $\varepsilon = 0$ turns into some as yet unknown solution, with period 2π, of the nonperturbed system. Putting $t = (1+\varepsilon\omega(\varepsilon))\tau$ we will obtain from (3.15.1) the equation

$$\frac{dx}{d\tau} = Ax + \varepsilon\omega Ax + (i + \varepsilon\omega)\varepsilon g(x). \tag{3.15.1'}$$

Put

$$\varphi = (\alpha_1 e^{iq_1\tau}, \bar{\alpha}_1 e^{-iq_1\tau}, \dots, \alpha_m e^{iq_m\tau}, \bar{\alpha}_m e^{-iq_m\tau}, 0, \dots, 0),$$

where $\pm iq_j$ are eigenvalues of A and let us seek $\omega(\varepsilon)$ and $\alpha(\varepsilon)$ so that $x = \varphi(\varepsilon, \tau) + \varepsilon y$ is a 2π-periodic solution of (3.15.1′).

Since $\varphi(\tau + h)$ is also a periodic solution of the nonperturbed system, then the number of essential real parameters which define φ is $2m - 1$ and not $2m$.

Remark 3.15.1 The representation $x = \varphi(\varepsilon, \tau) + \varepsilon y$ is not unambiguously defined. For example, putting $\varphi(\varepsilon, \tau) = \varphi(0, \tau) + \varepsilon\psi(\varepsilon, \tau)$, we get $x = \varphi(0, \tau) + \varepsilon(\psi(\varepsilon, \tau) + y)$, implying that y and consequently φ are defined up to an additive summand ψ. Therefore we may assume that, e.g., α_j's do not depend on ε and search for them and for initial values for y under the condition that x is 2π-periodic. We may normalize φ and y some other way. (For example, assuming $y(0) = 0$, then α_j will essentially depend on ε.)

The equation for y which follows from (3.15.1′) is

$$\frac{dy}{d\tau} = Ay + \omega A\varphi + \varepsilon\omega Ay + (1 + \varepsilon\omega)g(\varphi + \varepsilon y),$$

or, after expanding $g(\varphi + \varepsilon y)$ into a power series in ε,

$$\frac{dy}{d\tau} = Ay + (\omega A\varphi + g(\varphi)) + \varepsilon b_1(\tau, y) + \varepsilon^2 b_2(\tau, y) + \cdots. \tag{3.15.2}$$

The first order operator corresponding to (3.15.2) is $X_0 + \varepsilon B_1 + \varepsilon^2 B_2 + \cdots$ [we have put $\omega A\varphi + g(\varphi) = f(\tau) = \{f_1, f_{-1}, \dots, f_m, f_{-m}, \dots, f_n\}$], where

$$\begin{aligned} X_0 y_\ell &= iq_\ell y_\ell + f_\ell(\tau) \qquad (\ell = 1, \dots, m),\\ X_0 y_s &= \lambda_s y_s + f_s \qquad (s \geq p+1),\\ X_0 \tau &= 1. \end{aligned} \tag{3.15.3}$$

Setting $y_s = z_s + c_s$ for $s \geq p+1$ we get $X_0 z_s = \lambda_s z_s + \lambda_s c_s - \partial c_s/\partial\tau + f_s$. But λ_s are such that we may find 2π-periodic functions c_s which annihilate $\lambda_s c_s - \partial c_s/\partial\tau + f_s$. Similarly putting $y_s = \eta_s + d_s$ for $|s| \leq m$ we may find functions d_s with period 2π such that only resonance terms of corresponding f's will remain. After these substitutions, X_0 is reduced to the form

$$X_0 \eta_\ell = iq_\ell \eta_\ell + a_\ell^0 e^{iq_\ell \tau}, \qquad X_0 z_s = \lambda_s z_s, \qquad X_0 \tau = 1.$$

Eigenfunctions of X_0 are z_s and $e^{i\tau}$, and functions η_ℓ correspond to the eigenvalue iq_ℓ; i.e.,

$$X_0 \eta_\ell = iq_\ell \eta_\ell + a_\ell^0 e^{iq_\ell \tau}, \qquad X_0 e^{iq_\ell \tau} = iq_\ell e^{iq_\ell \tau}. \tag{3.15.4}$$

Now it is easy to demonstrate that X is reducible to the following normal form:

$$\begin{aligned} \tilde{X} z_s &= \lambda_s z_s + \varepsilon P_s^{(1)} + \varepsilon^2 P_s^{(2)} + \cdots,\\ \tilde{X} \eta_\ell &= iq_\ell \eta_\ell + (a_\ell^0 + \varepsilon a_\ell^1 + \cdots) e^{iq_\ell \tau} + \varepsilon r_\ell^0 + \varepsilon^2 r_\ell^1 + \cdots. \end{aligned} \tag{3.15.5}$$

Here $P_s^{(j)}$ vanish if $z_{p+1} = \cdots = z_n = 0$; functions r_ℓ^j correspond to eigenvalues iq_ℓ and vanish if $\eta_\ell = 0$ for all ℓ; functions $a_\ell^1, a_\ell^2, \dots$ depend on z_s's and on invariants of X_0 which vanish when $\eta_\ell = 0$ for all ℓ. Reasoning as above, set $z_s = 0$ and seek a 2π-periodic solution of the system

$$\frac{d\eta_\ell}{d\tau} = iq_\ell \eta_\ell + \overline{(a_\ell^0 + \varepsilon a_\ell^1 + \cdots) e^{iq_\ell \tau}} + \overline{\varepsilon r_\ell^0} + \cdots,$$

where the overbar means that in the corresponding expression $z_s = 0$. With the above remark taken into account we may assume that $\eta_\ell(0) = 0$. But then, clearly, $\eta_\ell \equiv 0$ and the branching equations are

$$\widehat{a_\ell^0 + \varepsilon a_\ell^1 + \cdots} = 0 \qquad \text{for } \ell = 1, \dots, p \tag{3.15.6}$$

where $\widehat{\ }$ means that in the corresponding expression $z_s = \eta_\ell = 0$.

3.16 Example: One Problem on Eigenvalues

Let us consider the equation

$$Au - \lambda u + \varepsilon B(u) = 0, \tag{3.16.1}$$

, where A is a linear operator with a discrete spectrum, B is nonlinear, and $B(0) = 0$. We suppose that u is an element of a space which might be just a vector space or a space of vector functions $u(x)$, where $x = (x_1, \dots, x_n)$, $u = (u_1(x), \dots, u_m(x))$. Usually we assume that $B(u)$ is a polynomial in u and (if u is a vector function) in its partial derivatives. Let φ_k, where $k = 1, 2, \dots$, be a system of eigenfunctions of A; i.e., $A\varphi_k = \lambda_k \varphi_k$, and let there be a simple method which enables one to express "any" vector u in the form $u = \sum_{k \geq 1} c_k \varphi_k$, (e.g., there is a scalar product and a system of biorthogonal functions ψ_k). Let $\lambda = \lambda_j$ be an eigenvalue of multiplicity $\ell < \infty$. Denote the corresponding eigenfunctions by $\eta_1, \dots, \eta_\ell$ retaining the notation φ for the rest. We do not suppose yet that A possesses Jordan cells.

Our problem is to find a solution ψ of (3.16.1), where $\lambda = \lambda_j + \omega$, $\omega \to 0$ as $\varepsilon \to 0$, and ψ passes into a linear combination of η's as $\varepsilon \to 0$. Let us write u in the form

$$u = \sum_{1 \leq s \leq \ell} \xi_s \eta_s + \sum c_k \varphi_k.$$

Then (3.16.1) may be rewritten in the form

$$z_s \equiv -\omega \xi_s + \varepsilon B_s(\xi, c) = 0, \qquad \text{where } s = 1, \dots, \ell, \tag{3.16.2}$$

$$G_k \equiv \tilde{\lambda}_k c_k - \omega c_k + \varepsilon B_k(\xi, c) = 0, \qquad \text{where } \tilde{\lambda}_k = \lambda_k - \lambda_j,\ k > \ell. \tag{3.16.2'}$$

We will be looking for a convenient method to write branching equations. Since $\tilde{\lambda}_k \neq 0$, then a branching equation arises when we exclude (with the needed accuracy) c_k from (3.16.2). There are several possibilities for approaching the problem. Let us list some.

(A) Let us seek c_k in the form $c_k = \varepsilon c_k^{(1)} + \varepsilon^2 c_k^{(2)} + \cdots$, supposing ω to be unknown. From (3.16.2′) we derive that

$$\begin{aligned} &\varepsilon c_k^{(1)} + \varepsilon^2 c_k^{(2)} + \cdots \\ &+ \frac{\varepsilon}{\tilde{\lambda}_k - \omega} \left(B_k(\xi, 0) + \sum \frac{\partial B_k(\xi, 0)}{\partial c_\mu} c_\mu + \sum \frac{1}{2} \frac{\partial^2 B_k(\xi, 0)}{\partial c_\mu \partial c_\nu} c_\mu c_\nu + \cdots \right) \\ &= 0. \end{aligned} \tag{3.16.3}$$

Inserting in (3.16.3) the decompositions for c_μ, we get

$$c_k^{(1)} = -\frac{1}{\tilde{\lambda}_k - \omega} B_k(\xi, 0), \qquad c_k^{(2)} = -\frac{1}{\tilde{\lambda}_k - \omega} \sum \frac{\partial B_k(\xi, 0)}{\partial c_\mu} c_\mu^{(1)}, \dots. \tag{3.16.4}$$

Substituting (3.16.4) in (3.16.2) gives the desired equations. Let, for instance, all components of B be homogeneous polynomials of degree 2 in variables ξ and c. Then (3.16.4) implies that

$$c_k^{(1)} = -\frac{1}{\tilde{\lambda}_k - \omega}\gamma_k^{1,\alpha,\beta}\xi_\alpha\xi_\beta,$$

$$c_k^{(2)} = -\frac{1}{\tilde{\lambda}_k - \omega}(\gamma_{k,\mu}^{2,\nu}\xi_\nu)c_\mu^{(1)}, \tag{3.16.5}$$

$$\vdots$$

i.e., $c_k^{(h)}$ are polynomials in ξ of degree $h+1$. Returning to (3.16.2) we come to equations

$$-\omega\xi_s + \varepsilon B_s(\xi,0) + \varepsilon^2\frac{\partial B_s(\xi,0)}{\partial c_\nu}c_\nu^{(1)}(\xi) + \cdots = 0, \qquad \text{where } s = 1,\ldots. \tag{3.16.6}$$

It is clear that various different possibilities arise. Let us discuss briefly the simple one when $B_s(\xi,0) \neq 0$ for all s. It is immediately clear that ω might be also taken in the form $\omega = \tilde{\omega}\varepsilon + \tilde{\omega}_2\varepsilon^2 + \cdots$. After this the main system to investigate will be $\tilde{\omega}\xi_s = B_s(\xi,0)$. If $\ell = 2$, then in principle the problem is simple, since putting $\xi_2/\xi_1 = \mu$ we obtain the cubic equation $\mu = B_2(1,\mu)/B_1(1,\mu)$ to define μ and then find also $\tilde{\omega}$. We will obtain a solution depending on an arbitrary parameter ξ_1, which is natural since we are seeking an "eigenfunction." The "eigenvalue" also may depend on ξ_1. Further approximations are obtained in a trivial way.

In the same example, let us see what will be changed when A possesses Jordan cells. The principal part of (3.16.6) can be rewritten in the form

$$-\omega\xi_1 + \varepsilon B_1(\xi,0) = 0, \tag{3.16.7}$$

$$-\omega\xi_2 + \xi_1 + \varepsilon B_2(\xi,0) = 0. \tag{3.16.7$'$}$$

From (3.16.7) and (3.16.7′) it follows that ξ_1 cannot be of order 1 since then $\omega \sim \varepsilon$ and (3.16.7′) leads to contradiction. Therefore, assuming $\xi_2 \sim 1$, we obtain from (3.16.7′) that $\xi_1 \sim \omega$ and from (3.16.7) that $\omega \sim \sqrt{\varepsilon}$. Therefore we may put $\xi_1 = \sqrt{\varepsilon}\tilde{\xi}_1$, $\omega = \sqrt{\varepsilon}\tilde{\omega}$. Then from (3.16.7) and (3.16.7′) equations of the first approximation arise in the form $-\tilde{\omega}\tilde{\xi}_1 + B_1(0,\tilde{\xi}_2;0) = 0, -\tilde{\omega}\xi_2 + \tilde{\xi}_1 = 0$ and two branches appear which depend on the parameter ξ_2. It is evident that the complete solution may be found in the form of expansions of $\tilde{\omega}$ and $\tilde{\xi}_1$ with respect to powers of $\varepsilon^{1/2}$.

Remark 3.16.1 Let us underline one circumstance which is important, though trivial. We should not hurry in guessing the functional dependence of all unknowns on the parameter. In our example the expansion of c_k with respect to powers of ε is evident, whereas the form of the expansion of ω is defined by the structure of A and should be deduced.

(B) Let us consider the variables (z_s, G_k) in (3.16.2) as a vector function defining a map F of the space H into itself. We seek a point $h \in H$ transformed into 0 under this transformation and tending as $\varepsilon \to 0$ to a point of the ℓ-dimensional subspace g defined by the equation $c = 0$. Since $F(0) = 0$ we may choose two transformations S_1 and S_2 of H, close to the identity as $\varepsilon \to 0$ and preserving 0; putting $S_1 F(S_2(h)) = F_1(h)$, solve the problem for the function F_1 while trying to select S_1, S_2 in a most convenient way. Setting $S_2 = E + \varepsilon\Gamma(\xi, c)$ and $S_1 = E - \varepsilon\Gamma(\xi, c)$, we will obtain from (3.16.2)

$$-\omega\xi + \varepsilon[-\omega\Gamma_s(\xi, c) - \Gamma_s(-\omega\xi, \tilde{\tilde{\lambda}}c) + B_s(\xi, c)] + \varepsilon^2[\cdot] = 0, \qquad (3.16.8)$$

$$\tilde{\tilde{\lambda}}_k c_k + \varepsilon[\tilde{\tilde{\lambda}}_k \Gamma_k(\xi, c) - \Gamma_k(-\omega\xi, \tilde{\lambda}c) + B_k(\xi, c)] + \varepsilon^2[\cdot] = 0. \qquad (3.16.8')$$

where $\tilde{\tilde{\lambda}}_k = \tilde{\lambda}_k - \omega$.

Choose Γ so that the coefficient of ε in (3.16.8′) vanishes at $c = 0$. Then we may continue the process and transform the left-hand sides of (3.16.8) and (3.16.8′) to the "normal" form so that all terms of (3.16.8′) vanish at $c = 0$. Then, putting $c = 0$, we get from (3.16.8) the branching equations for ξ. Suppose, e.g., that B_k's are again homogeneous forms of degree 2. Then from (3.16.8′), putting $\Gamma_k = \gamma_k^{\alpha\beta}\xi_\alpha\xi_\beta$, we get

$$(\tilde{\tilde{\lambda}}_k - \omega^2)\gamma_k^{\alpha\beta}\xi_\alpha\xi_\beta + b_k^{\alpha\beta}\xi_\alpha\xi_\beta = 0,$$

where $b_k^{\alpha\beta}\xi_\alpha\xi_\beta$ is the corresponding summand in B_k. Set other Γ to be 0.

(C) Finally, we may look at the left-hand sides of (3.16.2) as coefficients of the first order operator

$$X = z_s\frac{\partial}{\partial\xi_s} + G_k\frac{\partial}{\partial c_k} = -\omega\xi_s\frac{\partial}{\partial\xi_s} + (\tilde{\lambda}_k - \omega)c_k\frac{\partial}{\partial c_k} + \varepsilon X_1 + \cdots$$
$$= X_0 + \varepsilon X_1 + \cdots.$$

In such an interpretation we search for a fixed point of X. Then we may seek a transformation $\tilde{X} = e^{-\varepsilon S}Xe^{\varepsilon S}$, where $S = S_1 + \varepsilon S_2 + \cdots$ is a first order operator with 0 as a fixed point, pose the same problem for $\tilde{X}$, and look for an appropriate S.

All these methods are equivalent and the choice of them depends on the way the problem is reduced to Equations (3.16.2). For example, if the initial system of eigenfunctions is an orthonormal one, then the expansion $u = \sum \xi_s\eta_s + \sum c_k\varphi_k$ is a representation in the form $u = \sum \xi_s\eta_s + \psi$, where $\psi \perp \eta_s$ for all s. Since the branching equation contains explicitly only ξ_s, we should avoid, as far as possible, computations of Fourier coefficients of ψ. This will define the choice of the method (cf. [23], Ch. 10, Appendix).

Remark 3.16.2 It was supposed above that the spectrum of A is discrete. (In mathematical terminology A is usually supposed to be reducible to a completely continuous operator, which permits a rigorous justification of the computation procedure.) Most essential is the assumption of finite-dimensionality of the subspace of elements corresponding to the eigenvalue λ_j. Unfortunately, plenty of important problems do not satisfy this condition.

The next example illustrates Section 2.5 in a very simple situation as far as the construction of the canonical form is concerned.

3.17 Example: A. M. Lyapunov's Problem

In 1893 A. M. Lyapunov [31] considered the stability problem for the equilibrium point $x = y = 0$ of the system

$$\frac{dx}{dt} = y + Q_2(x,y) + Q_3(x,y) + \cdots,$$
$$\frac{dy}{dt} = R_2(x,y) + R_3(x,y) + \cdots, \qquad (3.17.1)$$

where Q_j and R_j are homogeneous polynomials of degree j in x and y.

L. G. Khazin [26], remaining in the frame of Lyapunov's ideas, simplified his investigations. We will consider this problem as an example of an application of the canonical form in the sense of Section 2.5 and the (generalized) shearing transformation.

(1) Reduction to the Canonical Form

For convenience's sake let us make the transformation $x \mapsto \varepsilon x$, $y \mapsto \varepsilon y$ so that the system (3.17.1) takes the form

$$\frac{dx}{dt} = y + \varepsilon Q_2(x,y) + \varepsilon^2 Q_3(x,y) + \cdots,$$
$$\frac{dy}{dt} = \varepsilon R_2(x,y) + \varepsilon^2 R_3(x,y) + \cdots. \qquad (3.17.1')$$

The corresponding operator is $X = X_0 + \varepsilon X_1 + \varepsilon^2 X_2 + \cdots$, where

$$X_0 = y\frac{\partial}{\partial x}, \qquad X_j = Q_{j+1}\frac{\partial}{\partial x} + R_{j+1}\frac{\partial}{\partial y} \qquad \text{for } j = 1, 2, \ldots.$$

Note that the leading operator X_0 is nilpotent. If we search for the operator S_j in the same form as for X_j, i.e., so that $S_j x$ and $S_j y$ are homogeneous polynomials of degree $j+1$ in x, y, then the system in new variables $x = Mx$ and $y = My$ will evidently have the same form as (3.17.1′). Due to Section 2.5, M may be obtained in the canonical form in the indicated sense. Let us show precisely what this means. In the notations of the Section 2.5 we have $x = e_{1/2}$ and $y = e_{-1/2}$. The normal basis in P_{N-1} is generated by the element $e_{1/2}^N$. In fact, the operator Z_+ from Section 2.5 satisfies $Z_+e_{1/2} = 0$ and $Z_+e_{-1/2} = e_{1/2}$, implying that the equation $Z_+p = 0$, where p is the sum of monomials corresponding to the same eigenvalues with respect to the operator Z_0, i.e., such that $Z_0e_s = se_s$, has the unique (up to a multiple) solution $e_{1/2}^{2r}$, where $r = N/2$.

Further, returning to formulas (2.5.6), we have $k = 1/2, r = N/2$, $\hat{e}_r = e_{1/2}^{2r}$, and m may be either $N - 1/2$ or $N + 1/2$.

For $m = (N-1)/2$ the index s might be either $-1/2$ or $1/2$ and

$$b_{-1/2}\left(\frac{N-1}{2}\right) = (N-1)!, \qquad b_{1/2} = \left(\frac{N-1}{2}\right) = N!,$$

respectively. For $m = (N+1)/2$ the index s might be only $-1/2$ and accordingly $b_{-1/2}(N + 1/2) = N!$.

Inserting the values obtained into (2.5.7) we find two generating elements of the normal basis in T_{N-1}; i.e.,

$$\begin{aligned} Y_{N-1/2}e_{-1/2} &= (N-1)!\hat{e}_{(N/2)-1}, \\ Y_{N-1/2}e_{1/2} &= N!\hat{e}_{N/2}, \\ Y^*_{N+1/2}e_{-1/2} &= N!\hat{e}_{N/2} \qquad (Y^*_{N+1/2}e_{1/2} = 0). \end{aligned}$$

Taking into account that $\hat{e}_{N/2} = x^N$, $\hat{e}_{(N/2)-1} = X_0\hat{e}_{N/2} = Ne_{1/2}^{N-1}e_{-1/2} = Nx^{n-1}y$ and dividing by $N!$, we get two operators

$$x^N\frac{\partial}{\partial x} + x^{N-1}y\frac{\partial}{\partial y}, \qquad x^N\frac{\partial}{\partial y} \tag{3.17.2}$$

which are generating elements of two chains of the normal basis of T_{N-1}. The index of the first of the operators (3.17.2) is $s = N - 1/2$ and that of the second one is $s = N + 1/2$. This answer might also be verified directly.

Thus we obtain

$$M_j = A_j\left(x^{j+1}\frac{\partial}{\partial x} + x^jy\frac{\partial}{\partial y}\right) + B_jx^{j+1}\frac{\partial}{\partial y},$$

where A_j and B_j are constants; hence, reduce the system (3.17.1′) to the form

$$\begin{aligned} \frac{dx}{dt} &= y + \varepsilon A_1x^2 + \varepsilon^2A_2x^3 + \cdots, \\ \frac{dy}{dt} &= \varepsilon(B_1x^2 + A_1xy) + \varepsilon^2(B_2x^3 + A_2x^2y) + \cdots. \end{aligned} \tag{3.17.3}$$

The concrete computation of coefficients A_j and B_j in terms of coefficients of the initial system will be omitted. Instead we show the relation between A_j and B_j and values introduced by A. M. Lyapunov (see Remark 3.17.2).

(2) The Model System

Our next aim is to single out in M a new leading operator. In M consider the summand

$$X_0 + \varepsilon^j B_j x^{j+1} \frac{\partial}{\partial y},$$

where $B_j \neq 0$. We have already mentioned the shearing transformation in accordance with (2.5.1), the fact that $x = e_{1/2}$, $y = e_{-1/2}$, and Remark 2.5.3:

$$x \mapsto \varepsilon^{-j/(4+j)} x, \qquad y \mapsto \varepsilon^{j/(4+j)} y, \tag{3.17.4}$$

so that, for a fixed j,

$$X_0 + \varepsilon^j B_j x^{j+1} \frac{\partial}{\partial y} \to \varepsilon^{2j/(4+j)} \left(X_0 + B_j x^{j+1} \frac{\partial}{\partial y} \right). \tag{3.17.5}$$

Under the transformation (3.17.4) we also get for any i that

$$\varepsilon^i M_i \to \varepsilon^{\frac{2j}{4+j}} \left\{ \varepsilon^{2(2i-j)/(4+j)} A \left(x^{i+1} \frac{\partial}{\partial x} + x^i y \frac{\partial}{\partial y} \right) + \varepsilon^{4(i-j)/(4+j)} B_i x^{i+1} \frac{\partial}{\partial y} \right\}. \tag{3.17.6}$$

Now let us express the result of the transformation (3.17.4) in the form $M \to \varepsilon^{2j/4+j} \tilde{M}$ and find under what conditions the leading summand in $\tilde{M}$ is correctly chosen; i.e., when in $\tilde{M}$ there are no negative powers of ε.

First, let j be odd: $j = 2n+1$, where $n = 0, 1, 2, \ldots$. Then, as is clear from (3.17.6), in $\tilde{M}$ there are no negative powers of ε when $A_i = 0$ for $i \leq n$, $B_i = 0$ for $i \leq 2n$.

Then, as is clear from (3.17.5) and (3.17.6), terms with the 0-th power of ε in $\tilde{M}$ are exhausted by the summand

$$X_0 + B_j x^{j+1} \frac{\partial}{\partial y}, \qquad \text{where } j = 2n+1. \tag{3.17.7}$$

The system corresponding to the operator (3.17.7),

$$\frac{dx}{dt} = y, \qquad \frac{dy}{dt} = B_{2n+1} x^{2n+2}, \tag{3.17.8}$$

is the new leading system.

Let us state from the very beginning that the system (3.17.8) considered as the model one for (3.17.3) implies the instability of the equilibrium (the proof is standard) so that our case, i.e., $B_{2n+1} \neq 0$, $A_i = 0$ for $i \leq n$, and $B_i = 0$ for $i \leq 2n$, is not considered in the sequel.

Now let j be even; i.e., $j = 2(n-1)$ for $n = 2, 3, \ldots$. Then there are no negative powers of ε in $\tilde{M}$ whenever $A_i = 0$ for $i < n-1$ and $B_i = 0$ for $i < 2(n-1)$. Then terms with the 0 power of ε in $\tilde{M}$ are

$$X_0 + B_{2n-2} x^{2n-1} \frac{\partial}{\partial y} + A_{n-1} \left(x^n \frac{\partial}{\partial x} + x^{n-1} y \frac{\partial}{\partial y} \right). \tag{3.17.9}$$

This is the new leading operator.

The highest perturbation term in $\tilde{M}$ is

$$\varepsilon^{2/n+1}\left\{A_n\left(x^{n+1}\frac{\partial}{\partial x}+x^n y\frac{\partial}{\partial y}\right)+B_{2n-1}x^{2n}\frac{\partial}{\partial y}\right\}. \tag{3.17.10}$$

Thus we have singled out the new leading operator and the principal perturbation term.

Remark 3.17.1 We have considered the possibility when

$$B_j x^{j+1}\frac{\partial}{\partial y}$$

is "added" to X_0. Then the operator

$$x^N\frac{\partial}{\partial x}+x^{N-1}y\frac{\partial}{\partial y}$$

also enters into the leading operator (3.17.9). Clearly, if we had started the discussion with the summand

$$X_0+\varepsilon^j A_j\left(x^{j+1}\frac{\partial}{\partial x}+x^j y\frac{\partial}{\partial y}\right)$$

we would have also arrived at (3.17.9).

Now put the artificially introduced parameter ε to be equal to 1, and denote coefficients as follows for brevity's sake: $B_{2n-2}=\beta_0$, $B_{2n+1}=\beta_1$, $\ldots$, $A_{n-1}=\alpha_0$, $A_n=\alpha_1,\ldots$ and write out terms corresponding to (3.17.9) and (3.17.10):

$$\begin{aligned}\boxed{\frac{dx}{dt}=y+\alpha_0x^n}&+\alpha_1x^{n+1}+\cdots\\ \boxed{\frac{dy}{dt}=\beta_0x^{2n-1}+\alpha_0x^{n-1}y}&+\beta_1x^{2n}+\alpha_1x^ny+\cdots,\end{aligned} \tag{3.17.11}$$

where $n\geq 2$, $\alpha_0^2+\beta_0^2\neq 0$ and the new leading system is arranged into the bounding box. The system (3.17.11) is the model one for (3.17.1). The criterion of (asymptotic) stability is deduced further only for the model system; its applicability for the complete system may be proved by standard means.

Remark 3.17.2 A. M. Lyapunov has considered the following model system

$$\begin{aligned}\boxed{\frac{dx}{dy}=y}&+dxy,\\ \boxed{\frac{dy}{dt}=ax^{2n-1}+bx^{n-1}y}&+a_1x^{2n}+b_1x^ny+cy^2+\cdots\end{aligned} \tag{3.17.12}$$

containing a larger number of parameters (six instead of four). If in this system we undertake the transformation

$$x \mapsto x + \frac{c+d}{2}x^2,$$
$$y \mapsto y + \alpha_0 x^n + cxy + (c\alpha_0 + \alpha_1)x^{n+1}$$

and put

$$\begin{aligned}(n+1)\alpha_0 &= b,\\ (n+2)\alpha_1 &= \frac{n-1}{2}b(c+d) + b_1,\\ \beta_0 - \alpha_0^2 &= a,\\ \beta_1 - 2\alpha_0\alpha_1 &= \frac{2n-1}{2}a(c+d) - ca + a_1,\end{aligned} \tag{3.17.13}$$

then up to accuracy $0(x^{n+2})$ in dx/dt and $0(x^{2n+1})$ in dy/dt, where $y = 0(x^n)$, we obtain (3.17.11).

Thus, it does not matter which of the systems (3.17.11) and (3.17.12) is taken as the model one (on this subject see Remark 3.17.3 later). Formulas (3.17.13) express α_0, β_0, α_1, and β_1 in terms of a, b, c, d, a_1, and b_1. The study of the leading system is not difficult [it is integrated explicitly; substituting $y = zx^n$ gives the Bernoulli equation for $z(x)$] and the necessary condition for stability is

$$\Delta \equiv (n-1)^2\alpha_0^2 + 4n\beta_0 < 0. \tag{3.17.14}$$

In particular, we should have $\beta_0 < 0$ and the condition $\Delta < 0$ gives two simple (necessary) Lyapunov's conditions $a < 0$ and $|b| \le 2\sqrt{n|a|}$; see formulas (3.17.13).

Besides, if n is odd, then the leading system itself can be considered as a model one: for $\alpha_0 < 0$ it is asymptotically stable, for $\alpha_0 > 0$ it is unstable, and $\alpha_0 = 0$ is not considered. If n is even, then for (3.17.14) the leading system is stable but not asymptotically stable.

The main problem is to obtain conditions of (asymptotic) stability of (3.17.11) for even n when (3.17.14) is verified.

(3) The Main Lyapunov Equation

Let k be a root of the equation $nk^2 - (n-1)\alpha_0 k - \beta_0 = 0$ [by (3.17.14) this equation has complex roots]. Set $\varphi = y + kx^n$. Then the leading system in (3.17.11) takes the form

$$\frac{d\varphi}{dt} = \lambda x^{n-1}\varphi, \qquad \frac{d\bar\varphi}{dt} = \bar\lambda x^{n-1}\bar\varphi,$$

where $\lambda = \alpha_0 + nk$. This implies that the integral of the leading system is $\bar\varphi^{\lambda}\varphi^{-\bar\lambda} = \text{const}$ which in real form is $H = \text{const}$, where we have

$$H = [ny^2 + (n-1)\alpha_0 yx^n - \beta_0 x^{2n}]x,$$

where

$$x = \exp\left(\frac{2(n+1)\alpha_0}{\sqrt{-\Delta}} \arctan \frac{2ny + (n-1)\alpha_0 x^n}{x^n\sqrt{-\Delta}}\right).$$

The function H is positive. Now let us make the change of variables $x \to x + \mathcal{A}x^2$, $y \to y + \mathcal{B}x^{n+1}$, choosing constants $\mathcal{A}, \mathcal{B}$ so that the model system (3.17.11) takes the form

$$\begin{aligned} \frac{d\varphi}{dt} &= (\lambda x^{n-1} + \Lambda x^n)\varphi + \cdots, \\ \frac{d\bar{\varphi}}{dt} &= (\bar{\lambda} x^{n-1} + \bar{\Lambda} x^n)\bar{\varphi} + \cdots, \end{aligned} \tag{3.17.11'}$$

with the same accuracy as (3.17.11). It is possible to do this. Constants $\mathcal{A}$, $\mathcal{B}$, Λ, and $\bar{\Lambda}$ are uniquely expressed in terms of α_0, β_0, α_1, and β_1.

Now we make the "final" passage to the model system (3.17.11′), dropping small perturbations. Due to the model system (3.17.11′) we have

$$\frac{dH}{dt} = hx^n H, \tag{3.17.15}$$

where H is described above and $h = 2(\bar{\lambda}\Lambda - \lambda\bar{\Lambda})/(\bar{\lambda} - \lambda)$. The function H is, certainly, not a Lyapunov function, since $hx^n H$ is sign positive for an even n but it is not sign definite. However, we may judge stability by the sign of h.

Let $h < 0$. Let us show that asymptotic stability occurs. Since H is positive definite and does not increase due to (3.17.15), the distance $r = (x^2 + y^2)^{1/2}$ of the origin from the depicting point on the phase plane either tends to 0, and then there is (asymptotic) stability, or it is bounded away from 0, i.e., $r \geq \delta > 0$. Let us show that the latter case is impossible. The polar angle due to (3.17.11) changes monotonically:

$$y\frac{dx}{dt} - x\frac{dy}{dt} = y^2 - \beta_0 x^{2n} + \cdots, \qquad \text{where } \beta_0 < 0,$$

is positive. Therefore

$$\int_0^t x^n\,dt = \int_0^t 2^n \cos^n\theta\,dt = \int_{\theta_0}^{\theta} 2^n \cos^n\theta \frac{d\theta}{\theta} \int_{\theta_0}^{\theta} r^n \cos^n\theta \frac{d\theta}{\theta} \to \infty$$

if $r \geq \delta > 0$ for even n. But then $H \to 0$ since

$$H = \exp\left(h \int_0^t x^n\,dt\right)$$

for $h < 0$, hence $r \to 0$, a contradiction. Similarly we prove instability for $h > 0$. The case $h = 0$ is not considered. The evaluation of h shows that its sign coincides with that of $\mathcal{D} = (n+1)^2\alpha_0\beta_1 + [n\alpha_0^2 - (n+2)(2n+$

$1)\beta_0]\alpha_1$, which is expressed in terms of Lyapunov's coefficients [see formulas (3.17.13)] as

$$\mathcal{D} = -a\left\{\left(c - (n+1)\frac{a_1}{a} - nd\right)b + (2n+1)b_1\right\}.$$

The condition $\mathcal{D} < 0$ is the main Lyapunov stability condition.

Remark 3.17.3 We can undertake a similar investigation starting from the model Lyapunov system (3.17.12), so, strictly speaking, it is impossible to consider (3.17.11) to be a better model system than (3.17.12). Nevertheless, authors believe that the canonical form in the sense of Section 2.5 is more convenient.

The following example is a "formal" illustration for the Section 2.5 when X_0 has two Jordan cells:

$$\begin{pmatrix} 0 & 1 & 0 & 0 \\ 0 & 0 & 0 & 0 \\ 0 & 0 & 0 & 1 \\ 0 & 0 & 0 & 0 \end{pmatrix}.$$

3.18 Example: Illustration for Section 2.5

Suppose the leading operator is

$$X_0 = y_1\frac{\partial}{\partial x_1} + y_2\frac{\partial}{\partial x_2}, \tag{3.18.1}$$

where $x = (x_1, y_1, x_2, y_2)$, so that, in the notations of Section 2.5, we have

$$x_1 = e^{(1)}_{1/2}, \qquad y_1 = e^{(1)}_{-1/2}, \qquad x_2 = e^{(2)}_{1/2}, \qquad y_2 = e^{(2)}_{-1/2}.$$

A normal basis in P_{N-1} may be chosen so that

$$\frac{\hat{e}^{(\alpha,\beta,\gamma)}_{\alpha+\beta}}{2} = x_1^\alpha x_2^\beta (x_1y_2 - x_2y_1)^\gamma, \tag{3.18.2}$$

where $\alpha \geq 0$, $\beta \geq 0$, and $\gamma \geq 0$ are integers such that $\alpha + \beta + 2\gamma = N$. Then elements (3.18.2) may be taken as generating elements of chains in a normal basis P_{N-1} (to each set α, β, γ one chain corresponds and the dimension of the corresponding subspace is $\alpha + \beta + 1$).

In fact, the polynomial (3.18.2), as it is easy to verify, is annihilated by Z_+ and corresponds to the eigenvalue $\alpha + \beta/2$ with respect to Z_0 (the function $x_1y_2 - x_2y_1$ is an invariant for all three operators $Z_- \equiv X_0, Z_0, Z_+$). Therefore (3.18.2) is an element of $P_{\alpha+\beta/2,N-1}$ [see (2.5.3)]. Further, it is easy to see that polynomials (3.18.2) with mutually distinct sets of α, β, and

γ are linearly independent and therefore may be taken as basic elements in spaces $P_{\alpha+\beta/2,N-1}$. Therefore (3.18.2) may be considered as generating elements of some chains of the normal basis (this is proved in Section 2.5). It remains to show that the number of elements of all these chains equals that of P_{N-1}. The dimension of the space of homogeneous polynomials of degree N in n variables is equal to $(N+n-1)!/(n-1)!\,N!$. In our case $n=4$, hence

$$\dim P_{N-1} = \frac{(N+3)!}{3!\,N!} = \frac{1}{6}(N+1)(N+2)(N+3).$$

Further, the number of different sets α, β, γ for a fixed γ, where γ runs from zero up to $[N/2]$, equals $N-2\gamma+1 = \alpha+\beta+1$, the number of elements in the chain generated by (3.18.2). The number of all elements of all chains is

$$\sum_{0\le\gamma\le[N/2]} (N-2\gamma+1)^2 = \frac{1}{6}(N+1)(N+2)(N+3).$$

Thus a normal basis in P_{N-1} is constructed.

Remark 3.18.1 As compared with the preceding example, it might seem at the first glance that we should take the generating elements of the normal basis in the form $x_1^{\alpha}x_2^{\beta}$, where $\alpha+\beta=N$. But it appears that the presence of the quadratic invariant $x_1y_2-x_2y_1$, which has no analogue in the preceding example, is an essential feature.

Generating elements of a normal basis of an operator in T_{N-1} can be constructed as in the above example. For such elements we might take

$$Y^{(\alpha,\beta,\gamma,i)}_{(\alpha+\beta-1)/2} = (\alpha+\beta)\hat{e}^{(\alpha,\beta,\gamma)}_{(\alpha+\beta)/2}\frac{\partial}{\partial x_i} + \hat{e}^{(\alpha,\beta,\gamma)}_{(\alpha+\beta-1)/2}\frac{\partial}{\partial y_i}$$

$$\tilde{Y}^{(\alpha,\beta,\gamma,i)}_{(\alpha+\beta+1)/2} = \hat{e}^{(\alpha,\beta,\gamma)}_{(\alpha+\beta+1)/2}\frac{\partial}{\partial y_i}, \tag{3.18.3}$$

where $i=1,2$, and

$$\begin{aligned}\hat{e}^{(\alpha,\beta,\gamma)}_{(\alpha+\beta)/2-1} &= X_0\hat{e}^{(\alpha,\beta,\gamma)}_{(\alpha+\beta)/2}\\ &= (\alpha y_1x_2+\beta x_1y_2)x_1^{\alpha-1}x_2^{\beta-1}(x_1y_2-x_2y_1)^{\gamma}.\end{aligned}$$

Now let $X = X_0+\varepsilon X_1+\varepsilon^2X_2+\cdots$ be, as in the preceding example, such that X_jx_i, X_jy_i for $i=1,2$, and $j=1,2,\ldots$ are homogeneous polynomials in x_1, x_2; y_1, y_2 of degree $j+1$. Then the operator $M=e^{-S}Xe^{S}$, by the choice of $S=\varepsilon S_1+\varepsilon^2S_2+\cdots$, is reducible to the form

$$M = X_0+\varepsilon M_1+\varepsilon^2M_2+\cdots, \tag{3.18.4}$$

where M_j are linear combinations of operators (3.18.3), where $\alpha+\beta+2\gamma = j+1$. What might be a new leading operator? Let us make the (generalized) shearing transformation (2.5.1) with the parameter $t = \varepsilon^{-k}$, where k will be fixed later,

$$x_i \to \varepsilon^{-k/2} x_i, \qquad y_i \to \varepsilon^{k/2} y_i \qquad \text{for } i = 1, 2. \tag{3.18.5}$$

Under the transformation (3.18.5), in (3.18.4) the following changes will occur. We will get $\varepsilon^k X_0$ instead of X_0 (see Remark 2.5.3), and operators $Y^{(\alpha,\beta,\gamma,i)}_{(\alpha+\beta-1)/2}$, $\tilde{Y}^{(\alpha,\beta,\gamma,i)}_{(\alpha+\beta+1)/2}$ which enter M_j will be factored by

$$\varepsilon^{k(1-\alpha-\beta)/2}, \qquad \varepsilon^{-k(1+\alpha+\beta)/2} \tag{3.18.6}$$

respectively [see (3.18.2), (3.18.3)]. Under the right choice of k in $\varepsilon^{-k} M = X_0 + \cdots = M$ there should be no negative powers of ε and terms without ε will constitute a new leading operator. Taking into account that $\alpha + \beta = j + 1 - 2\gamma$, let us rewrite factors (3.18.6) in the form

$$\varepsilon^{k(2\gamma-j)/2}, \qquad \varepsilon^{k(2\gamma-j-2)/2}. \tag{3.18.6'}$$

If k is correctly chosen, then

$$\begin{aligned} m(j,\gamma) &\equiv j + \frac{k}{2}(2\gamma - j) - k \geq 0, \\ \tilde{m}(j,\gamma) &\equiv j + \frac{k}{2}(2\gamma - j - 2) - k \geq 0 \end{aligned} \tag{3.18.7}$$

for all $Y^{(\alpha,\beta,\gamma,i)}_{(\alpha+\beta-1)/2}$ and $\tilde{Y}^{(\alpha,\beta,\gamma,i)}_{(\alpha+\beta+1)/2}$ which enter M_j with nonzero coefficients. Here we should have in mind that $\gamma < \frac{j+1}{2}$, where $\alpha + \beta > 0$ for $Y^{(\alpha,\beta,\gamma,i)}_{(\alpha+\beta-1)/2}$, and $\gamma \leq j + 1/2$, where $\alpha + \beta \geq 0$ for $\tilde{Y}^{(\alpha,\beta,\gamma,i)}_{(\alpha+\beta+1)/2}$ [see (3.18.3)].

Suppose the leading operator contains a summand from M_{j_0} and this summand is of the form

$$c_i \tilde{Y}^{(\alpha,\beta,\gamma_0,i)}_{\alpha+\beta-1/2}, \qquad \text{where } c_i = \text{const}, \ \gamma = \gamma_0.$$

Then $m(j_0, \gamma_0) = 0$, hence

$$k = \frac{2j_0}{j_0 + 2 - 2\gamma_0} \qquad \text{for } j_0 + 2 - 2\gamma_0 > j_0 + 2 - j_0 - 1 = 1. \tag{3.18.8}$$

For given k, conditions (3.18.7) take the form

$$(\gamma - 1)j_0 - (\gamma_0 - 1)j \geq 0, \qquad (\gamma - 2)j_0 - (\gamma_0 - 1)j \geq 0. \tag{3.18.9}$$

We should remember that conditions $m(j,\gamma) \geq 0$ and $\tilde{m}(j,\gamma) \geq 0$ are connected with operators $Y^{(\alpha,\beta,\gamma,i)}_{\alpha+\beta-1/2}$ and $\tilde{Y}^{(\alpha,\beta,\gamma,i)}_{\alpha+\beta+1/2}$ and values j, γ in $m(j,\gamma)$ and $\tilde{m}(j,\gamma)$ do not, generally speaking, coincide. If $\gamma_0 < 1$, i.e., $\gamma_0 = 0$, then (3.18.9) holds for $j \geq 2j_0$ and might fail for a finite number of terms ($j < 2j_0$). If coefficients in M of operators (3.18.3) for which (3.18.9) fails are zero then the value (3.18.8) for k is correctly chosen.

If $\gamma_0 \geq 1$, then (3.18.9) might be violated for any M_j and if the value (3.18.8) of k were correct an infinite number of conditions on coefficients M must be satisfied. Then in the new leading operator an infinite number of summands might enter. Similarly, if in the new leading operator a summand from M_{j_0} of the form

$$c_i \tilde{Y}^{(\alpha,\beta,\gamma_0,i)}_{(\alpha+\beta+1)/2}$$

enters, then

$$k = \frac{2j_0}{j_0 + 4 - 2\gamma_0}, \qquad j_0 + 4 - 2\gamma_0 \geq j_0 + 4 - j_0 - 1 = 3. \qquad (3.18.10)$$

For this k, conditions (3.18.7) take the form

$$(\gamma - 1)j_0 - (\gamma - 2)j_0 \geq 0, \qquad (\gamma - 2)j_0 - (\gamma - 2)j_0 \geq 0. \qquad (3.18.11)$$

If $\gamma_0 < 2$, i.e., $\gamma_0 = 0$ or $\gamma_0 = 1$, then (3.18.11) might fail only for a finite number of terms. If $\gamma_0 \geq 2$ then, as above, the correct choice of k in the form (3.18.10) corresponds to an infinite number of conditions on M.

Thus in "general position," i.e., when the perturbation contains the full set of operators (3.18.3), in the new leading operators only operators from (3.18.3) with $\gamma = 0$ or $\gamma = 1$ may enter, and then only a finite number of them. Generally speaking, the new leading operator might contain an infinite number of summands. For example, if each of the M_j's lacks

$$Y^{(\alpha,\beta,0,i)}_{(\alpha+\beta-1)/2}, \qquad \tilde{Y}^{(\alpha,\beta,0,i)}_{(\alpha+\beta+1)/2}, \qquad \tilde{Y}^{(\alpha,\beta,1,i)}_{(\alpha+\beta+1)/2},$$

then [see (3.18.9)], in the new leading operator from each M_j, summands with $Y^{(\alpha,\beta,1,i)}_{(\alpha+\beta-1)/2}$ and $\tilde{Y}^{(\alpha,\beta,2,i)}_{(\alpha+\beta+1)/2}$ will enter. Then, due to (3.18.8), $k = 2$ and $\gamma_0 = 1$.

In conclusion, note that for $M_1 \not\equiv 0$ we always have a "sensible" problem. Let us consider this case in more detail.

In M_1 only operators $Y^{(\alpha,\beta,0,i)}_{(\alpha+\beta-1)/2}$, $\tilde{Y}^{(\alpha,\beta,0,i)}_{(\alpha+\beta+1)/2}$, where $\alpha + \beta = 2$, and $\tilde{Y}^{(0,0,1,i)}_{1/2}$ may enter. Suppose the coefficient in M_1 of (at least one of) $\tilde{Y}^{(\alpha,\beta,0,i)}_{(\alpha+\beta+1)/2}$ is nonzero. Then putting $j_0 = 1$ and $\gamma_0 = 0$, we have

$$k = \frac{2}{5} \qquad \text{[see (3.18.10)]},$$

$$m(j,\gamma) = \frac{2}{5}(2j + \gamma - 1) > 0,$$

$$\tilde{m}(j,\gamma) = \frac{2}{5}(2j + \gamma - 2) > 0 \qquad \text{for } j > 1,$$

hence only these operators (in addition to X_0) will enter the new leading operator.

Thus one of the possible new leading operators (the leading operator in $\tilde{M}$) is

$$y_1\frac{\partial}{\partial x_1}+y_2\frac{\partial}{\partial x_2}+(\tilde{a}_{11}x_1^2+\tilde{a}_{12}x_2^2+\tilde{a}_{13}x_1x_2)\frac{\partial}{\partial y_1}$$
$$+(\tilde{a}_{21}x_1^2+\tilde{a}_{22}x_2^2+\tilde{a}_{23}x_1x_2)\frac{\partial}{\partial y_2} \tag{3.18.12}$$

if at least one of $\tilde{a}_{mn}$ is nonzero.

Now let coefficients $\tilde{a}_{mn}$ in M_1 of all $\tilde{Y}^{(\alpha,\beta,0,i)}_{(\alpha+\beta+1)/2}$'s be zero and of (at least one of) $Y^{(\alpha,\beta,0,i)}_{(\alpha+\beta-1)/2}$, where $\alpha+\beta=2$, be nonzero. Then putting [see (3.18.8)] $j_0=1$ and $\gamma_0=0$ we have

$$\begin{aligned} k &= \tfrac{2}{3} \\ m(j,\gamma) &= \tfrac{2}{3}(j+\gamma-1)>0 \qquad \text{for } j>1, \\ \tilde{m}(j,\gamma) &= \tfrac{2}{3}(j+\gamma-2)\geq 0 \qquad \text{for } j>1 \text{ or } j=1,\ \gamma\geq 1. \end{aligned}$$

In the new leading operator $\tilde{Y}^{(0,0,1;i)}_{1/2}$ from M_1 will also enter as will $\tilde{Y}^{(\alpha,\beta,0,i)}_{(\alpha+\beta+1)/2}$ from M_2, where $\alpha+\beta=3$. Thus the second of the possible new leading operators in $\tilde{M}$ is

$$\begin{aligned} &(y_1+2a_{11}x_1^2+2a_{12}x_2^2+2a_{13}x_1x_2)\frac{\partial}{\partial x_1} \\ &\quad+(y_2+2a_{21}x_1^2+2a_{22}x_2^2+2a_{23}x_1x_2)\frac{\partial}{\partial x_2} \\ &\quad+\{2a_{11}x_1y_1+2a_{12}x_2y_2+a_{13}(y_1x_2+x_1y_2) \\ &\qquad+b_1(x_1y_2-x_2y_1) \\ &\qquad+\tilde{c}_{11}x_1^3+\tilde{c}_{12}x_2^3+\tilde{c}_{13}x_1^2x_2+\tilde{c}_{14}x_1x_2^2\}\frac{\partial}{\partial y_1} \\ &\quad+\{2a_{21}x_1y_1+2a_{22}x_2y_2+a_{23}(y_1x_2+x_1y_2) \\ &\qquad+b_2(x_1y_2-x_2y_1) \\ &\qquad+\tilde{c}_{21}x_1^3+\tilde{c}_{22}x_2^3+\tilde{c}_{23}x_1^2x_2+\tilde{c}_{24}x_1x_2^2\}\frac{\partial}{\partial y_2}, \end{aligned} \tag{3.18.13}$$

where at least one of a_{mn} and b_m is nonzero. It is clear that it is not necessary to consider separately the case when the coefficient of $\tilde{Y}^{(0,0,1;i)}_{1/2}$ in M is nonzero. Formulas (3.18.12) and (3.18.13) give all possible new leading operators when $M_1\not\equiv 0$.

In the following problem we consider an example of a resonance for variable eigenvalues. This example might be attached to the reconstruction problem (see Chapter 4) but due to its simplicity it is placed here.

3.19 Example: Fast Rotation of a Solid Body

In this example we will turn to one of the classical problems of analytical mechanics: that of motion of a mass solid body with one fixed point.

(1) Euler–Poisson Equations

Suppose a Cartesian coordinate system with 0 as its origin is rigidly attached to the body and in this system 0_{xyz} vectors Ω, K, Γ, and G define the absolute angular velocity of the body, its kinetic momentum, the vertical axis, and the center of mass, respectively. Then

$$\frac{dK}{dt} + [\Omega, K] = mg[G, \Gamma], \qquad \frac{d\Gamma}{dt} + [\Omega, \Gamma] = 0, \tag{3.19.1}$$

where mg is the weight of the body. [The left-hand sides of (3.19.1) express absolute velocities of variations of K and Γ; the first of the equations (3.19.1) expresses the kinetic momentum theorem and the second one the condition of stability of the vertical axis in the absolute space.]

The kinetic momentum and the angular velocity are related by the equation $K = \hat{T}\Omega$, where $\hat{T}$ is a (symmetric) inertial tensor, so that (3.19.1) is a closed system of equations with respect to K and Γ (or Ω and Γ), where $\hat{T}$ and G are constants. Axes of the frame 0_{xyz} are usually principal axes of inertia of the body with respect to 0: i.e., $\hat{T}$ is a diagonal operator.

Setting in (3.19.1)

$$\begin{aligned} \Omega &= \{p, q, r\}, & K &= \{Ap, Bq, Cr\}, \\ \Gamma &= \{\gamma, \gamma', \gamma''\}, & G &= \{x_0, y_0, z_0\}, \end{aligned} \tag{3.19.2}$$

where A, B, C are inertia moments of the body with respect to the principal axes 0_{xyz}, we obtain equations of motion (the Euler–Poisson equations) in the standard form and standard notations [20].

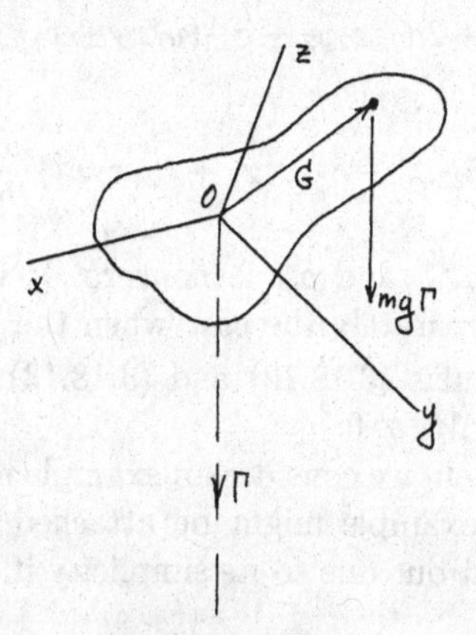

FIGURE 3.5.

Equations (3.19.1) have three mutual first integrals (of energy, of area, and the so-called geometric integral):

$$(K\Omega) - 2mg(G\Gamma) = \text{const}, \qquad (K\Gamma) = \text{const}, \qquad \Gamma^2 = 1. \tag{3.19.3}$$

It is known that to integrate (3.19.1) in quadratures it suffices to find one more, a fourth, common integral. [To define the position of the body in the absolute space, i.e., to define the dependence of the Euler angles on time, it is insufficient to integrate (3.19.1). However it suffices to perform one extra quadrature; see [20].]. Such an integral is found only in three famous cases (under certain restrictions on parameters A, B, C, X_0, Y_0, Z_0): those of Euler, Lagrange, and Kovalevskaya.

In what follows the Euler case will be important. This is the case of motion by inertia, where the fixed point coincides with the center of mass:

$$G = 0, \tag{3.19.4}$$

which is (formally) equivalent to the case of motion without gravity ($mg = 0$). The fourth integral in this case is

$$K^2 = \text{const}. \tag{3.19.5}$$

Then Equations (3.19.1) are integrated in elliptic Jacobi functions, or in elementary ones if the body has in addition an axis of dynamical symmetry, say $A = B$. The geometric interpretation of motion by inertia due to Poinsot (see, e.g., [2]) is also well known.

(2) Rapid Rotation. The Formulation of the Problem

We will speak about *rapid rotation* if

$$\frac{mg}{|K|}\frac{|G|}{|\Omega|} \ll 1; \tag{3.19.6}$$

i.e., when the kinetic energy is large compared with the potential energy. [Due to the existence of the energy integral (3.19.3), condition (3.19.6) holds if it holds at the initial moment.] A small parameter can be introduced into (3.19.1) by the formal transformation

$$\Omega \to \Omega/\varepsilon \tag{3.19.7}$$

so that in the new scale the body will perform $\sim 1/\varepsilon$ rotations during the time $t \sim 1$.

Under the transformation (3.19.7), Equations (3.19.1) will take the form ($K \to K/\varepsilon$):

$$\varepsilon\frac{dK}{dt} + [\Omega, K] = \varepsilon^2 mg[G, \Gamma], \qquad \varepsilon\frac{d\Gamma}{dt} + [\Omega, \Gamma] = 0 \tag{3.19.8}$$

where Ω, K, Γ, and G are taken in the form (3.19.2), all quantities except ε are assumed to be ~ 1, and $\varepsilon \ll 1$.

The system (3.19.8) will be referred to as *perturbed*, whereas the system

$$\varepsilon \frac{dK}{dt} + [\Omega, K] = 0, \qquad \varepsilon \frac{d\Gamma}{dt} + [\Omega, \Gamma] = 0 \tag{3.19.9}$$

with the same initial values as for (3.19.8) is called a *nonperturbed* one. Thus a nonperturbed system describes the Euler motion by inertia.

The well-known method of depicting a rapidly rotation solid body is to consider it as rotating around the kinetic momentum vector (approximately according to Poinsot rotation) while this vector is slowly precessing around the vertical axis; for instance, such is the motion of the top ($A = B$, $x_0 = y_0 = 0$) that rotates rapidly around a symmetry axis. In the accepted scale, the velocity of the top's precession is slow ($\sim \varepsilon$), hence the action of gravity becomes manifest [i.e., we observe a finite deviation from the motion by inertia due to (3.19.9)] after a large period of time ($\sim 1/\varepsilon$).

However, some extraordinary (in a sense) situations are possible when gravity tells already after some time $t \sim 1$. We will discuss here these "earlier than precession" mechanical effects. We will only consider the case when the body possesses a unique axis of dynamical symmetry. Let $A = B \neq C$ (the case of the total dynamical symmetry ($A = B = C$) is easy to integrate exactly [20]).

Remark 3.19.1 The case being considered is perhaps the most interesting from the mechanical viewpoint.

For $A > B > C$ (we do not discuss the case when inertia momenta can be ε-close) rotations of the body in the Euler motion around large and small inertia axes are stable, whereas the rotation around the middle axis is unstable. If at the initial moment the angular velocity is not situated close to separatrices [2] with the point of intersection on the middle axis, then the Euler motion during time $t \sim 1$ is preserved with accuracy $0(\varepsilon)$ in accordance with [3]. Otherwise, gravity will tell on the motion during time $t \sim 1$. This case, more difficult than that studied below, will not be considered.

Thus we will investigate (under the condition $A = B$) the perturbed system (3.19.8) during the time $0 \leq t \leq 1$ and be interested in cases when its solution, with accuracy up to $0(1)$ included, differs from that of the nonperturbed system (3.19.9).

(3) Remark on Accuracy. Resonances

Since the order of perturbation is $\sim \varepsilon^2$ [see (3.19.8)], then at first glance it seems that during the time $t \sim 1$ it cannot attribute a $0(1)$ change into the solution. In fact, this is not so, due to variability of eigenvalues. Let us clarify this. As we will show a little later, the system (3.19.8) may be reduced to the form

$$\varepsilon \frac{d\varphi_\nu}{dt} = i\lambda_\nu \varphi_\nu + \varepsilon^m F_\nu(\lambda, \varphi, \varepsilon),$$
$$\varepsilon \frac{d\lambda_\mu}{dt} = \varepsilon^m G_\mu(\lambda, \varphi, \varepsilon) \tag{3.19.10}$$

(the leading operator is diagonal, $\lambda_\nu \neq \text{const}$), where $m = 2$.

From (3.19.10) it is manifest that for $t \sim 1$ we may neglect the perturbation only when $m > 2$. In fact, the error in computing λ_ν is $0(\varepsilon^{m-1})$ which gives the error $0(\varepsilon^{m-2})$ in computing φ_ν. [The rigorous arguments consist of the straightforward passage from (3.19.10) to the corresponding integral equations.]

It is also clear that if a change of variables close to identity gives $G_\nu(\lambda, \varphi, \varepsilon) = 0(\varepsilon)$, i.e., $\varepsilon\, d\lambda_\nu/dt = 0(\varepsilon^{m+1})$, then the solution will be close to the solution of the nonperturbed system for $t \lesssim 1$.

Hence the question is if it is possible for (3.19.8) to solve equations

$$M_2\lambda_\nu \equiv ([X_0, S_2] + X_2)\lambda_\nu = X_0(S_2\lambda_\nu) + X_2\lambda_\nu = 0, \tag{3.19.11}$$

where λ_ν are eigenvalues of the nonperturbed problem. The system (3.19.8) can be written in the form

$$\begin{aligned} \varepsilon \tfrac{dp}{dt} &= \alpha qr - \varepsilon^2 \zeta\gamma', & \varepsilon \tfrac{d\gamma}{dt} &= r\gamma' - q\gamma'', \\ \varepsilon \tfrac{dq}{dt} &= -\alpha rp + \varepsilon^2[\zeta\gamma - (1-\alpha)\xi\gamma''], & \varepsilon \tfrac{d\gamma'}{dt} &= p\gamma'' - r\gamma, \\ \varepsilon \tfrac{dr}{dt} &= \varepsilon^2 \xi\gamma', & \varepsilon \tfrac{d\gamma''}{dt} &= q\gamma - p\gamma', \end{aligned} \tag{3.19.12}$$

where $A = B = C/(1-\alpha)$, $\xi = mgx_0/c$, $\zeta = mgz_0/A$, and $y_0 = 0$, because for $A = B$, axes of 0_{xyz} may be chosen so that the center of mass belongs to the plane 0_{xz}. To the system (3.19.12) corresponds the operator $X_0 + \varepsilon^2 X_2$, where

$$X_0 = \alpha r\left(q\frac{\partial}{\partial p} - p\frac{\partial}{\partial q}\right) + (r\gamma' - q\gamma'')\frac{\partial}{\partial \gamma} + (p\gamma'' - r\gamma)\frac{\partial}{\partial \gamma'} + (q\gamma - p\gamma')\frac{\partial}{\partial \gamma''}, \tag{3.19.13}$$

$$X_2 = -\zeta\gamma'\frac{\partial}{\partial p} + [\zeta\gamma - (1-\alpha)\xi\gamma'']\frac{\partial}{\partial q} + \xi\gamma'\frac{\partial}{\partial r}.$$

The invariants of X_0 are r, $k = p\gamma + q\gamma' - (1-\alpha)r\gamma''$, and $k^2 = p^2 + q^2 + (1-\alpha)^2 r^2$, and its eigenfunctions are

$$\varphi_{1,2} = q \pm ip, \qquad \psi_{1,2} = -k^2\gamma'' + (1-\alpha)rk \pm ik(q\gamma - p\gamma'), \tag{3.19.14}$$

corresponding to eigenvalues $\pm i\alpha r$ and $\pm ik$ respectively. In the domain $\omega \neq 0$, where

$$\omega^2 = p^2 + q^2 \tag{3.19.15}$$

(ω is an invariant of X_0), we may pass to the basic system r, k (3.19.14) in which X_0 possesses diagonal form and the form of the system is given by (3.19.10).

However, computations are easier to perform in initial variables. We are only interested in the possibility of solving Equations (3.19.11); i.e., the representability of $X_2 r$ and $X_2 k$ in the form $X_0 f$, since eigenvalues are $\pm i\alpha r$ and $\pm ik$.

Since

$$X_2 r = \xi\gamma', \qquad X_2 \frac{k^2}{2} = X_0\{(1-\alpha)\xi\gamma + \zeta\gamma''\} - \alpha(1-\alpha)\xi r\gamma'$$

due to (3.19.13), the problem is reduced to the representation of γ' in the form $X_0 f$. It is not difficult to compute

$$X_0^2\gamma' = -2\alpha r X_0\gamma - (k^2 - \alpha^2 r^2)\gamma' + \frac{k}{\alpha r} X_0 p,$$

implying that γ' is representable in the form $X_0 f$ when $r \neq 0$ and $k^2 - \alpha^2 r^2 \neq 0$, and (3.19.11) is solvable.

Thus if at the initial moment one of the quantities

$$r, \quad k^2 - \alpha^2 r^2 \tag{3.19.16}$$

is small [quantities (3.19.16) cannot be small simultaneously since $|\Omega| \sim 1$], namely, if either $|r| \leq \varepsilon$ or $|k^2 - \alpha^2 r^2| \lesssim \varepsilon$, then the proximity of the solution of the perturbed system to that of the nonperturbed one cannot be guaranteed. On the other hand, if at the initial moment $|r| \gg \varepsilon$ and $|k^2 - \alpha^2 r^2| \gg \varepsilon$, then during time $t \lesssim 1$ we will observe motion by inertia (up to small quantities). This is evident from the earlier arguments on errors caused by neglecting the perturbation.

Thus we might be interested only in cases when at the initial moment the vector of angular velocity lies ε-close to the plane $z = 0$, i.e., the equatorial plane of the inertia ellipsoid, or to the cone $x^2 + y^2 + (1 - 2\alpha)z^2 = 0$, $\alpha > 1/2$ [see (3.19.16)], where $k^2 = p^2 + q^2 + (1-\alpha)^2 r^2$.

From here on we will consider only the first case (when the body, unlike the top, rotates quickly *across* the axis of symmetry). The second case is more difficult but does not differ in principle from the first one. It is considered, e.g., in [7].

Note that the cases mentioned [when one of quantities (3.19.16) is small] have the evident meaning of resonances in the Euler motion by inertia.

Remark 3.19.2 Above we have assumed that ω [see (3.19.15)] is not small [$\omega = 0$ is a singular point for the transformation (3.19.14)]. If at the initial moment ω is small, i.e., p and q are small, then this is the case of a rapid rotation along the axis of symmetry. It can be considered assuming from the beginning that $p \sim 1$, $q \sim 1$ and $r \sim 1/\varepsilon$ in the system (3.19.1).

In the case for $t \lesssim 1$ the Euler motion is preserved and the (first) nontrivial effect of gravity, precession, arises at large times ($\sim 1/\varepsilon$).

(4) Rapid Rotation Across the Axis of Symmetry

Let us perform in (3.19.12) or, which is the same, in (3.19.13), the transformation

$$r \to \varepsilon r. \tag{3.19.17}$$

Since (3.19.17) is singular at $\varepsilon = 0$, the operator $X = X_0 + \varepsilon^2 X_2$ is *reconstructed*; i.e., it turns into $X = \tilde{X}_0 + \varepsilon \tilde{X}_1 + \varepsilon^2 \tilde{X}_2$, where

$$\begin{aligned} \tilde{X}_0 &= -q\gamma'' \frac{\partial}{\partial \gamma} + p\gamma'' \frac{\partial}{\partial \gamma'} + (q\gamma - p\gamma')\frac{\partial}{\partial \gamma''}, \\ \tilde{X}_1 &= \alpha r \left(q\frac{\partial}{\partial p} - p\frac{\partial}{\partial q}\right) + r\gamma' \frac{\partial}{\partial \gamma} - r\gamma' \frac{\partial}{\partial \gamma} + \xi\gamma' \frac{\partial}{\partial r}, \\ \tilde{X}_2 &= -\zeta\gamma' \frac{\partial}{\partial p} + [\zeta\gamma - (1-\alpha)\xi\gamma'']\frac{\partial}{\partial q}. \end{aligned} \tag{3.19.18}$$

The operator $\tilde{X}_0$ is quasilinear since p and q are its invariants and its eigenfunctions correspond to $\pm i\omega$, where $\omega^2 = p^2 + q^2$. The procedure of reduction of X to the normal form with respect to $\tilde{X}_0$ is the standard one. Here we prefer to make use of the commutation relations.

It is easy to obtain $\omega^2 \tilde{X}_1' + \tilde{X}_1''' = 0$, where $X_1' \equiv [\tilde{X}_0, X_1]$, implying

$$M_1 = \tilde{X}_1 + \frac{1}{\omega^2}\tilde{X}_1''.$$

Further, by arguments like those in (3) we verify whether $M_2\omega = 0$, since the eigenvalues are $\pm\omega$. Due to (3.19.18) we have

$$\tfrac{1}{2}\tilde{X}_2\omega^2 = \zeta(q\gamma - p\gamma') - (1-\alpha)\xi\gamma'' q = \tilde{X}_0(\zeta\gamma'' + (1-\alpha)\xi\gamma)$$

and $\tilde{X}_0\omega = \tilde{X}_1\omega = 0$. Putting $\frac{1}{2}S_2\omega^2 = -\zeta\gamma'' - (1-\alpha)\xi\gamma$ we get $M_2\omega = 0$; hence we may neglect terms in ε^2. It is convenient to express the system corresponding to $\tilde{X}_0 + \varepsilon M_1$ in variables

$$p,\ q,\ r,\ \gamma'',\quad u = p\gamma + q\gamma',\quad v = q\gamma - p\gamma', \tag{3.19.19}$$

so that

$$\begin{aligned} \tilde{X}_0 &= v\frac{\partial}{\partial \gamma''} - \omega^2 \gamma'' \frac{\partial}{\partial v}, \\ M_1 &= \alpha r \left(q\frac{\partial}{\partial p} - p\frac{\partial}{\partial q}\right) + \frac{\xi}{\omega^2} qu \frac{\partial}{\partial r}. \end{aligned}$$

Thus the motion of the body is described in the case being considered (in the principal order) by the system of equations

$$\begin{aligned} \frac{dp}{dt} &= \alpha qr, \\ \frac{dq}{dt} &= -\alpha rp, \\ \frac{dr}{dt} &= \frac{\xi}{\omega^2} qu, \\ \frac{du}{dt} &= 0, \\ \varepsilon \frac{dv}{dt} &= -\omega^2 \gamma'', \\ \varepsilon \frac{d\gamma''}{dt} &= v, \end{aligned} \tag{3.19.20}$$

where u and v are defined by (3.19.19). Besides obvious integrals

$$\begin{aligned} \omega^2 = p^2 + q^2 = \text{const}, \qquad u = \text{const}, \\ v^2 + \omega^2 \gamma''^2 = \text{const}, \end{aligned}$$

which are approximate corollaries of the three classical integrals of motion, the system (3.19.20) possesses a new fourth integral (*adiabatic invariant*)

$$r^2 - \frac{2\xi}{\alpha\omega^2} up = \text{const}$$

and therefore can be integrated in quadratures.

A simple investigation [7] shows that the motion described by (3.19.20) is as follows: pendulum-like oscillations of the body around the symmetry axis (nonlinear oscillations of the physical pendulum with the speed ~ 1) are imposed on the Euler inertial motion. In the second case, when at the initial moment $k^2 - \alpha^2 r^2$ is small, a similar effect is caused by gravity.

The following example is interesting for its "composite" asymptotics constructed separately in different subdomains separated by the singular surface on which asymptotic solutions are matched by continuity.

3.20 Example: The Langer Problem ([28])

Let us consider the equation

$$\varepsilon \frac{dp}{dx} = ixp + \varepsilon(a(x)p + b(x)\bar{p}) \qquad (\varepsilon > 0), \tag{3.20.1}$$

where $a(x)$ and $b(x)$ are given (smooth) complex-valued functions of the real argument x which runs over a finite segment containing $x = 0$; hereafter, a bar stands for complex conjugation. Since x is an eigenvalue of the leading problem, which vanishes in the considered domain, the standard reduction to the normal form in the whole domain is impossible to perform ($x = 0$ is a resonance 0-point; more exactly, $x = 0$ is the singular surface consisting of 0-points) if we apply by analogy the notion from Section 3.10. But here x is a variable, not a parameter, and arguments of Section 3.10 on reduction to the uniform normal form are inapplicable, at least directly. This problem is in the class of the reconstruction problems which are discussed in the next chapter. For these problems it is natural to make use of transformations singular at $\varepsilon = 0$ which lead to a "reconstruction" of the operator (system). In this problem such a transformation [see (3.20.2)] is easy from simple scale considerations.

Put

$$\varepsilon = \delta^2, \qquad x = \delta\tau, \tag{3.20.2}$$

and pass to the extended system following from (3.20.1), due to (3.20.2) and considered (which is important to remember) for $|\tau| \lesssim 1/\delta$:

$$\frac{dp}{d\tau} = i\tau p + \delta(a(x)p + b(x)\bar{p}), \qquad \frac{dx}{d\tau} = \delta. \tag{3.20.3}$$

To the system (3.20.3) corresponds the operator $X = X_0 + \delta X_1$, where

$$X_0 x = 0, \qquad X_0\tau = 1, \qquad X_0 p = i\tau p, \tag{3.20.4}$$

$$X_1 x = 1, \qquad X_1\tau = 0, \qquad X_1 p = a(x)p + b(x)\bar{p}. \tag{3.20.5}$$

The operator X_0 possesses an invariant

$$q = p\exp\left(-\frac{i}{2}\tau^2\right). \tag{3.20.6}$$

In variables x, τ, and q, where x and q are invariants of X_0 and τ corresponds to 0, operators X_0 and X_1 will take (due to (3.20.4), (3.20.5), and (3.20.6)) the form

$$X_0 x = 0, \qquad X_0\tau = 1, \qquad X_0 q = 0, \tag{3.20.7}$$

$$X_1 x = 1, \qquad X_1\tau = 0, \qquad X_1 q = a(x)q + b(x)\bar{q}e^{-i\tau^2}. \tag{3.20.8}$$

Let us find the operators of the first approximation; i.e., S_1 and $M_1 = [X_0, S_1] + X_1$. First, it is clear that we may put

$$S_1 x = S_1\tau = 0 \tag{3.20.9}$$

so that

$$M_1 x = 1, \qquad M_1\tau = 0. \tag{3.20.10}$$

Further,

$$M_1 q = \frac{\partial}{\partial\tau}(S_1 q) + a(x)q + b(x)\bar{q}e^{-i\tau^2}.$$

It is clear that we may put

$$M_1 q = a(x)q, \tag{3.20.11}$$

defining $S_1 q$ from the equation

$$\frac{\partial}{\partial\tau}(S_1 q) = -b(x)\bar{q}e^{-i\tau^2}.$$

Further, we will consider domains $x \geq 0$ $(\tau \geq 0)$ and $x \leq 0$ $(\tau \leq 0)$; i.e., construct asymptotics for $0 \leq x \lesssim 1$ and for $0 \geq x \gtrsim -1$ separately, requiring, naturally, the solution to be smooth at $x = 0$. For $\tau \geq 0$ put

$$S_1 q = b(x)\bar{q}\int_\tau^\infty e^{-i\zeta^2}\,d\zeta, \tag{3.20.12}$$

and for $\tau \leq 0$ put

$$S_1 q = -b(x)\bar{q}\int_{-\infty}^\tau e^{-i\zeta^2}\,d\zeta. \tag{3.20.12$'$}$$

Thus we impose the condition $S_1 q|_{|\tau|=\infty} = 0$ which will be essential in what follows. Now let us find operators of the second approximations; i.e., S_2 and $M_2 = [X_0, S_2] + \frac{1}{2}[M_1 + X_1, S_1]$. As above, it is clear that we may put $S_2 x = S_2\tau = 0$ so that $M_2 x = M_2\tau = 0$. Further, due to (3.20.7) - (3.20.12),

$$\begin{aligned} M_2 q = {} & \frac{\partial}{\partial\tau}(S_2 q) + \frac{1}{2}b(x)\overline{b(x)}(e^{i\tau^2}I(\tau) - e^{-i\tau^2}\overline{I(\tau)})q \\ & + \{b'(x) + b(x)[\overline{a(x)} - a(x)]\}I(\tau)\bar{q}, \end{aligned}$$

where, due to (3.20.12) and (3.20.12′),

$$\begin{aligned} I(\tau) &= \int_\tau^\infty e^{-i\zeta^2}\,d\zeta \qquad (\tau \geq 0), \\ I(\tau) &= -\int_{-\infty}^\tau e^{-i\zeta^2}\,d\zeta \qquad (\tau \leq 0), \end{aligned} \tag{3.20.13}$$

and $b'(x) = db(x)/dx$. Due to the choice of $I(\tau)$, we have

$$I(\tau) \sim \frac{e^{-i\tau^2}}{|\tau|} \qquad \text{as } |\tau| \to \infty$$

[recall once more that $I(\tau)$ is chosen separately for domains $\tau \geq 0$ and $\tau \leq 0$]. We may now "kill" the term with $\bar{q}$ and, more generally, get

$$\begin{aligned} M_2 q &= m_2(x,\tau)q, \\ m_2(x,\tau) &\equiv \tfrac{1}{2}b(x)\overline{b(x)}[e^{i\tau^2}I(\tau) - e^{-i\tau^2}\overline{I(\tau)}], \end{aligned}$$

obtaining $S_2 q$ like (3.20.12) and (3.20.12′); i.e., under the condition $S_2 q|_{|\tau|=\infty} = 0$ (and separately for domains $\tau \le 0$ and $\tau \ge 0$), and

$$S_2 q = \{-b'(x) + b(x)[a(x) - \overline{a(x)}]\}\left(\tau I(\tau) + \frac{i}{2}e^{-i\tau^2}\right)\bar{q}.$$

We get $M_2 q = m_2(x,\tau)q$ and, similarly, $M_3 q = m_3(x,\tau)q$, $M_4 q = m_4(x,\tau)q$, ..., so that $Mq = \delta m(x,\tau;\delta)q \equiv \delta[a(x) + \delta m_2(x,\tau) + \delta^2 m_3(x,\tau) + \cdots]q$; i.e., we get the system

$$\frac{dq}{d\tau} = \delta m(x,\tau,\delta)q, \qquad \frac{dx}{dt} = \delta$$

or

$$\frac{dq}{dx} = m\left(x, \frac{x}{\delta}, \delta\right) q.$$

We get

$$q^* = ce^{\theta(x,\delta)}, \quad \theta(x,\delta) \equiv \int_0^x m(\xi, \xi/\delta, \delta)\, d\xi, \tag{3.20.14}$$

where c is a (complex) constant and the notation q^* reminds us that (3.20.14) is the answer in new variables.

Remark 3.20.1 We have extended the system. Since $Sx = S\tau = 0$, $M_j x = M_j \tau = 0$ for $j \ge 2$ and $(X_0 + \delta M_1)(x - \delta\tau)|_{x=\delta\tau} = 0$ for $(X_0 + \delta M_1)(x - \delta\tau) \equiv 0$, we did not surpass the class of extended operators.

Returning to the old variables and restricting ourselves for simplicity's sake to accuracy $0(\delta) = 0(\sqrt{\varepsilon})$, we get

$$q = c\exp\int_0^x a(\xi)\, d\xi - \delta\bar{c}b(x)I(\tau)\exp\int_0^x \overline{a(\xi)}\, d\xi + 0(\delta^2),$$

and, according to (3.20.6) and (3.20.2), we have

$$p = e^{ix^2/2\varepsilon}\left(c\exp\int_0^x a(\xi)\, d\xi - \sqrt{\varepsilon}\bar{c}b(x)I(x/\sqrt{\varepsilon})\exp\int_0^x \overline{a(\xi)}\, d\xi\right) + 0(\varepsilon).$$

Thus, due to (3.20.13),

$$\begin{aligned} p = \exp\left(\frac{-ix^2}{2\varepsilon}\right)\Bigg(&c^{(+)}\exp\int_0^x a(\xi)\, d\xi \\ &- \sqrt{\varepsilon}\bar{c}^{(+)}b(x)\int_{x/\sqrt{\varepsilon}}^{\infty} e^{-i\zeta^2}\, d\zeta \exp\int_0^x \overline{a(\xi)}\, d\xi\Bigg) + 0(\varepsilon) \end{aligned} \tag{3.20.15}$$

for $x \ge 0$ and

$$\begin{aligned} p = \exp\left(\frac{-ix^2}{2\varepsilon}\right)\Bigg(&c^{(-)}\exp\int_0^x a(\xi)\, d\xi \\ &+ \sqrt{\varepsilon}\bar{c}^{(-)}b(x)\int_{-\infty}^{x/\sqrt{\varepsilon}} e^{-i\zeta^2}\, d\zeta \exp\int_0^x \overline{a(\xi)}\, d\xi\Bigg) + 0(\varepsilon) \end{aligned} \tag{3.20.15′}$$

for $x \leq 0$.

Putting $x = 0$ we obtain the relation between constants

$$c^{(+)} - \sqrt{\varepsilon c^{(+)}} b(0) \frac{\sqrt{\pi}}{2} e^{-i\pi/4} - c^{(-)} - \sqrt{\varepsilon} b(0) \bar{c}^{(-)} \frac{\sqrt{\pi}}{2} e^{-i\pi/4} = 0(\varepsilon),$$

implying

$$\begin{aligned} c^{(+)} &= A_0 + \sqrt{\varepsilon}\left(A_1 + \bar{A}_0 \frac{\sqrt{\pi}}{2} e^{-i\pi/4} b(0)\right) + 0(\varepsilon), \\ c^{(-)} &= A_0 + \sqrt{\varepsilon}\left(A_1 - \bar{A}_0 \frac{\sqrt{\pi}}{2} e^{-i\pi/4} b(0)\right) + 0(\varepsilon), \end{aligned} \tag{3.20.16}$$

where $A_0 + \sqrt{\varepsilon} A_1 + \cdots$ is an arbitrary constant. Conditions (3.20.16) are the matching conditions for asymptotics (3.20.15) and (3.20.15′).

4

Reconstruction

4.1 Introduction

In various problems we must employ variable transformations degenerate at $\varepsilon = 0$. One such example is the case of a nilpotent X_0 considered in Sections 1.5 and 2.5, where a special (shearing) transformation reconstructs X so that another operator, different from X_0, becomes the leading one.

Situations when reconstruction becomes necessary are often encountered in applications and in this chapter we discuss problems of this kind.

Two types of reconstruction problems whose description follows suit will not be sharply distinguished.

The First Types of Problems

We have accepted above as an axiom the statement that the reduction of the operator $X(M)$ to the normal form with respect to $X_0(M_j^{(\mu_j)} = 0)$ exhausts the means of the perturbation theory based on the "complete knowledge" of X_0.

Remark 4.1.1 It is true though that a normal form is not unique and as the very example of a nilpotent X_0 shows (Section 2.5) the question of the "best" form of M is rather important.

However, the normal form of M with respect to X_0 automatically provides neither the "complete knowledge" of M nor the necessity for the exact study of M without taking into account the smallness of ε. Therefore the question arises: what should we understand to be the perturbation theory for the operator which is already in the normal form with respect to X_0? In other words: which operator in M ought to be considered as the leading one (X_0 does not fit any more by definition) in order that the problem of reduction to the normal form with respect to this new leading operator arises, in a sense, naturally? (For example, is it possible, as in the linear problem for a diagonal X_0, to take $X_0 + \varepsilon M_1$ as a leading operator?) Without an (at least formal) answer to this question the perturbation theory remains logically incomplete.

Let us give one example when the answer is clear. Let $[X_0, M] = 0$; i.e., $M' = 0$. (This takes place when X_0 is diagonal and its eigenvalues are constants.) Let $\bar{S} = \varepsilon \bar{S}_1 + \varepsilon^2 \bar{S}_2 + \cdots$, where $\bar{S}_j' = 0$ and

$$\tilde{M} = e^{-\bar{S}} M e^{\bar{S}} = X_0 + \varepsilon e^{-\bar{S}}(M_1 + \varepsilon M_2 + \cdots)e^{\bar{S}} \qquad (\tilde{M}' = 0) \qquad (4.1.1)$$

so that

$$\tilde{M}_1 = M_1$$

and

$$\tilde{M}_2 = [M_1, \bar{S}_1] + M_2, \qquad \tilde{M}_3 = [M_1, \bar{S}_2] + M_3 + [M_2, \bar{S}_1] + \tfrac{1}{2}[[M_1, \bar{S}_1]\bar{S}].$$

In the last formulas all operators belong to the class of operators permutable with X_0. The operator M_1 plays the role of a leading operator; choosing $\bar{S}$ we can reduce $\tilde{M}$ to the normal form with respect to M_1 (clearly under certain conditions; see Theorem 2.4.3). Thus in this case M_1 can be called the new leading operator. Note also that $\tilde{M}$ is also normal with respect to $X_0 + \varepsilon M_1$. This operator also might be considered as a leading one, which is perhaps more logical (see below).

Remark 4.1.2 Since in this example we have not used singular transformations we do not consider it as a reconstruction problem.

If $M' \neq 0$ (which holds, generally speaking, even for a diagonal X_0, with variable eigenvalues and so much the more for a Jordan X_0), arguments (4.1.1) fail. In fact, if $\bar{S}_1' = 0$, then it is not possible, generally speaking, to reduce $\tilde{M}_2$ to the normal form with respect to M_1, since $\bar{S}_1$ belongs to a narrower class of operators ($\bar{S}_1' = 0$) than M_1 and M_2 do ($M_1' \neq 0$ and $M_2' \neq 0$). If $\bar{S}_1' \neq 0$, then (4.1.1) fails (it is impossible to eliminate X_0). The question of what is a new leading operator becomes nontrivial.

We can intuitively suppose that there exists a "principal" part $\zeta(\varepsilon)$ of the perturbation $\varepsilon M_1 + \cdots$, defined in a natural way, such that $X_0 + \zeta(\varepsilon)$ can be taken for a new leading operator [in the above example $\zeta(\varepsilon) = \varepsilon M_1$ and in the example with a nilpotent X_0, in Section 2.5, $\zeta(\varepsilon)$ is a sum of canonical operators of the highest "weight"]. We should expect as in the example in Section 2.5 that a reduction of $X_0 + \zeta(\varepsilon)$ to the Jordan form becomes possible only by a change of variables singular at $\varepsilon = 0$.

Problems associated with singular transformations where $\zeta(\varepsilon)$, i.e., the new leading operator $X_0 + \zeta(\varepsilon)$, will be referred to as reconstruction problems of the first type.

It ought to be said at the outset that the question of natural definition of $\zeta(\varepsilon)$ in the general case seems to be extremely difficult. We will formally answer it in Section 4.2 under certain assumptions on the perturbations; our main tool will be the shearing transformation of Section 2.5.

The Second Type of Problem

Let us give first an illustrative example. Consider the equation from Example 3.2: $\varepsilon y'' + a(x)y' + b(x)y = 0$ for $a(x) = x$ so that $a(0) = 0$ and $x \subset [0, 1]$.

As is clear from Example 3.2, operators M and S are singular at $x = 0$ and a reduction of X to the normal form near the surface $x = 0$ is impossible (and is possible in the domain $|x| \geq \alpha > 0$, where $\alpha = \text{const}$, for any $\alpha > 0$). It seems quite evident that near the surface $\alpha = 0$ the operator X_0 should be replaced by another, new, leading operator $X_0 + \hat{\zeta}(\varepsilon)$, where $\hat{\zeta}(\varepsilon)$ is a "principal" part of the perturbation such that $X_0 + \hat{\zeta}(\varepsilon)$ fits as a leading operator. Clearly, the question on $\hat{\zeta}(\varepsilon)$ is similar to the preceding one, but in principle is more difficult.

Generally, in problems of the second type there exist singular surfaces near which (some) coefficients of operators X, M, and S lose their smoothness (uniform boundedness). Most often these surfaces consist of resonance i-points in the sense of Section 3.10 (the notion of a resonance point from Section 3.10 is easy to generalize, since invariants of X_0 play the role of constants with respect to X_0). In this example the surface $x = 0$ consists of resonance 0-points: the operator $X_0 = z(\partial/\partial y) - xz(\partial/\partial z)$, diagonal at $x \neq 0$, becomes Jordan (nilpotent) at $x = 0$.

Methods of reconstruction of operators in problems of the second type will be given in Sections 4.3 and 4.4. The characteristic feature of problems of the second type is the necessity of matching asymptotics obtained far from and near to the singular surface. A method of matching will be indicated in Section 4.5. Several examples will be considered in detail in the end of the chapter.

4.2 New Leading Operators in the First Type Problems

Let

$$M = X_0 + \varepsilon M_1 + \varepsilon^2 M_2 + \cdots \equiv X_0 + \hat{M} \tag{4.2.1}$$

be an operator normal with respect to the (Jordan) operator X_0, let $\varphi_1, \ldots, \varphi_k$ be functions corresponding to eigenvalues $\lambda_1, \ldots, \lambda_k$, and let $\omega_1, \ldots, \omega_\ell$ be invariants with respect to X_0 such that (φ, ω) is an (extended) basic system $[\lambda_i = \lambda_i(\omega)]$. We will assume that the coefficients of operators M_j, i.e., functions $M_j\varphi_i$ and $M_j\omega_q$ are *polynomial in* φ with coefficients depending on ω. This is the *main formal* restriction. Then M is *normal* with respect to X_0 whenever $M_j\varphi_i$'s are combinations of monomials $\varphi_1^{m_1} \ldots \varphi_k^{m_k}$, where

$$m_1\lambda_1 + \cdots + m_k\lambda_k = \lambda_i \qquad (m_1 + \cdots + m_k \geq 1), \tag{4.2.2}$$

and $M_j\omega_q$ are combinations of monomials $\varphi_1^{n_1} \ldots \varphi_k^{n_k}$, where

$$n_1\lambda_1 + \cdots + n_k\lambda_k = 0, \tag{4.2.2$'$}$$

i.e., combinations of monomials corresponding to λ_i and 0 respectively. Suppose also that relations (4.2.2) and (4.2.2$'$) for all (integer) $m_1, \ldots, m_k, n_1,$

$\ldots, n_k$ for which these relations are possible can be satisfied only *identically* with respect to ω [see condition (2) of Theorem 2.4.3].

Consider the case of a diagonal X_0, i.e., $X_o\varphi_i = \lambda_i\varphi_i$, where λ_i are not, generally speaking, constants. First, note that the system corresponding to M, $d\varphi_i/d\tau = \lambda_i\varphi_i + \hat{M}\varphi_i$, $d\omega_q/d\tau = \hat{M}\omega_q, \ldots$, would have been reduced to the system $d\varphi_i/d\tau = \hat{M}\varphi_i$, $d\omega_q/d\tau = \hat{M}\omega_q$ after the change $\varphi_i \to \varphi_i \exp\int \lambda_i d\tau$ if we had found $\lambda_i(\tau)$. [In particular, if λ_i are constants $\varphi_i \to \varphi_i \exp \lambda_i\tau$ or if $\hat{M}\omega_q = a_q(\omega)$, then solving the system $d\omega_q/d\tau = a_q(\omega)$ we find $\lambda_i(\omega)$ as functions of τ.] This remark is based on the invariance of $\hat{M}\varphi_i$ and $\hat{M}\omega_q$ with respect to the transformation $\varphi_i \to \varphi_i \exp\int \lambda_i\, d\tau$, which follows from the assumption that M is normal and depends polynomially on φ. This clarifies what follows.

Let $\Omega_1, \ldots, \Omega_k$ be a system of invariants of X_0 satisfying resonance relations

$$m_1\Omega_1 + \cdots + m_k\Omega_k = \Omega_i, \qquad n_1\Omega_1 + \cdots + n_k\Omega_k = 0 \tag{4.2.2''}$$

which are satisfied by $\lambda_1, \ldots, \lambda_k$ in monomials that enter $\hat{M}\varphi_i$ and $\hat{M}\omega_q$. Let us perform a (singular) transformation

$$\varphi_i \to \varphi_i \exp\frac{\Omega_i}{\varepsilon}. \tag{4.2.3}$$

Then

$$X_0 \to X_0, \qquad \hat{M} \to \hat{M} - \sum_{1\le i\le k} \frac{1}{\varepsilon}(\hat{M}\Omega_i)\varphi_i\frac{\partial}{\partial\varphi_i}$$

due to (4.2.2), (4.2.2′), and (4.2.2″); i.e.,

$$M \to \sum_{1\le i\le k} (\lambda_i - M_1\Omega_i - \varepsilon M_2\Omega_i - \cdots)\varphi_i\frac{\partial}{\partial\varphi_i} + \hat{M}. \tag{4.2.4}$$

Now suppose that there exists a smooth (particular) solution of the equations

$$M_1\Omega_i = \lambda_i - \bar{\lambda}_i \qquad \text{for } i = 1, 2, \ldots, \tag{4.2.5}$$

where λ_i are suitable constants ("constant parts" of λ_i). The right-hand side of (4.2.4) takes the form

$$M^* = \sum_{1\le i\le k} \bar{\lambda}_i\varphi_i\frac{\partial}{\partial\varphi_i} + \varepsilon M_1^* + \varepsilon^2 M_2^* + \cdots,$$

where

$$M_1^* = M_1 - \sum_{1\le i\le k}(M_2\Omega_i)\frac{\partial}{\partial\varphi_i}, \qquad M_2^* = M_2 - \sum_{1\le i\le k}(M_3\Omega_i)\frac{\partial}{\partial\varphi_i}, \qquad \ldots,$$

which is normal with respect to

$$\sum_{1\le i\le k} \bar{\lambda}_i \varphi_i \frac{\partial}{\partial \varphi_i},$$

an operator with *constant* eigenvalues and (see Section 4.1) we can put $\zeta(\varepsilon) = \varepsilon M_1^*$.

Let us give a simple example. In Example 3.1 we have

$$M_1 = h(\omega)\left\{\left(\frac{5}{2}\varphi\frac{\partial}{\partial\varphi} + \frac{5}{2}\bar{\varphi}\frac{\partial}{\partial\bar{\varphi}}\right) + \omega\frac{\partial}{\partial\omega}\right\}.$$

Since

$$X_0 = i\omega\left(\varphi\frac{\partial}{\partial\varphi} - \bar{\varphi}\frac{\partial}{\partial\bar{\varphi}}\right),$$

put $\Omega_1 = -\Omega_2 = \Omega(\lambda_1 = -\lambda_2 = i\omega)$ where $\Omega = \Omega(\omega)$ and Equation (4.2.5) takes the form $\omega h g(\omega)\, d\Omega/d\omega = i\omega$. Putting $dH(\omega)/d\omega \equiv H'(\omega) = 1/h(\omega)$ we get $\Omega = iH(\omega)$. The transformation

$$\varphi \to \varphi \exp\frac{i}{\varepsilon}H, \qquad \bar{\varphi} \to \bar{\varphi}\exp\left(-\frac{i}{\varepsilon}H\right)$$

gives

$$M \to \varepsilon\left\{M_1 + i\frac{M_2\omega}{h(\omega)}\left(\varphi\frac{\partial}{\partial\varphi} - \bar{\varphi}\frac{\partial}{\partial\bar{\varphi}}\right)\right\} + 0(\varepsilon^2).$$

If we continue the computations of Example 3.1 we will easily find that $M_2\omega = 0$. Thus the new leading operator is M_1 (in new variables).

This method of reducing the problem to the case $\lambda_i = \text{const}$ can be made more versatile as follows. Put

$$\varphi_i \to \varphi_i \exp\frac{\Omega_i}{\varepsilon^\gamma} \qquad \text{for } \gamma > 0 \tag{4.2.3'}$$

instead of (4.2.3) so that

$$M \to \sum_{1\le i\le k}\left(\lambda_i - \frac{1}{\varepsilon^\gamma}\hat{M}\Omega_i\right)\varphi_i\frac{\partial}{\partial\varphi_i} + \hat{M} \tag{4.2.4'}$$

and let us seek a smooth (particular) solution $\Omega_i = \Omega_{i0} + \varepsilon\Omega_{i1} + \varepsilon^2\Omega_{i2} + \cdots$ of the equations

$$\frac{1}{\varepsilon^\gamma}\hat{M}\Omega_i = \lambda_i - \bar{\lambda}_i + 0(\varepsilon) \qquad (\bar{\lambda}_i = \text{const}) \tag{4.2.5'}$$

(for some γ). This generalization embraces not only the case $M_1 \equiv 0$ but also cases when for some reasons it is convenient to take, say, $M_1\Omega_{i0} = 0$, $M_2\Omega_{i0} + M_1\Omega_{i0} + M_1\Omega_{i1} = \lambda_i - \bar{\lambda}_i$, $\gamma = 2$, etc.

Now consider the case of a nilpotent X_0 ($\lambda_i = 0$). First suppose that perturbations are polynomial not only in φ but also in ω (in what follows this additional restriction will be dropped). Then we have the case already considered in Section 2.5. Recall briefly the contents of the Section 2.5. Let $Y = p_s(x)(\partial/\partial x_s)$ be an operator acting on functions in n variables $x = (x_1, \ldots, x_n)$ and let $p_s(x)$ be homogeneous polynomials of degree N in $x_1, \ldots, x_n$. The linear space of operators Y will be denoted by T_{N-1}. Let also X_0 be a nilpotent operator in T_0 and variables x constitute a canonical Jordan basis in T_0 with respect to X_0. Further let us enumerate symmetrically Jordan chains of the basis x; i.e.,

$$e_r,\ e_{r-1},\ \ldots,\ e_{-r+1},\ e_{-r}, \tag{4.2.6}$$

$$\tag{4.2.7}$$

where $X_0 e_s = e_{s-1}$ for $s = r, r-1, \ldots, -r+1$, and $X_0 e_{-r} = 0$ [here the index which denotes the number of the chain is omitted; i.e., $e_{r-q} = e_{i,r_i-q}$, where i is the number of the chain; r_i is an integer or half integer index and $\sum_i (2r_i + 1) = n$].

In T_{N-1}, introduce the operator of differentiation $DY \equiv [X_0, Y] = Y'$. The operator D is nilpotent in T_{N-1}, i.e., Y is bounded in height; $D^\mu Y = 0$ for some μ. In T_{N-1}, introduce a Jordan basis with respect to D and let us index its chains as in (4.2.6):

$$Y_m, Y_{m-1}, \ldots, Y_{-m+1}, Y_{-m}, \tag{4.2.8}$$

where $DY_s = Y_{s-1}$ and $DY_{-m} = 0$. The height of the generating element Y_m equals $2m+1$. Decomposing Y with respect to the basis we get

$$Y = \sum \sum_{-m \le s \le m} c_s Y_s,$$

where the external summation runs over the number of chains, we get [see (4.2.8)]

$$Y = \sum c_m Y_m + DZ \qquad \text{for } Z \subset T_{N-1}, \tag{4.2.9}$$

where $\sum c_m Y_m$ is a nonintegrable operator.

Remark 4.2.1 The formula (4.2.9) gives one of the solutions of the problem of how to represent Y in the form $Y = \bar{Y} + DZ$, where $\bar{Y}$ is a nonintegrable operator of minimum height. For any solution $\bar{Y} = \sum \sum_{-m \le s \le m} \bar{c}_s Y_s$ of this problem we have $\bar{c}_m = c_m$ and $\bar{c}_s = 0$ for $s > \mu - m - 1$, where μ is the height of $\bar{Y}$.

Now if we consider the operator $X = X_0 + \varepsilon X_1 + \varepsilon^2 X_2 + \cdots$, where X_j is the sum of operators of the form Y for different N, then by (4.2.9) we may get

$$M = e^{-S} X e^{S} = X_0 + \varepsilon M_1 + \varepsilon^2 M_2 + \cdots \tag{4.2.10}$$

so that M_j is the linear combination of elements of the form Y_m [generating elements of some chains (4.2.8) for different N].

Remark 4.2.2 It is clear (see Remark 4.2.2) that we simultaneously solve the problem of consecutive minimization of heights of operators M_j for $j = 1, 2, \dots$ (we find one solution of this problem).

Furthermore, in Section 2.5 we have proved the main Theorem 2.5.3 which states that in each T_{N-1} there exists a basic (4.2.8) such that the change of variables with a parameter

$$e_s \to t^s e_s \tag{4.2.11}$$

performed in all chains (4.2.6) yields

$$Y_s \to t^s Y_s \tag{4.2.12}$$

in all chains (4.2.8). Such a basis is called, in Section 2.5, a normal one.

Now consider the operator $X_0 + \Delta Y_s$, where Δ is a parameter, Y_s is an element of a normal basis, and $s \neq -1$. Note that the transformation (4.2.11) yields $X_0 \to t^{-1} X_0$. Then by (4.2.12) we get for $t = \Delta^{-1/(1+s)}$ that $X_0 + \Delta Y_s \to \Delta^{1/1+s}(X_0 + Y)$. In particular, for $s = m$ we have $X_0 + \Delta Y_m \to \Delta^{2/1+\mu}(X_0 + Y)$ where $\mu = 2m + 1$ is the height of Y_m with respect to X_0. The parameter $\Delta^{2/(1+\mu)}$ will be referred to as the *weight* of ΔY_m with respect to X_0. We will assume that M in (4.2.10) is constructed in a normal basis and denote by μ_j the maximum height of summands in M_j so that the maximal height of summands in M_j is $\varepsilon^{2j/(1+\mu_j)}$. Suppose that

$$k = \min_j \frac{2j}{1+\mu_j} > 0$$

exists and is reached. Then the transformation (4.2.11) for $t = \varepsilon^{-k}$ (a generalized shearing transformation) yields

$$M \to \varepsilon^k(\tilde{X}_0 + \varepsilon^{\nu_1}\tilde{X}_1 + \varepsilon^{\nu_2}\tilde{X}_2 + \cdots) \quad \text{for } 0 < \nu_1 < \nu_2 < \cdots, \tag{4.2.13}$$

where $\varepsilon^k \tilde{X}_0$ is the sum (in new variables) of X_0 and all operators of the maximal weight ε^k.

The formula (4.2.13) shows that $\tilde{X}_0$ can be (formally) considered as a new leading operator. For the linear problem, such a definition of the new leading operator turned out to be natural since each step of this kind is a regular approximation to the complete solution of the whole problem (see Section 1.5, Theorem 1.5.1).

Now note one aspect of computations not mentioned in Section 2.5. It is not difficult to note that if in the construction of M in (4.2.10) we had restricted ourselves with the requirement of the consecutive minimization of the height of operators M_j (see Remarks 4.2.2 and 4.2.3), then taking any solution of this problem and defining k, as earlier, to be $k = \min_j 2j/(1 + \mu_j)$, where μ_j is the height of M_j, we would have [after application of (4.2.11) for $t = \varepsilon^{-k}$] a formula similar to (4.2.13) with perhaps another k and $\tilde{X}_0$.

Remark 4.2.3 It seems that all such procedures are equivalent, but we lack a proof.

Any such procedure is formally applicable and liberates one from a cumbersome construction of a normal basis: the problem of consecutive minimization of heights of M_j can be solved in any basis and the transformation (4.2.11) for $t = \varepsilon^{-k}$ yields automatically the formula of the form (4.2.13). It goes without saying that this simplified method of computing also takes much time.

Remark 4.2.4 We have made only the first step as compared with the linear problem. Pitifully, as had been already said in Section 2.5, an analogue of Theorem 1.5.1 for a nonlinear case is not established.

Finally, note an evident circumstance. The dependence of coefficients of perturbation operators on elements of the form e_0, in particular, on generating elements of the form e_0, i.e., on "genuine" nonintegrable invariants (ω), might not be considered as polynomials since elements of the form e_0 do not suffer "dilatation"; see (4.2.11).

Let us come to the general case of a Jordan X_0. Let us split X_0 into diagonal and nilpotent parts: $X_0 = X_{0d} + X_{0n}$. Note the following evident facts:

1. $[X_{0d}; X_{on}] = 0$;
2. The operator M [(4.2.1)] is normal with respect to X_{0d};
3. The operator M is normal with respect to X_{0n}.

Without loss of generality we may assume that the problem of consecutive minimization of heights μ_{j0} of operators M_j with respect to X_{0n} is solved for M. Set

$$k_0 = \min_j \frac{2j}{1+\mu}$$

and let us perform a shearing transformation with $t = \varepsilon^{-k_0}$. Since X_{0d} is invariant with respect to transformations of the form (4.2.11) we get

$$M \to X_{0d} + \varepsilon^{k_0}\tilde{M}_1 + \varepsilon^{k_1}\tilde{M}_2 + \cdots \qquad \text{for } k_0 < k_1 < \cdots. \tag{4.2.14}$$

The right-hand side of (4.2.14) is normal with respect to the diagonal operator X_{0d} and the above-mentioned trick of constructing $\zeta(\varepsilon)$ is applicable to it.

Remark 4.2.5 The fact that in (4.2.14) powers of ε are rational is clearly unimportant.

To conclude this subsection let us give one more illustrative example. Consider the operator

$$X = X_0 + \varepsilon X_1 = -\rho(x)y\frac{\partial}{\partial z} + \varepsilon\left(z\frac{\partial}{\partial y} + \frac{\partial}{\partial x}\right), \tag{4.2.15}$$

where $\rho(x) \neq 0$. This operator arises in considering the well-known equations

$$\varepsilon y'' + \rho(x)y = 0. \tag{4.2.16}$$

We put $y' = z$ and pass to the system

$$\varepsilon y' = \varepsilon z, \qquad \varepsilon z' = -\rho(x)y$$

which is "incorrect" (whereas "correct" is to put $\sqrt{\varepsilon}\, y' = z$).

It is not difficult to verify that in (4.2.15) the operator X was (incidentally) normal from the start with respect to X_0. In fact, $X_1''' = 0$ (and the height of X_1, which equals 3, is the minimal height of the perturbation). However, this property does not by itself provide us with any progress in the study of (4.2.16). The problem requires a reconstruction.

Note that X_0 is a nilpotent operator and x is its invariant of the form e_0 which, as had been mentioned above, makes it possible to ignore in the reconstruction procedure the dependence on variables of this form and perform the shearing transformation

$$y \to t^{1/2}y, \qquad z \to t^{-1/2}z \qquad (y = e_{1/2},\ z = e_{-1/2}) \tag{4.2.17}$$

with (an as yet) arbitrary parameter t. Then

$$X \to t\left\{-\rho(x)y\frac{\partial}{\partial z} + \frac{\varepsilon^2}{t}z\frac{\partial}{\partial y} + \frac{\varepsilon}{t}\frac{\partial}{\partial x}\right\}.$$

This implies that the reconstruction is performed at $t = \sqrt{\varepsilon}$ [we diminish t starting with $t \sim 1$ until in $\{\quad\}$ a term of order ~ 1 which is different from $-\rho(x)y(\partial/\partial z)$ appears]. Thus the transformation (4.2.17) with $t = \sqrt{\varepsilon}$ gives us

$$X \to \sqrt{\varepsilon}\,Y,$$

where

$$Y = Y_0 + \sqrt{\varepsilon}\,Y_1 = -\rho(x)y\frac{\partial}{\partial z} + z\frac{\partial}{\partial y} + \sqrt{\varepsilon}\frac{\partial}{\partial x}.$$

We would have obtained Y under the "correct" passage to the system, too.

We might have discussed this somewhat differently. Since the heights of $\varepsilon z(\partial/\partial y)$ and $\varepsilon(\partial/\partial x)$ with respect to X_0 are 3 and 2 respectively, then their weights are $\varepsilon^{2/(1+3)} = \sqrt{\varepsilon}$ and $\varepsilon^{2/(1+2)} = \varepsilon^{2/3}$ respectively. The weight of the first operator is greater and we may take

$$-\rho(x)y\frac{\partial}{\partial z} + \varepsilon z\frac{\partial}{\partial y}$$

for the new leading operator. This operator is easy to study exactly: its eigenfunctions are $\varphi = \sqrt{\varepsilon}\,z + i\sqrt{\rho}\,y$ and $\bar{\varphi}$, corresponding to $i\sqrt{\rho}$ and $-i\sqrt{\rho}$, and x_0 is an invariant. In x, φ, and $\bar{\varphi}$ we have

$$X = \sqrt{\varepsilon}\,Y,$$

where

$$\begin{aligned} Y &= Y_0 + \sqrt{\varepsilon} Y_1 \\ &= i\sqrt{\rho}\left(\varphi\frac{\partial}{\partial\varphi} - \bar{\varphi}\frac{\partial}{\partial\bar{\varphi}}\right) + \sqrt{\varepsilon}\left\{\frac{\rho'}{4\rho}(\varphi-\bar{\varphi})\left(\frac{\partial}{\partial\varphi} - \frac{\partial}{\partial\bar{\varphi}}\right) + \frac{\partial}{\partial x}\right\}; \end{aligned}$$

i.e., we have obtained the "trivial" problem. Let us underline that in both arguments we have made use of the change of singularity at $\varepsilon = 0$. Let us come now to the second type problems.

4.3 The Second Type Problems. "Algebraic" Method of Reconstruction

Suppose that there is given a domain $\mathcal{D}$ and a singular surface E for the operator $X = X_0 + \varepsilon X_1 + \cdots$ and near E there is a smooth transformation of variables

$$x_i = a_i(y_1, \ldots, y_{n-1}, \xi), \qquad \text{where } i = 1, 2, \ldots, n, \tag{4.3.1}$$

such that in new variables $y_1, \ldots, y_{n-1}$, ξ, the surface E is defined by the equation $\xi = 0$; i.e., $x_i = a_i(y_1, \ldots, y_{n-1}, 0)$ are parametric equations for E. Suppose also that the transformation (4.3.1) satisfies the following conditions:

(a) In variables $y = (y_1, \ldots, y_n)$, where $y_n = \xi/\Delta$ and Δ is a parameter such that $\Delta > 0$, for small Δ, bounded y and $i = 0, 1, 2, \ldots$ operators X_i are of the form

$$X_i = \Delta^{\nu_i}(X_{i0} + \Delta^{\sigma_{i1}}\bar{X}_{i1} + \Delta^{\sigma_{i2}}\bar{X}_{i2} + \cdots),$$

where $0 < \sigma_{i1} < \sigma_{i2} < \cdots$.

(b) For $\Delta = \varepsilon^k$ and sufficiently small $x > 0$ all powers of ε in the operator

$$\Delta^{-\nu_0}X = (\bar{X}_{00} + \Delta^{\sigma_{01}}\bar{X}_{01} + \cdots) + \varepsilon\Delta^{\nu_1-\nu_0}(\bar{X}_{10} + \Delta^{\sigma_{11}}\bar{X}_{11} + \cdots) + \cdots \tag{4.3.2}$$

are *positive* and if the number of X_i's is unbounded, then $\lim_{i\to\infty}(k(\nu_i - \nu_0) + i) = \infty$.

(c) For $\Delta = \varepsilon^k$ and sufficiently small $k > 0$ the operator (4.3.2) is reducible to the normal form *with respect to* $\bar{X}_{00}$ satisfying all requirements of Section 4.2 for the (formal) construction of the new, as compared with $\bar{X}_{00}$, leading operator; i.e., for the indication of the corresponding $\zeta(\varepsilon)$.

(d) The new leading operator [or $\zeta(\varepsilon)$] which arises after the first type reconstruction in (4.3.2) (see the preceding condition) does not depend on k for sufficiently small $k > 0$.

Remark 4.3.1 The meaning of the condition (d) is simple: for a sufficiently large Δ (i.e., sufficiently "far" from E since Δ has the meaning of a "distance" from E), the operator X_0 should still remain fit to be a leading operator. Hence the principal part of $\bar{X}_{00} + \zeta(\varepsilon)$ should not be dependent on the perturbation if we reach E "not too quickly"; i.e., if k is sufficiently small. Reconstruction should begin for a certain "critical" value of k when (d) fails.

(e) There exists an upper bound $k_{cr} < \infty$ of values k satisfying the above conditions.

Remark 4.3.2 Usually at $k = k_{cr}$ the condition (d) is the first to fail (see Remark 4.3.1).

Now put $\Delta = \varepsilon^{k_{cr}}$ in (4.3.2). The obtained operator will be referred to as formally reconstructed *near* E.

Remark 4.3.3 For an operator formally reconstructed near E the first type reconstruction problem arises. Usually it is solved simultaneously with computing k_{cr} since in a "nice" problem the condition (c) holds also at $k = k_{cr}$.

Remark 4.3.4 We may come across the case $k_{cr} = \infty$. This actually means that the surface E is not singular.

The transformations (4.3.1) are individual for each problem. We cannot provide a general trick. Note only that often ζ can be introduced literally as a distance from E.

It is worth mentioning one more detail important in practice. Near E, when X_0 still "fits" ($k < k_{cr}$), the new (as compared with $\bar{X}_{00}$) leading operator is "simpler" than X_0, since a part of the information contained in X_0 is pumped in perturbation. Therefore the new leading operator obtained for $k = k_{cr}$ at the reconstruction will be "simpler" than X_0 plus the part of perturbation which arises at $k = k_{cr}$. At first glance it may seem that we should try to choose the transformation (4.3.1) so the "greatest" possible part of X_0 is "pumped into" the perturbation as we approach E. However, this approach may cause the effect that the solutions obtained "far" from ε, i.e., when $\Delta \sim 1$, and "near" E, i.e., when $\Delta = \Delta_{cr}$, are impossible to "match" in the domain $\Delta \sim \varepsilon^k$, where $0 < k < k_{cr}$. The other extremity is to make the transformation (4.3.1) so that the new (as compared with $\bar{X}_{00}$) leading operator for $k < k_{cr}$ is "the whole" X_0. Then the new leading operator obtained after reconstruction is X_0 plus the part of perturbation which fits until $k = 0$, so matching is not needed at all. But the new leading operator may turn out to be too complicated for investigation.

In practice we usually have to choose something in between these two extremities and the decisive argument in the choice of (4.3.1) might be only the possibility for matching which will be described in Section 4.5.

Let us give an illustrative example. (This example is considered from another viewpoint in the next section.)

Consider the equation

$$\varepsilon^2 u'' + xu = \varepsilon u^n \qquad \text{for an integer } n > 0. \tag{4.3.3}$$

Putting $\varepsilon u' = v$ we get the system $\varepsilon u' = v$, $\varepsilon v' = -xu + \varepsilon u^n$ to which the operator

$$X = X_0 + \varepsilon X_1 = v\frac{\partial}{\partial u} - xu\frac{\partial}{\partial v} + \varepsilon\left(u^n\frac{\partial}{\partial v} + \frac{\partial}{\partial x}\right) \tag{4.3.4}$$

corresponds. Here it is clear that $x = 0$ is a singular surface on which the operator X_0, diagonal for $x \neq 0$, becomes nilpotent. For the coordinates (4.3.1) we may take u, v, and x; i.e., just the initial variables. Now put $x = \Delta\tau$, so $\xi = x$, where τ is a new variable. We get

$$X = v\frac{\partial}{\partial u} - \Delta\tau u\frac{\partial}{\partial v} + \varepsilon u^n\frac{\partial}{\partial v} + \frac{\varepsilon}{\Delta}\frac{\partial}{\partial \tau}.$$

Conditions (a) and (b) are satisfied for $k < 1$ and $\bar{X}_{00}$ is $v(\partial/\partial u)$; i.e., a nilpotent operator. With respect to $\bar{X}_{00}$ the operator X has the canonical form in the sense of Section 2.5 so that conditions (c) and (d) are evidently verified. Heights of operators

$$-\Delta\tau u\frac{\partial}{\partial v}, \qquad \varepsilon u^n\frac{\partial}{\partial v}, \qquad \frac{\varepsilon}{\Delta}\frac{\partial}{\partial\tau}$$

with respect to $\bar{X}_{00}$ equal 3, $n+2$, and 1 respectively, their weights being $\Delta^{1/2}$, $\varepsilon^{2/(n+3)}$, and ε/Δ. The operator X_0 remains the leading operator for a sufficiently large Δ with the highest weight $\Delta^{1/2}$. The critical value of k is

$$k_{\text{cr}} = \min(\Delta^{1/2} = \varepsilon^{2/(n+3)}, \Delta^{1/2} = \varepsilon/\Delta, 1) \qquad \text{for } \Delta = \varepsilon^k.$$

For $n \leq 3$ we have $k_{\text{cr}} = 2/3$ and for $n \geq 3$ we have $k_{\text{cr}} = 4/(n+3)$. The reconstructed operator is

$$X = v\frac{\partial}{\partial u} - \varepsilon^{2/3}\tau u\frac{\partial}{\partial v} + \varepsilon u^n\frac{\partial}{\partial v} + \varepsilon^{1/3}\frac{\partial}{\partial\tau} \qquad \text{for } n \leq 3,$$

$$X = v\frac{\partial}{\partial u} - \varepsilon^{4/(n+3)}\tau u\frac{\partial}{\partial v} + \varepsilon u^n\frac{\partial}{\partial v} + \varepsilon^{(n-1)/(n+3)}\frac{\partial}{\partial\tau} \qquad \text{for } n \geq 3, \tag{4.3.5}$$

so that the new leading operator is

$$v\frac{\partial}{\partial u} - \varepsilon^{2/3}\tau u\frac{\partial}{\partial v} + \varepsilon^{1/3}\frac{\partial}{\partial\tau} \qquad \text{for } n \leq 3$$

and

$$v\frac{\partial}{\partial u} - \varepsilon^{4/(n+3)}\tau u\frac{\partial}{\partial v} + \varepsilon u^n\frac{\partial}{\partial v} \qquad \text{for } n \geq 3.$$

After the shearing transformation

$$u = \varepsilon^{-1/6} y, \qquad v = \varepsilon^{1/6} z \qquad (x = \varepsilon^{2/3}\tau,\ t = \varepsilon^{1/3}) \qquad \text{for } n \leq 3,$$

$$u = \varepsilon^{-1/n+3} y, \qquad v = \varepsilon^{1/(n+3)} z \qquad (x = \varepsilon^{4/(n+3)}\tau,\ t = \varepsilon^{2/(n+3)}) \quad \text{for } n \geq 3, \tag{4.3.6}$$

(here $u = e_{1/2}$ and $v = e_{-1/2}$) chosen in accordance with the weight $\Delta_{cr}^{1/2}$ of $-\Delta_{cr}\tau u(\partial/\partial v)$, we get the final result of the reconstruction in the form

$$X = \varepsilon^{1/3}\left\{ z\frac{\partial}{\partial y} - \tau y\frac{\partial}{\partial z} + \frac{\partial}{\partial \tau} + \varepsilon^{(3-n)/6} y^n \frac{\partial}{\partial z} \right\} \qquad \text{for } n < 3$$

$$X = \varepsilon^{2/(n+3)}\left\{ z\frac{\partial}{\partial y} - (\tau y - y^n)\frac{\partial}{\partial z} + \varepsilon^{(n-3)/(n+3)} \frac{\partial}{\partial \tau} \right\} \qquad \text{for } n > 3.$$

Remark 4.3.5 For $n = 3$ the problem loses a small parameter. the corresponding equation $d^2y/d\tau^2 + \tau y = y^3$ is the Painleve equation [20].

In conclusion, note that the mentioned method of reconstruction is purely formal and its shortcoming is that it does not take into account the position on E which trajectories approach and the behavior of trajectories near E which might give complementary information for the reconstruction. Arguments connected with the study of behavior of trajectories near E are considered in the following subsection.

4.4 "Trajectory" Method of Reconstruction

In what follows we give a method of reconstruction based on the study of asymptotic behavior near a singular surface of the trajectory described by approximate equations corresponding to M. We cannot characterize exactly the class of problems to which the method is applicable and try only to formulate the main considerations which allow us to obtain a solution in a number of cases.

A *singular surface* is a surface E_k of any dimension less than n which is the boundary of the set D such that in a neighborhood of each point x of D, operators X_i for $i = 0, \ldots, k$, and S_j, for $j = 1, \ldots, k$, are smooth but as $x \to E_k$ the uniform boundedness of some coefficient of at least one of these operators fails. The index k means that on E_k uniform boundedness is necessarily lost by some coefficients of operators with index k while X_j and S_j for $j < k$ are smooth in $D \cup E_k$.

First consider the example: Equation (4.3.3), $\varepsilon^2 u'' + xu = \varepsilon u^n$ from the preceding section to which the operator (4.3.4)

$$X = X_0 + \varepsilon X_1 = v\frac{\partial}{\partial u} - xu\frac{\partial}{\partial v} + \varepsilon\left(u^n\frac{\partial}{\partial v} + \frac{\partial}{\partial x}\right)$$

corresponds. Eigenfunctions of X_0 are $\varphi = v + i\sqrt{x}\,u, \bar{\varphi}$; $X_0\varphi = i\sqrt{x}\,\varphi$ and $X\varphi = i\sqrt{x}\,\varphi + \varepsilon[(2i)^{-n}(\varphi-\bar{\varphi})^n x^{-(1/2)n} + (\varphi-\bar{\varphi})(4x)^{-1}]$, $Xx = \varepsilon$. Consider the domain $x > 0$ and for simplicity's sake put $n = 2\ell$. The solution of the equation $[X_0, S_1] + X_1 = M_1$ is given by

$$S_1 x = 0, \qquad M_1 x = 1, \qquad M_1\varphi = \varphi/4x, \qquad M_1\bar{\varphi} = \bar{M}_1\bar{\varphi},$$

$$\begin{aligned} S_1\varphi &= \sum(-1)^p i \binom{n}{p} x^{-(n+1)/2}\varphi^p\varphi^{n-p}(2i)^{-n}(2p-(n+1))^{-1} \\ &\quad + \frac{1}{4}\bar{\varphi}(-2ix^{3/2})^{-1}, \\ S_1\bar{\varphi} &= \overline{S_1\varphi}. \end{aligned}$$

The differential equation corresponding to this normal form is

$$\varepsilon\frac{d\varphi^*}{dx} = \left(i\sqrt{x} + \frac{\varepsilon}{4x}\right)\varphi^*$$

and the singular surface E_1 is the surface $x = 0$. Consider the trajectory of the equation of the first approximation

$$\varphi^*(x) = \varphi_0^* x^{1/4}\exp\frac{2i}{3\varepsilon}x^{3/2}, \qquad \varphi_0^* = \text{const.} \tag{4.4.1}$$

It joins the point $(\varphi_0^*, \bar{\varphi}_0^*, 1)$ with the point $(0,0,0)$ on the surface E_1. Let us pose the following question: how small can x be so that the trajectory of the initial problem is still described (with accuracy diminishing as does x) by the trajectory (4.4.1)? Put $x = \Delta s$ and let us measure the "distance" of the point φ^* from E_1 by the value Δ, considering Δ small and $s \sim 1$. The behavior of φ^* can be described starting from (4.4.1) and putting $\varphi^* = \psi^*\Delta^{1/4}$, where $\psi^* \sim 1$ and ε is assumed to be fixed.

The nature of asymptotics being established, we naturally pass to coordinates $\varphi = \psi\Delta^{1/4}$, $\bar{\varphi} = \bar{\psi}\Delta^{1/4}$, $x = \Delta s$. In these coordinates we have

$$\begin{aligned} X\psi &= \Delta^{1/2}\left\{i\sqrt{s}\,\psi + \varepsilon\Delta^{-(n+3)/4}\left(\frac{1}{2i}\right)^n(\psi-\bar{\psi})^n s^{-n/2}\right. \\ &\qquad \left. + \Delta^{-3/2}\varepsilon(\psi-\bar{\psi})(4s)^{-1}\right\} \\ Xs &= \Delta^{1/2}\frac{\varepsilon}{\Delta^{3/2}}, \\ X &= \Delta^{1/2}\left(\tilde{X}_0 + \varepsilon\Delta^{-(n+3)/4}Y + \frac{\varepsilon}{\Delta^{3/2}}Z\right), \end{aligned} \tag{4.4.2}$$

where coefficients of $\tilde{X}_0$, Y, and Z are bounded in the domain $s \geq \alpha > 0$.

Now consider two possibilities: either $\Delta^{-(n+3)/4} > \Delta^{-3/2}$ or $\Delta^{-(n+3)/4} < \Delta^{-3/2}$; in other words, $n \lessgtr 3$. In the first case we leave the expression for

X in the form (4.4.2), and in the second one we will write X in the form $X = \Delta^{1/2}(\tilde{X} + \varepsilon\Delta^{-3/2}Z + \varepsilon\Delta^{-(n+3)/4}Y)$. Accordingly $\varepsilon S_1 = \delta\tilde{S}$, where $\delta = \varepsilon\Delta^{-(n+3)/4}$ or $\delta = \varepsilon\Delta^{-3/2}$ (for different $\tilde{S}$'s in both cases). Thus

$$e^{-\varepsilon S_1}(X_0 + \varepsilon X_1)e^{\varepsilon S_1} = \begin{cases} \Delta^{1/2}e^{-\delta\tilde{S}}[\tilde{X}_0 + \delta(Y + 0(\Delta))]e^{S\tilde{S}} & \text{for } n > 3 \\ \Delta^{1/2}e^{-\delta\tilde{S}}[\tilde{X}_0 + \delta(Z + 0(\Delta))]e^{S\tilde{S}} & \text{for } n < 3 \end{cases}, \tag{4.4.3}$$

Clearly if $\delta \ll 1$ then the change of variables defined by the operator $\delta\tilde{S} = \varepsilon S_1$ still fits. [It is not difficult to show that, applying Hausdorff's formula to (4.4.3), we get a formal power series in δ with coefficients being operators with bounded coefficients in the domain $s > \alpha > 0$].

Thus the answer to our question is as follows: we may use the operators $X_0 + \varepsilon M_1$ for approximate computation of the trajectory in the whole domain $|\psi| = 0(1)$, $s \geq \alpha > 0$, if $\delta \ll 1$ (evidently with an error which grows as Δ diminishes).

The critical value Δ_{cr} is $\Delta_{cr} = \varepsilon^{4/n+3}$ or $\Delta_{cr} = \varepsilon^{2/3}$ respectively. For $\Delta = \Delta_{cr}$ the new (reconstructed) leading operator $\tilde{X}_0 + Y$ or $\tilde{X}_0 + Z$ arises, which must be used in the study of the problem for $0 < s \leq 1$. Note that for $n = 3$ ($\Delta = \varepsilon^{2/3}$) the equation loses a small parameter; it turns into the known Painleve equation (see Section 4.3). It seems that in this case in the study of the operator on the whole segment $(0, \infty)$ the presence of a small parameter in the initial problem does not simplify it.

The above arguments are based on the following. Consider the formula

$$e^{-\varepsilon S_1}(X_0 + \varepsilon X_1 + \cdots)e^{\varepsilon S_1} = X_0 + \varepsilon M_1 + \varepsilon^2 N_2 + \cdots = \hat{X}. \tag{4.4.4}$$

Suppose (for the time being) that the singular surface is E_1 and *the trajectory of the equation corresponding* to $X_0 + \varepsilon M_1$ *approaches* E_1 *however closely.* Making use of the knowledge of the behavior of operators which enter the left-hand side of (4.4.4), on the trajectory (tube of trajectories) we try to define how near (the notion of the "distance" will be deciphered a little later) E_1 we can consider the right-hand side of (4.4.4) as a formal asymptotic series of the form $\nu(\tilde{X}_0 + \delta\tilde{M}_1 + \delta^2\tilde{M}_2 + \cdots)$, where δ and ν are small, depend on "distance" of the point of the trajectory from E_1 and on ε, and coefficients of all operators are bounded (only a finite number of summands and an estimate of the remainder are actually needed). The reader should keep in mind the considered example for the time being.

As long as is possible, a solution of the equation corresponding to the left-hand side of (4.4.4) can be approximated by the considered trajectory. We assume that as a result we can find a new reconstructed operator, appropriate near E_1.

This process is most often applicable under the following conditions:

(a) X_0 is a diagonal operator of the form

$$X_0 = \sum_k \lambda_k(e)\varphi_k\frac{\partial}{\partial\varphi_k} \qquad (X = X_0 + \varepsilon X_1 + \cdots).$$

(b) Besides, it is convenient to assume that there exists a basic system $\varphi_1, \ldots, \varphi_m, e_1, \ldots, e_{\ell-1}, \tau$, where $m + \ell = n$ and e_i and τ are invariants of X_0, such that x belongs to E_1 if and only if $\tau = 0$; i.e., the "distance" from E_1 is measured by the value of τ. (The word "convenient" in this sentence means only that under such an assumption the mainstream of our arguments is easy to clarify. The real problem might not satisfy any assumption made a priori, though the method—the indication of a direction of our attempts to solve it—may prove to be applicable.)

In these coordinates, equations of the first approximation (corresponding to $X_0 + \varepsilon M_1$) are of the form

$$\begin{aligned} \frac{d\varphi_i}{dt} &= (\lambda_i(e; h, \tau) + \varepsilon m_i(e; h, \tau))\varphi, \\ \frac{de_k}{dt} &= \varepsilon q_k(e, h, \tau), \\ \frac{d\tau}{dt} &= \varepsilon r(e; h, \tau), \end{aligned} \tag{4.4.5}$$

where h is a set of invariants of X_0 which are not functions of e.

(c) We assume that on each trajectory of (4.4.5) sufficiently near to the trajectory that starts from an initial point (x_0), asymptotic formulas

$$\begin{aligned} &\varphi_j = \tau^{p_j}(\tilde{\varphi}_{j0} + \tau^{\sigma_{j1}}\tilde{\varphi}_{j1} + \cdots), \quad \sigma_{j1} < \sigma_{j2} < \cdots \\ &e_\mu = \tau^{p'_\mu}(\tilde{e}_{\mu 0} + \tau^{\sigma'_{\mu 1}}\tilde{e}_{\mu 1} + \cdots), \quad \sigma'_{\mu 1} < \sigma'_{\mu 2} < \cdots, \ p'_\mu < p'_{\mu - 1} \end{aligned} \tag{4.4.6}$$

hold for sufficiently small $|\tau|$.

(d) Under the change of variables $\tau = \Delta S$, $\varphi_j = \Delta^{p_j}\psi_j$, $e_\mu = \Delta^{p'_\mu} g_\mu$, all operators X_k and S_1 take the form

$$\begin{aligned} X_0 &= \Delta^{\nu_0}(Y_0 + \Delta^{\nu_{01}} Y_{01} + \cdots), \\ X_k &= \Delta^{\nu_k}(Y_{k0} + \Delta^{\nu_{k1}} Y_{k1} + \cdots), \\ S_1 &= \Delta^{q_0}(\tilde{S}_0 + \Delta^{q_1}\tilde{S}_1 + \cdots), \end{aligned}$$

and operators Y and $\tilde{S}$ are smooth in the domain $s > \alpha > 0$, $\psi_j = 0(1)$, $g_\mu = 0(1)$.

Let us now rewrite (4.4.4) in new variables:

$$\begin{aligned} \Delta^{-\nu_0}\hat{X} = e^{-\varepsilon\Delta^{q_0}(\tilde{S}_0 + \Delta^{q_1}\tilde{S}_1 + \cdots)} \\ \times [(Y_0 + \Delta^{\nu_{01}} Y_1 + \cdots) + \varepsilon\Delta^{\nu_1 - \nu_0}(Y_{10} + \Delta^{\nu_{11}} Y_{11} + \cdots) \\ + \cdots] e^{\varepsilon\Delta^{q_0}(\tilde{S}_0 + \cdots)}. \end{aligned} \tag{4.4.7}$$

Put $\Delta = \varepsilon^k$, $k_s = (-q_0)^{-1}$ and let k grow monotonically in the domain $x > 0$.

Let $x_s > 0$. The operator εS_1 is small only if $0 < k < k_s$. If there exists k such that $0 < k < k_s$ and the exponent of ε in square brackets in (4.4.7) vanishes, then k_{cr} is the *smallest* [conceding it exists and a finite number of summands in (4.4.7) corresponds to it] of these numbers. Then the arguments used in the example show that the trajectory approximates a solution for all $k < k_{\text{cr}}$. Therefore for $k = k_{\text{cr}}$ there arises an essentially new situation and a new reconstructed operator is obtained if we put $\Delta = \varepsilon^{k_{\text{cr}}}$ in (4.4.7). The leading operator in the new problem is $Y_0 + Z$, where Z is the sum of all not small terms in parentheses (4.4.7).

If there are no such k then put $k_{\text{cr}} = k_s$. In this case the leading operator is Y_0. When $k_s < 0$, the previous arguments hold, except that if all powers of ε in the brackets in (4.4.7) are positive for all $k > 0$, then reconstruction does not appear.

Remark 4.4.1 The case when all powers in brackets in (4.4.7) are positive (k_s is arbitrary) is distinguished by the fact that the leading operator, which is Y_0, is not formally affected by a perturbation. (The perturbation effect tells on only the change of variables.)

Remark 4.4.2 If τ is an invariant, i.e., $r \equiv 0$ in (4.4.5), then the trajectory defined by (4.4.5) does not approach however closely the surface E_1 defined by $\tau = 0$. Therefore we must consider the further approximations.

A solution of the equation which arises after reconstruction should be an extension of the initial trajectory in a domain earlier inapproachable. The search for this solution is called the *matching problem* discussed below (in Section 4.5). For the time being note only that the use of only a first approximation might not secure the required accuracy or will not lead in the needed domain and the consideration of further terms of the normal form might be needed. (See also Remark 4.4.2.) In this case the reconstructed operator might turn out to be different from the one obtained earlier.

The investigation of the next approximation is carried out similarly: in the formula

$$e^{-\varepsilon S_1 - \varepsilon^2 S_2} X e^{\varepsilon S_1 + \varepsilon^2 S_2} = X_0 + \varepsilon M_1 + \varepsilon^2 M_2 + e^3 M_3 + \cdots$$

we compute operators S_1 and S_2, and, considering that

$$S_2 = \Delta^{\hat{S}_0}(\hat{S}_0 + \Delta^{\hat{S}_1}\hat{S}_1 + \cdots)$$

behaves like S_1, put k_s to be equal to the least of the positive numbers among $(-q_0)^{-1}$, $(-k_0)^{-1}$ (if there are positive among them), and later proceed as above.

Let us make an additional remark about the choice of coordinates. We start from the basic system φ and e. This system seems most natural but we can make use of the other systems. We should have in mind that different coordinate systems equivalent inside of the domain are not necessarily equivalent near the singular surface since in the vicinity of a singular surface the change of variables, which defines the passage from one coordinate system to another one, may turn out to be degenerate. Coordinate systems that we employ should be asymptotically "maximally different."

Let us elucidate this notion considering the condition $p'_\mu < p'_{\mu+1}$ in (4.4.6). If this condition fails [e.g., if it fails for variables $g_1 = e_1$, $g_\mu = e_1 + e_\mu$ $(\mu \geq 2)$, all of them having the same asymptotics], then the number of characteristic features of the behavior of the curve near E_1 will be smeared resulting in "bad" reconstruction.

Finally, note that as with any method of this kind, success is not guaranteed, and a "bad" reconstruction may occur. In this case it is worth studying the trajectories of the next approximation.

Let us consider briefly one simple example: the operator

$$X = ix^\alpha \varphi \frac{\partial}{\partial \varphi} - ix^\alpha \bar{\varphi} \frac{\partial}{\partial \bar{\varphi}} + \varepsilon x^{-1} g^n \left((\varphi + \bar{\varphi}) \frac{\partial}{\partial \bar{\varphi}} + (\varphi + \bar{\varphi}) \frac{\partial}{\partial \bar{\varphi}} \right) + \varepsilon \frac{\partial}{\partial x},$$

where $g = \varphi\bar{\varphi}$. We easily get

$$S_1 \varphi = \frac{x^{-1-\alpha}}{2i} g^n \bar{\varphi}, \qquad S_1 \bar{\varphi} = \bar{S}_1 \varphi, \qquad S_1(x) = 0.$$

The equation of the trajectory is

$$\frac{d\varphi}{dx} = i\frac{x^\alpha}{\varepsilon}\varphi + x^{-1} g^n \varphi,$$

where

$$\frac{dg}{dx} = 2g^{n+1}x^{-1} \quad \text{or} \quad g = \left(\frac{1}{D - n\ln x}\right)^{1/n}, \qquad \text{where } D = g^{-n}(1) > 0. \tag{4.4.8}$$

Let us try to find a solution in the segment $0 < x \leq 1$. Taking (4.4.8) into account, set $x = \Delta s$ and $\varphi = (-\ln\Delta)^{-1/2n}\psi$, getting

$$\Delta^{-\alpha} X = is^\alpha \psi \frac{\partial}{\partial \psi} - is^\alpha \bar{\psi} \frac{\partial}{\partial \bar{\psi}} + \frac{\varepsilon}{\Delta^{1+\alpha}} \tilde{\delta} \frac{1}{s} r^n (\psi + \bar{\psi}) \left(\frac{\partial}{\partial \psi} + \frac{\partial}{\partial \bar{\psi}} \right) + \frac{\varepsilon}{\Delta^{1+\alpha}} \frac{\partial}{\partial s},$$

$$\varepsilon S_1 \psi = \frac{\varepsilon}{\Delta^{1+\alpha}} \frac{\tilde{\delta} s^{-1-\alpha}}{2i} r^n \bar{\psi},$$

where $\tilde{\delta} = -1/\ln\Delta$ and $r = \psi\bar{\psi}$. Hence $\Delta_{\text{cr}}^{1+\alpha} = \varepsilon$. The reconstructed operator is

$$X_0 + \delta X_1 = is^\alpha \psi \frac{\partial}{\partial \psi} - is^\alpha \bar{\psi} \frac{\partial}{\partial \bar{\psi}} + \frac{\partial}{\partial s} + \frac{\delta}{s} r^n (\psi + \bar{\psi}) \left(\frac{\partial}{\partial \psi} + \frac{\partial}{\partial \bar{\psi}} \right),$$

where
$$\delta = \frac{1}{\ln \Delta_{cr}}.$$

Passing to new variables, $u = e^{-is^{\alpha+1}/(\alpha+1)}\psi$, $\bar{u}$, and s we get

$$X_0 + \delta X_1 = \frac{\partial}{\partial s} + \frac{\delta}{s} v^n \left[\left(u + \exp\left(-\frac{2is^{\alpha+1}}{\alpha+1} \right) \bar{u} \right) \frac{\partial}{\partial u} + \left(\bar{u} + \exp\left(\frac{2is^{\alpha+1}}{\alpha+1} \right) u \right) \frac{\partial}{\partial \bar{u}} \right]$$
$$(v = u\bar{u}). \tag{4.4.9}$$

Let us reduce the reconstructed operator to the normal form in the domain $s > h > 0$. From $[X_0, S_1]u + X_1 u = M_1 u$ we derive

$$\frac{d(S_1 u)}{ds} + \frac{v^n}{s} e^{-2i(s^{\alpha+1}/\alpha+1)} \bar{u} = 0 \quad \text{and} \quad M_1 u = \frac{v^n}{s} u. \tag{4.4.10}$$

It is convenient to introduce the following symbolic notations: write P_m for any polynomial of the form $c_p u^p \bar{u}^q$, where $p + q = m$, regardless of coefficients. Further, putting

$$s^{-1} e^{-2i(s^{\alpha+1}/\alpha+1)} = \mu(s), \qquad \nu(s) = \int_s^\infty e^{-2i(s^{\alpha+1}/\alpha+1)} \frac{ds}{s}$$

we get $S_1 u = \nu(s) P_{2n+1}$, $S_1 P_m = \nu(s) P_{m+2n}$. Now (4.4.10) yields

$$[X_0, S_1]u = \mu(s) P_{2n+1}, \qquad [X_0, S_1] P_m(u) = \mu(s) P_{2n+m}, \tag{4.4.11}$$

$$e^{-\delta S_1}(X_0 + \delta X_1) e^{\delta S_1} = \frac{\partial}{\partial s} + \delta \left(\frac{v^n}{s} u \frac{\partial}{\partial u} + \frac{v^n}{s} \bar{u} \frac{\partial}{\partial \bar{u}} \right) + \delta^2 \mu \nu P_{4n+1} + \delta^3 \mu \nu^3 P_{6n+1}. \tag{4.4.12}$$

We get similar results computing further approximations also.

We can make use of (4.4.11) in the domain (s, u) where $\nu(s)$ is bounded and $P_{2n\ell+1}$ are bounded for any ℓ. (As usual, we speak about an asymptotic formula.) But $\mu(s)$ and $\nu(s)$ are bounded in the whole domain $s \geq \alpha > 0$ and $P_{2n\ell+1}$ are bounded in the domain of bounded values of $|u|$. Solving the equation corresponding to (4.4.12), i.e., $du/ds = (\delta/s) v^n u$, we get

$$u = e^{i\theta} \frac{1}{(d - 2\delta n \ln s)^{1/2n}}, \tag{4.4.13}$$

where θ and d are constants.

The function u is bounded, roughly speaking, in the domain $\delta \ln s \sim 1$. Since $\Delta_{cr} s = x$ and $\delta = -(1/\ln \Delta_{cr})$, u is bounded on the trajectory in the whole domain $\Delta_{\text{cr}} \le x \le 1$ (for a corresponding value of α). Thus the reconstructed operator "serves" in the whole domain $\varepsilon^{1/1+\alpha} \le x \le 1$. In this domain the uniform asymptotics are arranged with respect to powers of δ, but the convergence of formulas improves as s increases and for $s \sim 1/\Delta_{\text{cr}}$, i.e., in the domain $1 \ge x > \gamma > 0$, the error of the first approximation becomes of order ε. We will not continue the solution further in the domain $0 \le x \le \varepsilon^{1/1+\alpha}$.

Remark 4.4.3 From the beginning we might seek the solution in the form of a series in δ. We prefer to retain in (4.4.12) the diagonal part in the coefficient of δ only from "aesthetic" considerations.

4.5 Matching

In problems with reconstruction there arises a problem of "matching" asymptotic developments obtained "far from" and "near" a singular surface. Asymptotics obtained by reducing the initial operator to the normal form in a domain, points of which are at a distance of order 1 from the singular surface ("far" from the singular surface), will be called exterior, whereas those obtained by reducing the reconstructed operator to the normal form (with respect to the new leading operator) in a finite domain of new variables (found as a result of reconstruction) are called inner ("near" the singular surface).

First let us give an account of certain simple facts and notions. Let δ be a small parameter and $M = \{\mu_1(\delta), \mu_2(\delta), \dots\}$ a sequence of functions satisfying $\mu_1(\delta) \gg \mu_2(\delta)$ for $\delta = o(1)$. [The functions $\mu_i(\delta)$ are not necessarily bounded as $\delta \to 0$.] Such a system will be called an *asymptotic* one.

The series $f \approx c_1\mu_1(\delta) + c_2\mu_2(\delta) + \cdots$ is called *M-asymptotic* for $f(\delta)$, if

$$\left| f(\delta) - \sum_{1\le p\le n} c_p\mu_p \right| \ll \mu_n(\delta)$$

for each $n = 1, 2, \dots$. Such an asymptotic development is clearly unique since $\sum_{k\ge1} a_k\mu_k(\delta) \approx 0$ implies $|a_1\mu_1| \ll \mu_1(\delta)$, i.e., $a_1 = 0$, etc.

Now consider two sequences of functions

$$M = \{\mu_1(\delta) \gg \mu_2(\delta) \gg \cdots\}; \qquad N = \{\nu_1(\delta) \gg \nu_2(\delta) \gg \cdots\},$$

and suppose they admit the ordering

$$M \cup N : \mu_1(\delta) \gg \mu_1(\delta) \gg \cdots \gg \mu_{i_1}(\delta) \gg \nu_1(\delta) \gg \cdots$$
$$\gg \nu_{j_1}(\delta) \gg \mu_{i_1+1}(\delta) \gg \cdots \gg \mu_{i_1}(\delta) \gg \nu_{j_1+1}(\delta),$$

the coincidence of some μ with corresponding ν not being excluded. Denote the set of coinciding functions by $D(M, N)$. The uniqueness theorem implies that if $f \approx \sum c_k\mu_k$ and $f \approx \sum c'_e\nu_e$, then coefficients c and c' which do not correspond to functions of $D(M,N)$ must vanish since both decompositions are developments into series with respect to functions from $M \cup N$, i.e., f is actually decomposable into a series with respect to functions from $D(M,N)$.

The sequence of bounded functions $\psi_1(\tau)$ and $\psi_2(\tau)$ is ∞-*independent* if $|a_1\psi_1(\tau) + \cdots + a_n\psi_n(\tau)| = o(1/\tau)$ implies $a_i = 0$, where $i = 1, \ldots, n$ for any n The requirement of boundedness can be considerably weakened further. A simple but important example of a ∞-independent sequence is $\{e^{ik\tau}\}$.

Now we turn to the matching problem. Suppose that for an exterior solution ("far from" the singular surface) a system of functions $\mu_i(\tau,\delta)$ is found such that for given arbitrary constants (c') the external solution χ_1 is developed into the asymptotic series $\chi_1 \simeq \sum a_i(c')\mu_i(\tau,\delta)$ for $\tau \gg 1$ (this system arises as the solution of the exterior problem). Suppose also that the inner solution ("near" the singular surface) in the domain $\tau \ll \delta^{-h}$ is developed into an asymptotic series of the form

$$\chi_2 \simeq \sum_i b_i(c'')\mu_i(\tau,\delta)$$

with respect to the same functions. This is possible since we assume that χ_2 coincides with χ_1 in the domain $\Omega : 1 \ll \tau \ll \delta^{-h}$ called the *domain of matching.* (From the beginning, note that in concrete problems exterior and inner solutions are obtained by different methods and one has to first express formally different asymptotics of both solutions in terms of the same system of functions μ.) Further, suppose that for any $\tau = \delta^{-\alpha} \subset \Omega$ functions $\mu_i(\delta^{-\alpha},\delta)$ constitute an asymptotic system and for any two different admissible values $\alpha = \alpha_1$ and $\alpha = \alpha_2$, systems $M : \mu_i(\delta^{-\alpha_1},\delta)$ and $N : \mu_j(\delta^{-\alpha_2},\delta)$ generate the systems M and N just described. Finally, suppose that the product of any two functions from M belongs to M and the same holds for N.

Let us write the matching condition of χ_1 and χ_2 in the matching domain for $\alpha = \alpha_1$ and $\alpha = \alpha_2$. We get

$$\sum_{i\geq 1}(a_i(c') - b_i(c''))\mu_i(\delta^{-\alpha_1},\delta) = 0,$$
$$\sum_{i\geq 1}(a_i(c') - b_i(c''))\mu_i(\gamma^{-\alpha_2},\delta) = 0. \tag{4.5.1}$$

In this problem constants c' are given, constants c'' must be found, and they should be decomposable into an asymptotic series with respect to some functions in δ. Due to the last assumption we can assume that c'' decomposes into an asymptotic series with respect to $\mu_i(\delta^{-\alpha_1},\delta)$ and get via (4.5.1) recursive relations for the definition of coefficients in the decomposition. Similarly, we can assume a decomposability of c'' with respect to $\mu_i(\delta^{-\alpha_2},\delta)$. However, (4.5.1) and the uniqueness theorem show that c''

actually splits into an asymptotic series with respect to functions from $D(M, N)$. Supposing c'' decomposes into asymptotic series with respect to this system we obtain from (4.5.1) the desired recurrences.

Now we consider how many computations are needed to get the given accuracy δ^ν. This question is always subject to discussion since the bulk of computations increases extremely rapidly as accuracy rises. Suppose we have to define a solution of the interior problem in the domain $\tau \sim 1$ up to δ^ν. It clearly means that constants c'' should be defined with accuracy δ^ν, but, generally speaking, does *not* mean that a solution should be defined in the whole matching domain with this accuracy.

In fact, in defining c'' we may start from any value of α in (4.5.1). First, take $\alpha = \alpha_1$ to be near to $-h$. Since c'' should be defined with accuracy δ^ν, we can compute the *exterior* solution in the domain $\delta^{-\alpha_1} \lesssim \tau$ with this accuracy; i.e., practically without entering the matching domain (clearly, the interior solution should then be defined in the domain $\tau \lesssim \delta^{-\alpha_1}$ with accuracy δ^ν). On the other hand, for calculations we can similarly make use of a value of α close to 0. Note also that in computations employing (4.5.1) for an intermediate admissible value $\alpha = \alpha_3$ we have to use only the functions $\mu_i(\delta^{-\alpha_2}, \delta)$ which exceed δ^ν both for inner and exterior developments. But our constants do not depend on the chosen value of α, yielding the following recipe for computational procedure.

We should compare with accuracy δ^ν an exterior solution for a value of α however close to $-h$ and exclude from the obtained asymptotic development with respect to functions $\mu_i(\delta^{-\alpha}, \delta)$ terms which for a certain admissible α (which is now considered a variable in the whole admissible domain) become less than δ^ν. Perform a similar operation over an interior solution considering, first, α close to 0. Equate the obtained expressions and, considering c'' developed into a series with respect to functions from $D(M, N)$, find the unknown coefficients.

Let us note that these computations are admissible if we do not insist on accuracy δ^ν in Ω, where the accuracy is worse.

One more remark: suppose that the exterior and interior developments considered are of the form (for admissible α)

$$\chi_1 \simeq \left(\sum_{k,e} a_{e,k}\mu_k(\delta^{-\alpha}, \delta)\right)\psi_e(\delta^{-\alpha}) = \sum_k \left(\sum_e a_{e,k}\psi_e\right)\mu_k(\delta^{-\alpha}, \delta)$$

$$\simeq \sum_k h_k(\delta^{-\alpha})\mu_k(\delta^{-\alpha}, \delta),$$

$$\chi_2 \simeq \sum_k \hat{h}_k(\delta^{-\alpha})\mu_k(\delta^{-\alpha}, \delta),$$

and let $h_k(\delta^{-\alpha})$ and $\hat{h}_k(\delta^{-\alpha})$ be sums of a *finite* number of summands, for any k. Now if $\psi_e(\tau)$ is ∞-independent, then inside of the matching domain, $\hat{h}_k$ and h_k should coincide for all k. This remark (following directly from the imposed conditions) can shorten the calculations.

Concluding this chapter we give several examples. The first of them, Example 4.6, is intended mostly to demonstrate the matching technique and is discussed in a rather detailed fashion: in a sense, it has a sound expository value, though it is somewhat cumbersome and not very simple. In the last example, 4.13 (relaxation oscillations in a Van der Pol oscillator) also considered in detail, the matching technique is illustrated once again. Other examples are devoted to different modifications of considerations of reconstruction making use of particular features of the concrete problem.

4.6 Example: Illustration for 4.5

Consider the equation

$$\varepsilon^2 u'' + p(x)u = \varepsilon u^2, \qquad 0 \le x \le a \sim 1, \tag{4.6.1}$$

where $p(0) = 0$, $p'(0) = 1$ and $p(x) > 0$ for $x > 0$ with the initial conditions $u(a) \sim 1$ and $\varepsilon u'(a) \sim 1$.

This equation is quite similar to (4.3.3). We will not repeat it for our arguments concerning reconstruction given in Sections 4.3 and 4.4 for Equation (4.3.3) and will immediately make use of the reconstruction transformation (4.3.6) for $n = 2$. Put

$$\delta = \varepsilon^{1/6}, \qquad x = \delta^4 \tau, \qquad u = \delta^{-1} y \tag{4.6.2}$$

and, developing $p(x)$ into a Taylor series, we get the reconstructed equation

$$\frac{d^2 y}{d\tau^2} + \tau y = \delta y^2 - \left(\delta^4 \frac{p''(0)}{2} \tau^2 + \cdots\right) y \tag{4.6.3}$$

(note that here we should, generally speaking, assume $\tau \ll \delta^{-4}$).

Now consider various methods of constructing the exterior (in the domain $x \sim 1$) and the interior (in the domain $\tau \sim 1$) asymptotics.

Exterior Asymptotics

The exterior asymptotics can be constructed making use of the standard reduction of the corresponding operator (4.6.1) to the normal form. [Computations are similar to those given in 4.4 for (4.3.3).] Then initial computations suggest that the solution is of the form (4.6.5) given below, so the search for asymptotics can be considerably shortened if from the beginning we seek them in this form. Set

$$\zeta = \exp\left(\frac{i}{\varepsilon} \int_0^x \sqrt{p(\eta)}\, d\eta\right). \tag{4.6.4}$$

Let us seek a solution of (4.6.1) in the form

$$u = u_0(\zeta, \bar{\zeta}, x) + \varepsilon u_1(\zeta, \bar{\zeta}, x) + \cdots, \tag{4.6.5}$$

where

$$u_j = u_j(\zeta, \bar{\zeta}, x) = g_j(x) + w_j(\zeta, x) + \overline{w_j(\zeta, x)} \qquad [g_0(x) \equiv 0],$$
$$w_j = w_j(\zeta, x) = A_{j,1}(x)\zeta + A_{j,2}(x)\zeta^2 + \cdots + A_{j,j+1}(x)\zeta^{j+1}$$

(hereafter, the overbar stands for complex conjugation). Substituting (4.6.5) into (4.6.1) and differentiating the polynomial w_j with respect to ζ according to the rule

$$\varepsilon w_j' = i\sqrt{\rho}\,\zeta \frac{\partial w_j}{\partial \zeta} + \varepsilon \frac{\partial w_j}{\partial x} \qquad [\rho = \rho(x)],$$

and repeating this procedure $[\varepsilon^2 w_j'' = \varepsilon(\varepsilon w_j)']$, we will get, due to (4.6.4) first, that $\varepsilon^2 u'' + \rho u = \varepsilon F_1 + \varepsilon^2 F_2 + \cdots$ is a series of the same form as (4.6.5); i.e., F_j is a polynomial in ζ, $\bar{\zeta}$ of the same form as u_j, and second, we can write the series $\varepsilon u^2 = \varepsilon G_1 + \varepsilon^2 G_2$ in the same form taking into account that $\zeta\bar{\zeta} = 1$.

Equation (4.6.1) will be satisfied if $F_j - G_j = 0$ identically with respect to ζ, $\bar{\zeta}$. In the result, we get, as it not difficult to verify, a recursive system of equations for coefficients g_j and $A_{j,k}$. Differential equations arise, actually, only for $A_{j,1}$ (other coefficients are found without integration) and are of the form $4\rho A_{j,1}' + \rho' A_{j,1} = f_j(x)$, where $f_0(x) \equiv 0$ and $f_j(x)$ are found recursively [the equation for $A_{j,1}$ arises from $F_{j+1} - G_{j+1} = 0$ but $f_j(x)$ does not depend on u_{j+1}].

Putting $A_{01} = c\rho^{-1/4}$, where c is an (arbitrary) complex constant, for subsequent $A_{j,1}$ we can take partial solutions since it is possible, clearly, to satisfy initial conditions.

Let us give the result for the first three terms of the decomposition (4.6.5) which will be used later in a concrete example of matching:

$$\begin{aligned} u = \frac{1}{\sqrt{\rho}}\Bigg\{ & c\zeta + \bar{c}\bar{\zeta} \\ & + \varepsilon\left(\frac{2c\bar{c}}{\rho^{5/4}} + i\left(h' - \frac{\rho'}{8\rho^{3/2}}\right)(c\zeta - \bar{c}\bar{\zeta}) - \frac{1}{3\rho^{5/4}}(c^2\zeta^2 + \bar{c}^2\bar{\zeta}^2)\right) \\ & + \varepsilon^2\Bigg[\left(\frac{\rho' h}{8\rho^{3/2}} - \frac{h^2}{2} - \frac{11\rho'^2}{128\rho^3} + \frac{11c\bar{c}}{18\rho^{3/2}} + \frac{\rho''}{16\rho^2}\right)(c\zeta + \bar{c}\bar{\zeta}) \\ & + i\left(\frac{23\rho'}{36\rho^{11/4}} - \frac{2h}{3\rho^{5/4}}\right)(c^2\zeta^2 - \bar{c}^2\bar{\zeta}^2) \\ & + \frac{1}{12\rho^{5/2}}(c^3\zeta^3 + \bar{c}^3\bar{\zeta}^3)\Bigg] + 0\left(\frac{\varepsilon^3}{\rho^{9/2}}\right)\Bigg\}, \end{aligned} \qquad (4.6.6)$$

where h stands for a function $h(x)$ satisfying $h' \equiv a(x) = -\rho'^2/32\rho^{5/2} - 5c\bar{c}/3\rho^2$. Since

$$a(x) \equiv -\frac{\rho'^2}{32\rho^{5/2}} - \frac{5c\bar{c}}{3\rho^2}$$

$$= -\frac{1}{32}\left(\frac{1}{x^{5/2}} + \frac{3}{4}\rho''(0)x^{-3/2}\right) - \frac{5c\bar{c}}{3}\left(\frac{1}{x^2} - \frac{\rho''(0)}{x}\right) + 0\left(\frac{1}{\sqrt{x}}\right)$$

$$\equiv b(x) + 0\left(\frac{1}{\sqrt{x}}\right)$$

for a small x, it will be convenient to put

$$h = \frac{1}{48x^{3/2}} + \frac{3\rho''(0)}{64\sqrt{x}} + \frac{5c\bar{c}}{3}\left(\frac{1}{x} + \rho''(0)\ln x\right) + \int_0^x [a(\eta) - b(\eta)]\, d\eta. \quad (4.6.7)$$

Let us add that the error of the m-term approximation (4.6.5) for a small ρ (small x) is not difficult to estimate. It constitutes $0(\varepsilon^m/\rho^{3m/2})$ with respect to $\rho^{-1/4}$; hence, the exterior asymptotics are extendable into the domain of small x only if $x \gg \varepsilon^{2/3} = \delta^4$ (which corresponds to the reconstruction).

Inner Asymptotics

The simplest way of constructing the inner asymptotics is, evidently, to seek the form

$$y = y_0 + \delta y_1 + \delta^2 y_2 + \cdots \qquad [y_j = y_j(\tau)], \qquad (4.6.8)$$

where the principal term y_0 satisfies the leading (Airy) equation

$$\frac{d^2 y_0}{d\tau^2} + \tau y_0 = 0. \qquad (4.6.9)$$

[This method of constructing inner asymptotics is equivalent to the reduction of the operator $Y = Y_0 + \delta Y_1 + \cdots$, where

$$Y_0 = z\frac{\partial}{\partial y} - \tau y\frac{\partial}{\partial z} + \frac{\partial}{\partial \tau}$$

and $z = dy/d\tau$; corresponding to (4.6.3) to the simplest normal form, namely, to the leading operator Y_0 itself.] Let us give computations and note certain details.

Let us take a convenient fundamental system of solutions of the Airy equation (4.6.9). Put $\alpha = \alpha(\tau)$ for a solution of (4.6.9) with the asymptotics

$$\alpha = \frac{1}{\sqrt[4]{\tau}} e^{(2i/3)\tau^{3/2}}\left(1 + \sum_{n\geq 1} \frac{(-i)^n c_n}{\tau^{3n/2}}\right),$$

where

$$c_n = \frac{1}{48^n n!}\prod_{k=1}^{n}(6k-5)(6k-1) \qquad (4.6.10)$$

as $\tau \to \infty$ (on the Airy functions see, e.g., [4]), from which, as is known, we can recover initial values $\alpha(0)$ and $\alpha'(0)$. Making use of any of the standard fundamental systems it is not difficult to compute that

$$\alpha(0) = \frac{\sqrt[3]{3}}{\sqrt{3}\pi}\Gamma\left(\frac{1}{3}\right)e^{-\pi i/12}, \qquad \alpha'(0) = -\frac{\sqrt[3]{9}}{\sqrt{3}\pi}\Gamma\left(\frac{2}{3}\right)e^{-5\pi i/12}$$

where $\Gamma(x)$ is the Euler gamma function.

Take α and $\bar{\alpha}$ as a fundamental system. The Wronski determinant is

$$\bar{\alpha}\frac{d\alpha}{d\tau} - \alpha\frac{d\bar{\alpha}}{d\tau} = 2i. \tag{4.6.11}$$

Let us seek a solution of (4.6.3) by the variation of constants method. Put $y = \mathcal{B}\alpha + \bar{\mathcal{B}}\bar{\alpha}$, $\alpha(d\mathcal{B}/d\tau) + \bar{\alpha}(d\bar{\mathcal{B}}/d\tau) = 0$. Then by (4.6.11) Equation (4.6.3) takes the form

$$\frac{d\mathcal{B}}{d\tau} = \frac{i}{2}\left[-\delta(\mathcal{B}\alpha + \bar{\mathcal{B}}\bar{\alpha})^2\bar{\alpha} + \left(\delta^4\frac{p''(0)}{2}\tau^2 + \cdots\right)(\mathcal{B}\alpha + \bar{\mathcal{B}}\bar{\alpha})\bar{\alpha}\right]. \tag{4.6.12}$$

Now, putting

$$\mathcal{B} = B + \delta\mathcal{B}_1(\tau) + \delta^2\mathcal{B}_2(\tau) + \cdots, \qquad B = \text{const}, \tag{4.6.13}$$

so that

$$y_0 = B\alpha + \bar{B}\bar{\alpha} \tag{4.6.14}$$

and $y_j = \mathcal{B}_j(\tau)\alpha + \bar{\mathcal{B}}_j(\tau)\bar{\alpha}$, it is not difficult to compute any approximation in the decomposition (4.6.8), taking partial solutions for $\mathcal{B}_j$, since B is an arbitrary complex constant.

Making use of (4.6.10), we can find the domain of applicability of (4.6.8).

For $\rho(x) \equiv x$ the linear asymptotics (4.6.8) can be used for all $\tau > 0$ since the $\mathcal{B}_j$'s decrease as $\tau \to +\infty$ (if partial solutions are chosen accordingly). It is easy to notice that for $\rho(x) \equiv x$ the right-hand side of (4.6.12) only depends on $\alpha, \bar{\alpha}$ and $\mathcal{B}, \bar{\mathcal{B}}$ and is cubic in $\alpha, \bar{\alpha}$. Therefore, for odd j, the $\mathcal{B}_j$ contain in the asymptotics as $\tau \to +\infty$ only terms with oscillating multiples since $\exp[\pm(2i/3)\tau^{3/2}]$ are multiplied an odd number of times. This accounts for the boundedness of all $\mathcal{B}_j$'s, as is easy to compute, though $\mathcal{B}_j$ for even j contain nonoscillating summands in the asymptotics as $\tau \to +\infty$. Integrating each time from τ to $+\infty$, we get that $\mathcal{B}_{2\nu-1}$ and $\mathcal{B}_{2\nu}$ decrease as $\tau^{-\nu-1/4}$ and $\tau^{-\nu}$, respectively.

If $\rho(x) \not\equiv x$ and $\alpha\bar{\alpha} \simeq \tau^{-1/2}$, (4.6.12) yields that, beginning with $\mathcal{B}_4 \sim \tau^{5/2}$ [if $\rho''(0) \neq 0$], $\mathcal{B}_j$ will grow as $\tau \to +\infty$. A straightforward investigation (which we will omit) shows that $\mathcal{B}_{4\nu}$'s grow as $\tau^{5\nu/2}$ for $\rho''(0) \neq 0$ (which will be assumed hereafter for the sake of simplicity) and $\mathcal{B}_{4\nu+1}$, $\mathcal{B}_{4\nu+2}$, and $\mathcal{B}_{4\nu+3}$ grow no faster than $\tau^{5\nu/2-5/4}$, $\tau^{5\nu/2-1}$, and $\tau^{5\nu/2-9/4}$ respectively, so that (4.6.8) is applicable when $\delta^4\tau^{5/2} \ll 1$.

Note an evident shortcoming of this method of constructing inner asymptotics: the matching domain is considerably "narrower" than the domain of applicability of the reconstructed equation: $1 \lesssim \tau \ll \delta^{-4}$, or, more exactly, than of the maximally "wide" matching domain $1 \ll \tau \ll \delta^{-4}$ on which we might have counted.

The use of (4.6.8) requires approximately $4/\gamma_{\text{cr}} = 2.5$ times more approximations (if the accuracy is high) in the exterior asymptotics for a matching procedure as compared with the inner asymptotics for which $\gamma_{cr} = 4$ (a little later we will show how this can be constructed).

We can improve the process of construction of the inner asymptotics by putting, in (4.6.12), $\mathcal{B} = \mathcal{A}e^{i\theta}$, where $\theta = \delta^4\tilde{\partial}_1(\tau) + \delta^8\tilde{\partial}_2(\tau)$ and $\tilde{\partial}_j(\tau)$ are real, so that

$$\frac{d\mathcal{A}}{d\tau} = \frac{i}{2}\Big[- \delta(\mathcal{A}e^{i\theta}\alpha + \mathcal{A}\bar{e}^{-i\theta}\bar{\alpha})^2 e^{-i\theta}\bar{\alpha} + \left(\delta^4\frac{\rho''(0)}{2}\tau^2 + \cdots\right) e^{-2i\theta}\bar{\mathcal{A}}\bar{\alpha}^2$$
$$+ \left(\left(\delta^4\frac{\rho''(0)}{2}\tau^2 + \cdots\right)\alpha\bar{\alpha} - 2\left(\delta^4\frac{d\tilde{\partial}_1}{d\tau} + \cdots\right)\right)\mathcal{A}\Big].$$

Now we can search for $\mathcal{A}$ in the form $\mathcal{A}_0 + \delta\mathcal{A}_1 + \cdots$, where $\mathcal{A}_0 = \text{const}$, ignoring the dependence of $e^{i\theta}$ on δ, the choice of $\tilde{\partial}_j$ being subject to applicability of the asymptotics obtained in the widest domain. Omitting the investigation, we note that $\gamma_{cr} = 4$ can be obtained.

Remark 4.6.1 It is interesting that from the matching considerations we can immediately indicate the answer for θ:

$$\theta = \int_0^\tau \left(\sqrt{\rho(\delta^4\eta)/\delta^4} - \sqrt{\eta}\right) d\eta,$$

where the quadrature is computed by a formal development into the Taylor series in $\sqrt{\tau}$ at $\sqrt{\tau} = 0$. This is clear since in the matching domain $1 \ll \tau \ll \delta^{-4}$ the exponential terms should be ζ^k and $\bar{\zeta}^k$ (up to multiples).

The use of "improved" inner asymptotics saves in the outer approximations but introduces complications in computations; we cannot say which process is preferable.

In the example of matching given below we will restrict ourselves to a low accuracy of computations $[0(\delta^6) = 0(\varepsilon)$ included$]$ for $u = \delta^{-1}y$ and employ the simplest inner asymptotics (4.6.8) ($y \simeq y_0 + \delta y_1 + \cdots + \delta^7 y_7$).

Let us choose particular solutions for y_j, where $j \geq 1$ [arbitrary constants enter y_0 defined by (4.6.14)]. Since the answer for this approximation is rather cumbersome, formulas to follow are somewhat unnatural due to our wish to shorten notation.

Denote by $z = z_0 + \delta z_1 + \cdots$, where $z_0 \equiv y_0$, the solution of an auxiliary equation $z'' + \tau z = \delta z^2$ [the case $\rho(x) \equiv x$] and introduce additional functions r_0, r_1, and r_2, solutions of

$$r_0'' + \tau r_0 = -\tau z_0^2 \equiv f_0(\tau),$$
$$r_1'' + \tau r_1 = 2(r_0 - \tau z_1)z_0 \equiv f_1(\tau),$$
$$r_2'' + \tau r_2 = 2(r_1 - \tau z_2)z_0 + (2r_0 - \tau z_1)z_1 \equiv f_2(\tau).$$

Then, as it is easy to verify by a direct substitution in (4.6.3), we can put

$$y \simeq z + \frac{\rho''(0)}{2}\left\{\frac{\delta^4}{5}\left(\tau^2\frac{dz}{d\tau} - \tau z\right) + \delta^5 r_0 + \delta^6 r_1 + \delta^7 r_2\right\} \qquad (4.6.15)$$

[where z is computed up to $\delta^7 z_7$ included, and terms $\tau^2(dz/d\tau) - \tau z$, respectively, up to $\delta^3 z_3$ included].

For $z_1, \ldots; r_0, r_1$, and r_2 we choose the following particular solutions:

$$z_j = \frac{i}{2} \sum_{0 \le k \le j-1} \int_\tau^\infty K(\tau,\eta) z_k(\eta) z_{j-1-k}(\eta)\, d\eta,$$

where

$$K(\tau,\eta) = \alpha(\tau)\bar{\alpha}(\eta) - \bar{\alpha}(\tau)\alpha(\eta), \quad z_0 \equiv y_0 = B\alpha + \bar{B}\bar{\alpha}, \tag{4.6.16}$$

$$\begin{aligned} r_0 &= \frac{i}{2} \int_\tau^\infty K(\tau,\eta) f_0(\eta)\, d\eta, \\ r_1 &= \frac{10iB\bar{B}}{3}(B\alpha - \bar{B}\bar{\alpha}) \ln\tau \\ &\quad + i \int_\tau^\infty \left\{ \frac{10B\bar{B}}{3}[B\alpha(\tau) - \bar{B}\bar{\alpha}(\tau)]\frac{1}{\eta} + K(\tau,\eta)\frac{f_1(\eta)}{2} \right\} d\eta \end{aligned} \tag{4.6.17}$$

[in the computation of r_1 the asymptotics in $K(\tau,\eta)(f_1(\eta)/2)$ as $\eta \to +\infty$ are taken into account],

$$r_2 = \frac{i}{2} \int_\tau^\infty K(\tau,\eta) f_2(\eta)\, d\eta. \tag{4.6.18}$$

Note that in the sequel it is important only to fix the choice of the particular solutions for the y_j's, where $j \ge 1$, and the above choice is unimportant.

Matching

Restricting ourselves with approximation (4.6.15) for y we find y with accuracy $0(\delta^7)$ included and error $0(\delta^8)$ for $\tau \sim 1$. To compute $y = \delta u$ with the same accuracy via formulas for the exterior asymptotics for $\tau \sim \delta^{-8/5} (x \sim \delta^{12/5})$ it suffices to make use of the approximation (4.6.6). But the error will constitute $0(\delta^{38/5})$, which is greater than δ^8. Therefore, by the matching principle, making use of (4.6.6) we should believe that B will be found with the error $\sim \delta^{38/5}$.

Passing to computations, we first find the asymptotics of y_j for large τ. For brevity's sake, introduce

$$\begin{aligned} w_k &= B^k e^{(2ki/3)\tau^{3/2}} + \bar{B}^k e^{-(2ki/3)\tau^{3/2}}, \\ \hat{w}_k &= i\left(B^k e^{(2ki/3)\tau^{3/2}} - \bar{B}^k e^{-(2ki/3)\tau^{3/2}}\right). \end{aligned} \tag{4.6.19}$$

Due to (4.6.10) and (4.6.19) we have

$$y_0 \simeq \frac{w_1}{\tau^{1/4}} - \frac{5\hat{w}_1}{48\tau^{7/4}} - \frac{385 w_1}{4608\tau^{13/4}}, \tag{4.6.20}$$

where the dropped terms are $\sim \tau^{-19/4}$, which for $\tau \sim \delta^{-8/5}$ constitutes $\sim \delta^{38/5}$, and, by the matching principle, these terms can be ignored. Further, making use of (4.6.16) and (4.6.10) and $z_0 \equiv y_0$, $y_1 = z_1$, $y_2 = z_2$, and $y_3 = z_3$ we find (integrating by parts) that

$$\begin{aligned} y_1 &\simeq \frac{1}{\tau^{3/2}}\left(2B\bar{B} - \frac{w_2}{3}\right) + \frac{5\hat{w}_2}{8\tau^3}, \\ y_2 &\simeq \frac{5B\bar{B}\hat{w}_1}{\tau^{5/4}} + \frac{1}{12\tau^{11/4}}\left(\frac{113B\bar{B}}{12}w_1 + w_3\right), \\ y_3 &\simeq -\frac{10B\bar{B}\hat{w}_2}{9\tau^{5/2}}, \end{aligned} \tag{4.6.21}$$

where omitted terms are $\sim \tau^{-9/2}$, $\tau^{-17/4}$, and τ^{-4}, respectively, so that in $\delta y_1 + \delta^2 y_2 + \delta^3 y_3$ they can be ignored as above. Besides, we have z_4, z_5, z_6, $z_7 \sim \tau^{-9/4}$, $\tau^{-7/2}$, $\tau^{-13/4}$, and $\tau^{-9/2}$, respectively, and terms $\delta^4 z_4 + \cdots + \delta^7 z_7$ can be also ignored.

From (4.6.17) and (4.6.18) we similarly find that

$$\begin{aligned} r_0 &\simeq \frac{1}{\sqrt{\tau}}\left(-2B\bar{B} + \frac{w_2}{3}\right) + 0\left(\frac{1}{\tau^2}\right), \\ r_1 &\simeq \frac{10B\bar{B}\hat{w}_1}{3\tau^{1/4}}\ln\tau + 0\left(\frac{\ln\tau}{\tau^{7/4}}\right), \\ r_2 &\simeq 0\left(\frac{\ln\tau}{\tau^{3/2}}\right), \end{aligned}$$

and, taking (4.6.15), (4.6.20), and (4.6.21) into account, we get

$$\begin{aligned} y_4 &\simeq \frac{\rho''(0)}{2}\left(\frac{\tau^{9/4}\hat{w}_1}{5} - \frac{11\tau^{3/4}w_1}{48} + \frac{187\hat{w}_1}{4608\tau^{3/4}}\right), \\ y_5 &\simeq \frac{\rho''(0)}{2}\left(-\frac{2\tau\hat{w}_2}{15} + \frac{1}{\sqrt{\tau}}\left(-3B\bar{B} + \frac{w_2}{4}\right)\right), \\ y_6 &\simeq \frac{\rho''(0)}{2}\left(-\frac{B\bar{B}\tau^{5/4}w_1}{3} - \frac{427B\bar{B}\hat{w}_1}{720\tau^{1/4}} + \frac{10B\bar{B}\hat{w}_1}{3\tau^{1/4}}\ln\tau + \frac{\hat{w}_3}{20\tau^{1/4}}\right), \\ y_7 &\simeq \frac{2\rho''(0)B\bar{B}w_2}{9}, \end{aligned} \tag{4.6.22}$$

where omitted terms are $\sim \tau^{-9/4}$, τ^{-2}, $(\ln\tau)\tau^{-7/4}$, and $(\ln\tau)\tau^{-3/2}$ and can be ignored in $\delta^4 y_4 + \cdots + \delta^7 y_7$.

Now it is clear with respect to which functions u_j should be developed in the matching domain $\delta^{12/5} = \varepsilon^{2/5} \gg x \gg \delta^4 = \varepsilon^{2/3}$.

For small x ($\varepsilon^{2/5} \gg x \gg \varepsilon^{2/3}$) we have [see (4.6.4)]

$$\zeta \simeq e^{(2i/3\varepsilon)x^{3/2}}\left(1 + \frac{i\rho''(0)}{10\varepsilon}x^{5/2} + 0\left(\frac{x^5}{\varepsilon^2}\right)\right),$$

where terms of the development (into the power series) not written explicitly do not affect the result of the following computation.

Expanding coefficients of powers of ζ in (4.6.6) with respect to negative and positive powers of x and $\ln x$, and inserting ζ, we get the expansion of u_0, εu_1, and $\varepsilon^2 u_2$ with respect to functions of the form $\varepsilon^\nu \zeta_0^m \bar{\zeta}_0^k x^\mu (\ln x)$, where $mk = 0$ and

$$\zeta_0 = \exp\left(\frac{2i}{3\varepsilon} x^{3/2}\right).$$

Here $|\zeta_0| = 1$ and $(\ln x)^\lambda$ grows when $x \sim \delta^\gamma$, slower than any power of ε. Therefore, we should only pay attention to $\varepsilon^\nu x^\mu$. By the matching principle, each summand of this kind can be omitted if it can reach only values less than the matching error $\delta^{-1}\delta^{38/5} = \delta^{33/5} = \varepsilon^{11/10}$ (recall that $u = \delta^{-1} y$). If $\mu > 0$ and $\nu + \mu(2/3) \geq 11/10$ and also if $\mu < 0$ and $\nu + \mu(2/5) \geq 11/10$ and, finally, if $\mu = 0$ and $\nu \geq 11/10$, we omit a summand, $\varepsilon^\nu x^\mu$.

Introducing for brevity's sake

$$\begin{aligned} w_k &= c^k e^{(2ki/3)\tau^{3/2}} + \bar{c}^k e^{-(2ki/3)\tau^{3/2}}, \\ \hat{w}_k &= i(c^k e^{(2ki/3)\tau^{3/2}} - \bar{c}^k e^{-(2ki/3)\tau^{3/2}}), \end{aligned} \tag{4.6.23}$$

the result of the above procedure after the substitution $x = \delta^4 \tau$ is

$$\begin{aligned} \delta u_0 &\simeq \frac{w_1}{\tau^{1/4}} + \delta^4 \frac{\rho''(0)}{2}\left(\frac{\tau^{9/4}\hat{w}_1}{5} - \frac{\tau^{3/4} w_1}{4}\right), \\ \delta\varepsilon u_1 &\simeq -\frac{5\hat{w}_1}{48\tau^{7/4}} + \delta \frac{1}{\tau^{3/2}}\left(2c\bar{c} - \frac{w_2}{3}\right) + \delta^2 \frac{5c\bar{c}\hat{w}_1}{3\tau^{5/4}} \\ &\quad + \delta^4 \frac{\rho''(0)}{2}\left(\frac{\tau^{3/4} w_1}{48} + \frac{11\hat{w}_1}{192\tau^{3/4}}\right) \\ &\quad + \delta^5 \frac{\rho''(0)}{2}\left(-\frac{\tau^2 \hat{w}}{15} - \frac{3c\bar{c}}{\sqrt{\tau}} + \frac{w_2}{2\sqrt{\tau}}\right) \\ &\quad + \delta^6 \frac{\rho''(0)}{2}\left(-\frac{c\bar{c}\tau^{5/4} w_1}{3} + \left(\frac{10c\bar{c}}{3\tau^{1/4}} \ln(\delta^4\tau) - \frac{5c\bar{c}}{12\tau^{1/4}}\right)\hat{w}_1\right) \\ \delta\varepsilon^2 u_2 &\simeq -\frac{385 w_1}{4608\tau^{13/4}} + \delta \frac{5\hat{w}_2}{8\tau^3} + \delta^2 \frac{1}{12\tau^{11/4}}\left(\frac{113c\hat{c}}{12} w_1 + w_3\right) \\ &\quad - \delta^3 \frac{10c\bar{c}\hat{w}_2}{9\tau^{5/2}} - \delta^4 \frac{\rho''(0)}{2} \frac{77\hat{w}_1}{4608\tau^{3/4}} - \delta^5 \frac{\rho''(0)}{2} \frac{w_2}{4\sqrt{\tau}} \\ &\quad + \delta^6 \frac{\rho''(0)}{2}\left(\frac{113c\bar{c}\hat{w}_1}{720\tau^{1/4}} + \frac{\hat{w}_3}{20\tau^{1/4}}\right) + \delta^7 \frac{2c\bar{c}\rho''(0) w_2}{9}, \end{aligned}$$

or, summing,

$$\begin{aligned} y &\simeq \delta(u_0 + \varepsilon u_1 + \varepsilon^2 u_2) \\ &\simeq \frac{w_1}{\tau^{1/4}} - \frac{5\hat{w}_1}{48\tau^{7/4}} - \frac{385 w_1}{4608\tau^{13/4}} + \delta\left(\frac{1}{\tau^{3/2}}\left(2c\bar{c} - \frac{w_2}{3}\right) + \frac{5\hat{w}_2}{8\tau^3}\right) \end{aligned}$$

$$+\delta^2\left(\frac{5c\bar{c}\hat{w}_1}{3\tau^{5/4}}+\frac{1}{12\tau^{11/4}}\left(\frac{113c\bar{c}}{12}w_1+w_3\right)\right)-\delta^3\frac{10c\bar{c}\hat{w}_2}{9\tau^{5/2}}$$
$$+\frac{\rho''(0)}{2}\left\{\delta^4\left(\frac{\tau^{9/4}\hat{w}_1}{5}-\frac{11\tau^{3/4}w_1}{98}+\frac{187\hat{w}_1}{4608\tau^{3/4}}\right)\right.$$
$$+\delta^5\left(-\frac{2\tau\hat{w}_2}{15}+\frac{1}{\sqrt{\tau}}\left(-3c\bar{c}+\frac{w_2}{4}\right)\right)$$
$$+\delta^6\left(-\frac{c\bar{c}\tau^{5/4}w_1}{3}-\frac{187c\bar{c}\hat{w}_1}{720\tau^{1/4}}+\frac{10c\bar{c}(\ln\tau+4\ln\delta)}{3\tau^{1/4}}\hat{w}_1+\frac{\hat{w}_3}{20\tau^{1/4}}\right)$$
$$\left.+\delta^7\frac{2c\bar{c}\rho''(0)w_2}{9}\right\}. \tag{4.6.24}$$

Taking into account that for $B = C$ formulas (4.6.19) and (4.6.23) coincide and comparing (4.6.24) with (4.6.20), (4.6.21) and (4.6.22), we see that $B \simeq C(1 + i\delta^6 q)$, where $q = 0(\ln\delta)$. Substituting $B \simeq C(1 + i\delta^6 q)$, and the possibility of omitting terms reaching values less than $\delta^{38/5}$ being taken into account, we get for $y_0, y_1, \ldots, y_7$ formulas (4.6.20), (4.6.21), and (4.6.22), with the replacements $\hat{w}_k \to w_k$, $w_k \to \hat{w}_k$ and increment $\delta^6 q\hat{w}_1/\tau^{1/4}$ in y_0. Then we get q. Finally,

$$B \simeq C\left(1+\varepsilon\frac{i\rho''(0)}{6}\left(1+\frac{20}{3}\ln\varepsilon\right)c\bar{c}\right).$$

The reconstruction of the operator near the singular surface enables one to extend the trajectory into the domain contracting to the singular surface as $\varepsilon \to 0$ faster than the domain into which the exterior asymptotics can be extended is contracted, but does not exclude the appearance of new singular surfaces (in new variables). An example of such a situation is given by the already known Equation (4.3.3) in the case $n > 3$.

4.7 Example: Appearance of a New Singularity

Returning to Equation (4.3.3), $\varepsilon^2 u'' + xu = \varepsilon u^n$, where $x \geq 0$, $n > 3$, to which the system $\varepsilon u' = v$, $\varepsilon v' = -xu + \varepsilon u^n$ and the operator

$$X = X_0 + \varepsilon X_1 = v\frac{\partial}{\partial u} - xu\frac{\partial}{\partial v} + \varepsilon\left(u^n\frac{\partial}{\partial v} + \frac{\partial}{\partial x}\right)$$

correspond, let us recall the reconstruction transformation (4.3.6) found in Section 4.3 in the case $n > 3$:

$$u = \varepsilon^{-1/n+3}y, \qquad v = \varepsilon^{1/n+3}z, \qquad x = \varepsilon^{4/n+3}\tau,$$

and, accordingly, the reconstructed operator X is $\varepsilon^{2/n+3}Y$, where

$$Y = Y_0 + \varepsilon Y_1 = z\frac{\partial}{\partial y} - (\tau y - y^n)\frac{\partial}{\partial z} + \delta\frac{\partial}{\partial\tau},$$
$$\delta = \varepsilon^{n-3/n+3}. \tag{4.7.1}$$

Making use of the reconstructed operator we can extend the trajectory into the previously inaccessible domain $\tau \sim 1$, but unlike the previous problem ($n < 3$) a new singular surface $\tau = 0$ arises, as we will presently show. Let us pass to new variables

$$p = \tau^{-1/n-1} y, \qquad q = \tau^{-n+1/2(n-1)} z, \qquad s = \frac{2}{3}\tau^{3/2}, \tag{4.7.2}$$

in which the operator (4.7.1) is of the form

$$Y = \left(\frac{3}{2}s\right)^{1/3} Z,$$

where

$$\begin{aligned} Z &= Z_0 + \delta Z_1 \\ &= q\frac{\partial}{\partial p} - (p - p^n)\frac{\partial}{\partial q} + \delta\left\{\frac{\partial}{\partial s} - \frac{2}{3(n-1)s}\left(p\frac{\partial}{\partial p} + \frac{n+1}{2}q\frac{\partial}{\partial q}\right)\right\}. \end{aligned} \tag{4.7.3}$$

Here, the leading operator Z_0 possesses an invariant

$$g = q^2 + p^2 - \frac{2}{n+1}p^{n+1}. \tag{4.7.4}$$

The leading system is autonomous; its phase trajectories in the plane (p, q) are closed if $0 < g < n - 1/n + 2$ and $|p| < 1$. A loop of the separatrix $g = n - 1/n + 1$ passing through one or two (depending on the parity of n) points of unstable equilibrium $q = 0$, $p = +1$ or -1, bounds the domain of close trajectories. At the initial "moment" the depicting point is near the origin [since we consider the problem with initial values $u(x_0) \sim 1$ and $\varepsilon u'(x_0) = v(x_0) \sim 1$, this corresponds to $p \sim \varepsilon^{1/n-1}$ and $q \sim \varepsilon^{1/n-1}$]; i.e., belongs to the inside of the separatrix loop. The reduction of Z to the normal form (Z serves both the domain $s \sim 1$ and the domain of large s) is complicated by the nonlinearity of the leading oscillator, but is simple in principle. We will give only the equation which describes the evolution of the phase trajectory (the dependence of y on s).

Putting $M_1 = [Z_0, S_1] + Z_1$ we get for $S_1 g$, taking (4.7.4) and (4.7.3) into account, the equation

$$Z_0(S_1 g) = \frac{2}{3s}\left(\frac{n+1}{n-1}g - p^2\right) + M_1 g, \tag{4.7.5}$$

where $M_1 g$ is a function in invariants g and s of Z_0. Now set

$$W(g, p) = g - p^2 + \frac{2}{n+1}p^{n+1} \tag{4.7.6}$$

and denote by $p_1(g)$ the absolute minimum of the negative root of the equation $w(g,p)=0$ and by $p_2(g)$ the minimal positive root of this equation (for $0<g<n-1/n+1$).

Finally, put $S_1 g = A(s,g,p)q$, $M_1 g = (2/3s)F(g)$. Then, taking $q^2 = W(q,p)$ into account [see (4.7.4) and (4.7.6)], along with

$$Z_0(S_1 g) = W\frac{\partial A}{\partial p} + \frac{A}{2}\frac{\partial W}{\partial p} = \sqrt{W}\frac{\partial}{\partial p}(\sqrt{W}\,A),$$

we get (4.7.5) in the form

$$\sqrt{W}\frac{\partial}{\partial p}(\sqrt{W}\,A) = \frac{2}{3s}\left(\left(\frac{n+1}{n-1}g - p^2\right) + F(g)\right) \equiv Q(s,g;p).$$

The solution of this equation which has no singularities at $p=p_1(g)$ and is unique:

$$A = \frac{1}{\sqrt{W(g,p)}}\int_{p_1(g)}^{p}\frac{Q(s,g;\eta)}{\sqrt{W(g,\eta)}}\,d\eta.$$

In order for this solution to have no singularities at $p=p_2(g)$ we should choose $F(g)$ from the condition that A vanishes at $p=p_2(g)$. We will get

$$F(g) = -\int_{p_1(g)}^{p_2(g)}\frac{\frac{n+1}{n-1}g-\eta^2}{\sqrt{W(g,\eta)}}\,d\eta\left[\int_{p_1(g)}^{p_2(g)}\frac{d\eta}{\sqrt{W(g,\eta)}}\right]^{-1}. \tag{4.7.7}$$

S_1 and M_1 can be completely found (we have only found $S_1 g$ and $M_1 g$) from the condition $M_1'' = 0$, which reduces to the equation of the type (4.7.5). Recall that for variable eigenvalues, which is the case in this problem due to the nonlinearity of the leading oscillator, we have, generally speaking, $-M_1' \neq 0$.

The equation for g in the leading order is

$$\frac{dg}{ds} = \frac{2F(g)}{3s}. \tag{4.7.8}$$

From (4.7.8) and (4.7.7) it follows that as s diminishes, the phase trajectory expands, nearing the loop of the separatrix $g=n-1/n+1$. However, for $s\sim\delta$ the perturbation theory fails: the operator (4.7.3) is reconstructed and a further study of the trajectory is, seemingly, impossible without exact integration of the problem [putting $s=\delta\sigma$ we lose a small parameter in (4.7.3)].

Remark 4.7.1 The dilatation $u=\varepsilon^{1/3(n-1)}\hat{u}$, $x=\varepsilon^{2/3}\hat{x}$ "kills" a small parameter in the initial equation and, as it is easy to verify, as a result of $s=\delta\sigma$ and previous transformations, we have obtained just these dilatations. Therefore, the deduction that the perturbation theory is inapplicable for $x\lesssim\varepsilon^{2/3}$ seems trivial. However, it is not exactly true: for $n<3$ the problem is completely solved, the reason being that in this case, $n<3$, the trajectory does not enter the domain $u\sim\varepsilon^{-1/3(n-1)}$ for $x\sim\varepsilon^{2/3}$ (whereas for $n\geq 3$ it does enter).

In some problems where the resonance surface does not consist of 0-points we can do without reconstruction and matching, making use of the uniform normal form presented in Section 3.10. Let us give a simple example (cf. the solution given below with [39]).

4.8 Example: Passing Through a Resonance

Consider the equation

$$\ddot{x} + \omega^2(\varepsilon t)x = \varepsilon a \cos t, \qquad a = \text{const}, \tag{4.8.1}$$

where $0 < \omega(\varepsilon t) < 2$ and $\omega(\varepsilon t)$ can be equal to 1. The system is

$$\frac{dx}{dt} = y, \qquad \frac{dy}{dt} = -\omega^2(\varepsilon t)x + \varepsilon a \cos t.$$

Introduce also variables $\psi = e^{it}$, $\bar{\Psi} = e^{-it}$, $\tau = \varepsilon t$:

$$\frac{dx}{dt} = y, \qquad \frac{dy}{dt} = \frac{a}{2}\omega^2(\tau)x + \varepsilon\frac{a}{2}(\psi + \bar{\psi}),$$

$$\frac{d\tau}{dt} = \varepsilon, \qquad \frac{d\psi}{dt} = i\psi, \qquad \frac{d\bar{\psi}}{dt} = -i\psi.$$

The operator is

$$\begin{aligned} X &= X_0 + \varepsilon X_1 \\ &= y\frac{\partial}{\partial x} - \omega^2(\tau)x\frac{\partial}{\partial y} + i\left(\psi\frac{\partial}{\partial \psi} - \bar{\psi}\frac{\partial}{\partial \bar{\psi}}\right) + \varepsilon\left\{\frac{\partial}{\partial \tau} + \frac{a}{2}(\psi + \bar{\psi})\frac{\partial}{\partial y}\right\}. \end{aligned}$$

The operator X_0 possesses eigenfunctions $\varphi = y + i\omega(\tau)x$ and $\bar{\varphi} = y - i\omega(\tau)x$ corresponding to $\pm i\omega(\tau)$ and eigenfunctions ψ and $\bar{\psi}$ such that

$$X_0\varphi = i\omega(\tau)\varphi, \qquad X_0\psi = i\psi, \qquad X_0\tau = 0. \tag{4.8.2}$$

Then

$$\begin{aligned} X_1\varphi &= \frac{\omega'}{2\omega}(\varphi - \bar{\varphi}) + \frac{a}{2}(\psi + \bar{\psi}), \\ X_1\psi &= 0, \\ X_1\tau &= 1 \left(\omega' = \frac{d\omega}{d\tau}\right). \end{aligned} \tag{4.8.3}$$

The surfaces $\tau = \tau_0$, where $\omega(\tau_0) = 1$, consist of resonance 1-points.

Let us reduce X to the uniform normal form; i.e., to the normal form with respect to X_0:

$$\hat{X}_0\varphi = i\varphi, \qquad \hat{X}_0\psi = i\psi, \qquad \hat{X}_0\tau = 0. \tag{4.8.4}$$

From (4.8.3) and (4.8.4) it is immediately clear that we can put

$$M_1\tau = 1, \qquad M_1\varphi = \frac{\omega'}{2\omega}\varphi + \frac{a}{2}\psi, \qquad M_1\psi = 0.$$

The solution of these equations is also evident due to (4.8.2):

$$S_1\tau = 0, \qquad S_1\varphi = \frac{i\omega'}{4\omega^2}\bar{\varphi} - \frac{ia}{2(1+\omega)}\bar{\psi}, \qquad S_1\psi = 0.$$

Restricting ourselves for simplicity's sake to this approximation we get that the principal term of the asymptotics is described by the equation

$$\frac{d\varphi}{dt} = i\omega\varphi + \varepsilon\left(\frac{\omega'}{2\omega}\varphi + \frac{a}{2}\psi\right), \qquad \text{where } \psi = e^{it},$$

the solution of which,

$$\varphi = \sqrt{\omega(\varepsilon t)}\, e^{i\int_0^t \omega(\varepsilon\eta)\,d\eta}\left(c + \frac{\varepsilon a}{2}\int_0^t \frac{1}{\sqrt{\omega(\varepsilon\eta)}} \exp i\left(\eta - \int_0^\eta \omega(\varepsilon\xi)\,d\xi\right) d\eta\right),$$

$$c = \text{const},$$

is the answer for $0 \leq t \leq 1/\varepsilon$, with accuracy $0(\varepsilon)$.

We can similarly seek higher approximations (with subsequent return to the old variables via e^{-S}).

Remark 4.8.1 This example is, certainly, very simple and it is not always possible to integrate the uniform normal system in the final form without considering asymptotics "far from" and "near" the resonance surface.

However, the obtaining of the uniform form is always useful.

Sometimes it is possible by artificial tricks to find a reconstruction transformation and a generalization of a normal form of an operator in such a way that the resulting asymptotics fit in the whole domain. The simplest examples of this type we have already seen above. Let us give one more.

4.9 Example: WKB-Type Problem

Consider the equation

$$\frac{d^2y}{dx^2} + [\lambda^2\rho(x) + g(x)]y = 0, \tag{4.9.1}$$

where $\rho(x) = x^n h^2(x)$, $n > 0$, $h(x) \neq 0$, $-\infty \leq x \leq +\infty$, $h(x)$ and $g(x)$ are smooth, and λ is a large parameter. We seek asymptotics of a general solution. This is one of the simplest WKB-problems ([39]) with the so-called turning point ($x = 0$). [A complicated problem is, e.g., the equation $y'' + [\lambda^2\rho(x) + \lambda m(x) + g(x)]y = 0$, where $m(x) = 0(x^{n/2-1})$ as $x \to 0$.]

Preliminary Transformations and Reconstruction

Putting $y' = \lambda z$, we pass to the system,

$$\frac{1}{\lambda}\frac{dx}{dt} = \frac{1}{\lambda}, \qquad \frac{1}{\lambda}\frac{dy}{dt} = z, \qquad \frac{1}{\lambda}\frac{dz}{dt} = -\rho(x)y - \frac{1}{\lambda^2}g(x)y,$$

to which the operator

$$Z = Z_0 + \frac{1}{\lambda}Z_1 + \frac{1}{\lambda^2}Z_2 = z\frac{\partial}{\partial y} - \rho(x)y\frac{\partial}{\partial z} + \frac{1}{\lambda}\frac{\partial}{\partial x} - \frac{1}{\lambda^2}g(x)y\frac{\partial}{\partial z} \tag{4.9.2}$$

corresponds. Since $\rho(x) = x^n h^2(x)$, then, clearly, $x = 0$ is a singular surface. The reconstruction in this problem can be made quite analogously to (4.6.1), but, putting $x = \Delta\tau$, $z = \Delta^{n/4}v$, $y = \Delta^{-n/4}u$, etc., we will be forced to develop $h^2(\Delta\tau)$ into a Taylor series which leads with necessity to matching (in contradiction with our aim).

Let us try to find a reconstruction such that the new leading operator includes the old one. For this let us first perform a shearing transformation (in accordance with $y = e_{1/2}$, $z = e_{-1/2}$) depending on x (an invariant of Z_0):

$$y = \sqrt{a(x)}\,\tilde{y}, \qquad z = \frac{\tilde{z}}{\sqrt{a(x)}}. \tag{4.9.3}$$

Then

$$Z = \frac{1}{a(x)}\tilde{Z} = \frac{1}{a(x)}\left\{\tilde{z}\frac{\partial}{\partial\tilde{y}} - \rho(x)a^2(x)\tilde{y}\frac{\partial}{\partial\tilde{z}} + \frac{1}{\lambda}\left(a(x)\frac{\partial}{\partial x} - \frac{a'(x)}{2}\tilde{y}\frac{\partial}{\partial\tilde{y}} + \frac{a'(x)}{2}\tilde{z}\frac{\partial}{\partial\tilde{z}}\right) - \frac{1}{\lambda^2}g(x)a^2(x)\tilde{y}\frac{\partial}{\partial\tilde{z}}\right\}.$$

We can note that, putting

$$\tilde{z} = \tilde{\tilde{z}} + \frac{a'(x)}{2\lambda}\tilde{y}, \tag{4.9.4}$$

we will get

$$\tilde{Z} = \tilde{\tilde{z}}\frac{\partial}{\partial\tilde{y}} - \rho(x)a^2(x)\tilde{y}\frac{\partial}{\partial\tilde{\tilde{z}}} + \frac{1}{\lambda}a(x)\frac{\partial}{\partial x} + \frac{1}{\lambda^2}\left(\frac{a'^2(x)}{4} - \frac{a(x)a''(x)}{2} - g(x)a^2(x)\right)\tilde{y}\frac{\partial}{\partial\tilde{\tilde{z}}}$$

so that the perturbation operator in $\tilde{Z}$ is somewhat simplified. This simplification is not principal and the usage of (4.9.4) is connected only with convenience of exposition.

Now let us introduce the new variable

$$\tilde{x} = \int_0^x \frac{d\eta}{a(\eta)} \qquad (\tilde{Z}\tilde{x} = \tfrac{1}{\lambda}),$$

and put $\rho(x)a^2(x) = \tilde{x}^h$; i.e., $1/a(x) = (x/\tilde{x})^{n/2}h(x)$. Then

$$\tilde{x} = \left(\frac{n+2}{2}\int_0^x \eta^{n/2}h(\eta)\,d\eta\right)^{2/n+2} \tag{4.9.5}$$

and $\tilde{Z}$ takes the form

$$\tilde{Z} = \tilde{z}\frac{\partial}{\partial\tilde{y}} - \tilde{x}^n\tilde{y}\frac{\partial}{\partial\tilde{z}} + \frac{1}{\lambda}\frac{\partial}{\partial\tilde{x}} - \frac{1}{\lambda^2}\tilde{g}(\tilde{x})\tilde{y}\frac{\partial}{\partial\tilde{z}}. \tag{4.9.2'}$$

Replacements (4.9.3)–(4.9.5) are equivalent to the well-known Langer transformation [38] which shows that in (4.9.1) it is possible to put $h(x) \equiv 1$ without loss of generality; cf. (4.9.2) and (4.9.2′).

In what follows we will assume

$$\rho(x) = x^n \qquad [h(x) \equiv 1]. \tag{4.9.6}$$

Now [in order not to develop $g(x)$] introduce an auxiliary variable "doubling" x:

$$\xi = x \tag{4.9.7}$$

and due to (4.9.6) and (4.9.7) consider the extended operator (4.9.2):

$$Z_p = z\frac{\partial}{\partial y} - \xi^n y\frac{\partial}{\partial z} + \frac{1}{\lambda}\left(\frac{\partial}{\partial\xi} + \frac{\partial}{\partial x}\right) - \frac{1}{\lambda^2}g(x)y\frac{\partial}{\partial z}. \tag{4.9.2''}$$

Here, the singular surface is $\xi = 0$. Putting $\xi = \Delta\tau$, $z = \Delta^{n/4}v$, $y = \Delta^{-n/4}u$ we find $\Delta_{\text{cr}} = \lambda^{-2/n+2}$; i.e., the critical value of Δ at which the reconstruction is performed.

Thus, putting

$$\lambda^{-2/n+2} = \varepsilon, \qquad \xi = \varepsilon\tau, \qquad z = \varepsilon^{n/4}v, \qquad y = \varepsilon^{-n/4}u, \tag{4.9.8}$$

we get $Z_p = \varepsilon^{1/2}X$, where

$$\begin{aligned} X &= X_0 + \varepsilon X_1 + \varepsilon^2 X_2 \\ &= v\frac{\partial}{\partial u} - \tau^h u\frac{\partial}{\partial v} + \frac{\partial}{\partial\tau} + \varepsilon\frac{\partial}{\partial x} + \varepsilon^2 g(x)u\frac{\partial}{\partial v} \end{aligned} \tag{4.9.9}$$

is the reconstructed operator.

Reduction to the Normal Form

Our goal is now to obtain the normal form M of X applicable (and subject to effective investigation) for $-\infty \leq \tau \leq +\infty$. Let us make several preliminary remarks which follow from the study of (4.9.9).

1. Operators $X_0 = v(\partial/\partial u) - \tau^n u(\partial/\partial v) + (\partial/\partial\tau)$ and $X_1 = \partial/\partial x$ commute and we can put $S_1 = 0$, $M_1 = X_1$.

2. We can put $S_j x = 0$, $M_j x = 0$ for $j \geq 1$ and in the construction of S_j and M_j the variable x plays the role of a parameter. We can also put $S_j\tau = M_j\tau = 0$.

3. Since Xu and Xv are linear functions in u and v we can take $S_j u$, $S_j v$, $M_j u$, and $M_j v$ to be linear (homogeneous) functions in u and v with coefficients depending on τ (and on x as a parameter).

4. This class of operators in which S_j and M_j are sought can be made narrower considering operators of the form

$$Y = (Av + Bu)\frac{\partial}{\partial u} + (-Bv + Cu)\frac{\partial}{\partial v}, \tag{4.9.10}$$

where A, B, C are functions in τ (and a parameter x).
In fact, $Y_3 = [Y_1, Y_2]$, where

$$Y_i = (A_i v + B_i u)\frac{\partial}{\partial u} + (-B_i v + C_i u)\frac{\partial}{\partial v} \qquad \text{for } i = 1, 2,$$

is an operator of the same form:

$$\begin{aligned} Y_3 &= (A_3 v + B_3 u)\frac{\partial}{\partial u} + (-B_3 v + C_3 u)\frac{\partial}{\partial v}, \\ A_3 &= 2(A_1 B_2 - A_2 B_1), \\ B_3 &= A_2 C_1 - A_1 C_2, \\ C_3 &= 2(B_1 C_2 - B_2 C_1). \end{aligned}$$

Furthermore,

$$\begin{aligned} [X_0, Y] = &((A' + 2B)v + (B' - \tau^n A - C)u)\frac{\partial}{\partial u} \\ &+ (-(B' - \tau^n A - C)v + (C' + 2\tau^n B)u)\frac{\partial}{\partial v}, \\ &\left(F' = \tfrac{\partial F}{\partial \tau}\right) \end{aligned} \tag{4.9.11}$$

is an operator of the form (4.9.10). Finally, X_2 is an operator of the same form.

5. Requiring the boundedness (smoothness) of coefficients A, B, and C of operators S_j for $j \geq 2$ for all $-\infty \leq \tau C + \infty$, we will have in the equation

$$M = [X_0, S] + \bar{Y} \qquad (M = M_j,\ S = S_j,\ \bar{Y} = \bar{Y}_j) \tag{4.9.12}$$

arising in the j-th step the operator $\bar{Y}$, recursively known, with bounded (smooth) coefficients. This follows from the general formulas (1.2.3) and the boundedness of the coefficients of X_2.

Thus, we will be interested in the following question: how to recover from the operator $\bar{Y}$ given by its bounded (smooth) coefficients $\bar{A}$, $\bar{B}$, and $\bar{C}$ an operator S of the class (4.9.10) with bounded (smooth) coefficients so that M, satisfying (4.9.12), is in a sense a simplest one (running ahead we can say that we will not manage to obtain boundedness of M in height, hence we speak about a generalization of a normal form).

First, consider the equation

$$[X_0, Y] + \bar{Y} = 0 \tag{4.9.13}$$

without requiring boundedness of coefficients of Y. The system of equations for A, B, and C which follows from (4.9.13) and (4.9.11) after excluding B and C,

$$B = -\tfrac{1}{2}(A' + \bar{A}), \qquad C = -\tfrac{1}{2}(A'' + \bar{A}') - \tau^n A + \bar{B}, \tag{4.9.14}$$

reduces to one-third order linear equation for A:

$$A''' + 4\tau^n A' + 2n\tau^{n-1} A = F(\tau), \tag{4.9.15}$$

$$F(\tau) = -\bar{A}'' - 2\tau^n \bar{A} + 2\bar{B}' + 2\bar{C}.$$

Since (4.9.15) is linear, it suffices to study the homogeneous equation $F(\tau) = 0$. Clearly, this equation should be closely connected with the equation

$$u'' + \tau^n u = 0 \tag{4.9.16}$$

to which the leading system (corresponding to X_0) reduces. In fact, if $\alpha = \alpha(\tau)$, $\bar{\alpha} = \bar{\alpha}(\tau)$ is a fundamental system of solutions of (4.9.16), then, as is subject to a straightforward verification, $\beta = \alpha\bar{\alpha}$, $\gamma = \alpha^2$, $\bar{\gamma} = \bar{\alpha}^2$ is the fundamental system of solutions of the homogeneous equation (4.9.15) with the Wronski determinant being $W(\beta, \gamma, \bar{\gamma}) = -2w^3(\alpha, \bar{\alpha})$ ($=$ const) (in general, the product of any two solutions of (4.9.16) is a solution of a homogeneous equation (4.9.15); see [24]).

Concerning (4.9.16), it is known that it is a form of the Bessel equation of index $\nu = 1/n + 2$ [putting $u = \sqrt{\tau}\, w$, $t = 2\nu\tau^{1/2\nu}$ and $\nu = 1/n + 2$ we get $t^2\ddot{w} + t\dot{w} + (t^2 - \nu^2)w = 0$].

Thus, integration of (4.9.15) by the usual method of variation of constants is not a problem. Note also that the recovery of a particular solution of (4.9.15) from its known asymptotics for large τ offers no principal difficulties since the asymptotics of Bessel functions are known [14].

Note also that the homogeneous equation (4.9.15) has a particular solution $\beta = \beta(\tau)$ with asymptotics, as $\tau \to +\infty$,

$$\beta \simeq \frac{1}{\tau^{n/2}}\left(1 + \sum_{k\geq 1} \frac{a_k}{\tau^{k(n+2)}}\right), \qquad \text{where } a_k \text{ are constants.} \tag{4.9.17}$$

Remark 4.9.1 The function $\beta = \beta(\tau)$ also satisfies the equation $2\beta\beta'' - \beta'^2 + 4\tau^n\beta^2 = 4$. It follows from the identity

$$0 = 2\beta(\beta''' + 4\tau^n\beta' + 2n\tau^{n-1}\beta) = (2\beta\beta'' - \beta'^2 + 4\tau^n\beta^2)'$$

(i.e., $2\beta\beta'' - \beta'^2 + 4\tau^n\beta^2 = \text{const}$); hence the value of the constant is clear from (4.9.17).

There exists a solution β^* of the homogeneous equation (4.9.15) with the asymptotics (4.9.17) as $\tau \to -\infty$ but it does not coincide (even up to a multiple) with β. This is a well-known fact [14].

In this connection since the following construction of M employs (4.9.17) we want to cite a reservation: we will reduce $X(M)$ to the normal form separately in domains $\tau \geq 0$ and $\tau \leq 0$ and in what follows for definiteness' sake we will speak about the domain $\tau \geq 0$ (for $\tau \leq 0$ the construction is completely analogous) having in mind that at $\tau = 0$ ($x = 0$) asymptotics are to be "matched" by continuity as, e.g., in Langer's problem 3.20 (a generalization of the normal form in this problem is also analogous to 3.20).

With the help of β let us construct

$$H = \left(\beta v - \frac{\beta'}{2}u\right)\frac{\partial}{\partial u} + \left(\frac{\beta'}{2}v - \left(\frac{\beta''}{2} + \tau^n\beta\right)u\right)\frac{\partial}{\partial v}, \tag{4.9.18}$$

permutable with X_0 [it is a solution of (4.9.13) for $\bar{Y} = 0$; see formulas (4.9.14)], which will play the leading part in the construction of M.

Returning to (4.9.12) let us first solve the auxiliary equation (4.9.13), choosing for A a particular solution which grows no faster than $a\tau$, where $a = \text{const}$. The existence of such a solution is clear from (4.9.15) and assumptions on the smoothness of $\bar{A}$, $\bar{B}$, and $\bar{C}$.

Now put

$$S = Y - \chi(\tau)H, \tag{4.9.19}$$

where $\chi = \chi(\tau)$ is chosen from the boundedness (smoothness) condition on the function $\tau^n(A - \chi\beta)$. Coefficients in S are [see (4.9.14) and (4.9.18)]

$$A - \chi\beta, \qquad -\tfrac{1}{2}(A - \chi\beta)' - \tfrac{1}{2}\chi'\beta - \tfrac{1}{2}\bar{A},$$
$$-\tfrac{1}{2}(A - \chi\beta)'' - \tfrac{1}{2}\chi''\beta - \chi'\beta' - \tfrac{1}{2}\bar{A}' - \tau^n(A - \chi\beta) + \bar{B}. \tag{4.9.20}$$

Since $\tau^n(A - \chi\beta)$ is bounded, $\beta = 0(\tau^{-n/2})$, and χ grows no faster than $\tau^{1+n/2}$, it is easy to see that coefficients of (4.9.20) in S are bounded (smooth).

By (4.9.19), (4.9.13), and (4.9.12) we will have

$$M = -\chi'(\tau)H \qquad (M = M_j,\ \chi = \chi_j).$$

Remark 4.9.2 Coefficients in M grow according to our estimates (in fact, more slowly) but at any rate, their growth is uniformly bounded in j (the largest coefficient is that of $u(\partial/\partial v)$, and it grows no faster than τ^n) so that the series $\varepsilon^2 M_2 + \cdots$ has an asymptotic meaning.

Thus, X can be reduced in the domain $0 \le \tau \le +\infty$ (and, similarly, in the domain $-\infty \le \tau \le 0$) to the normal for

$$X_0 + \varepsilon\frac{\partial}{\partial x} + \varepsilon^2 k(\tau, x, \varepsilon)H \tag{4.9.21}$$

[the parameter x will enter $k(\tau, x, \varepsilon)$ through $g(x)$, $g'(x)$, $g''(x)$, ...].

Then, since $S\tau = Sx = H\tau = Hx = 0$, we are still in the class of extended operators. We can again put [see (4.9.7) and (4.9.8)] $x = \varepsilon\tau$, considering instead of the extended operator (4.9.21) the operator

$$v\frac{\partial}{\partial u} - \tau^n u\frac{\partial}{\partial v} + \frac{\partial}{\partial \tau} + \varepsilon^2 k(\tau, \varepsilon)H$$

to which the system

$$\frac{du}{d\tau} = v + \varepsilon^2\hat{k}(\tau,\varepsilon)\left(\beta v - \frac{\beta'}{2}u\right)$$
$$\frac{dv}{d\tau} = -\tau^n u + \varepsilon^2\hat{k}(\tau,\varepsilon)\left(\frac{\beta'}{2}v - \left(\frac{\beta''}{2} + \tau^n\beta\right)u\right), \tag{4.9.22}$$

where $\hat{k}(\tau,\varepsilon) = k(\tau, \varepsilon\tau, \varepsilon)$, corresponds.

Integrating the System in the New Variables

Choose a fundamental system of solutions $\alpha = \alpha(\tau)$, $\bar{\alpha} = \bar{\alpha}(\tau)$ of (4.9.16) so that $\beta = \alpha\bar{\alpha}$ and let w be the Wronski determinant $\alpha\bar{\alpha}' - \alpha'\bar{\alpha} = w$ (= const).

Now put $u = p\alpha + \bar{p}\bar{\alpha}$, $v = p\alpha' + \bar{p}\bar{\alpha}'$, where p and $\bar{p}$ are new variables. Making use of the relations $\alpha\bar{\alpha}' - \alpha'\bar{\alpha} = w, \alpha'' + \tau^n\alpha = 0$, $\bar{\alpha}'' + \tau^n\bar{\alpha} = 0$,

$$\alpha\bar{\alpha} = \beta, \qquad \alpha'\bar{\alpha}' = \frac{\beta''}{2} + \tau^n\beta, \qquad 2\beta\beta'' - \beta'^2 + 4\tau^n\beta^2 = 4$$

(see Remark 4.9.1), Equations (4.9.22) can be written in variables p and $\bar{p}$ in the form

$$\frac{dp}{d\tau} = \frac{2\varepsilon^2}{w}\hat{k}(\tau,\varepsilon)p, \qquad \frac{d\bar{p}}{d\tau} = -\frac{2\varepsilon^2}{w}\hat{k}(\tau,\varepsilon)\bar{p},$$

and can be integrated immediately.

Remark 4.9.3 Integrability of (4.9.22) is explained by the permutability of X_0 and H, thanks to which these operators have a common invariant

$$\left(\frac{\beta''}{2} + \tau^n \beta\right) u^2 + \beta v^2 - \beta' uv \qquad (= -w^2 p\bar{p}),$$

and the system (4.9.22) has the integral $p\bar{p} = \text{const}$.

Remark 4.9.4 We have considered a formal construction of the asymptotics of the general solution. Since exponential asymptotics are possible here (for $x < 0$ and odd n), we ought to have in mind the well-known problem of a "one-way type of matching formulas" [39], which we do not deal with here. This remark also applies to Example 4.12.

It has been said above that a characteristic feature of the reconstruction problems is the necessity of using a change of variables (singular at $\varepsilon = 0$). In certain cases, the reconstruction of an operator can be performed by nondegenerate changes, so that the singularity is hidden by implicitness of a solution. The following example is of this kind (though it can be investigated by a "canonical" way). This example is also interesting since for a certain formulation of the problem, reconstruction needs to be performed an "infinite" number of times.

4.10 Example: Lighthill's Problem [38]

$$(x + \varepsilon y)\frac{dy}{dx} + (2 + x)y = 0, \quad y(1) = \frac{1}{\ell}. \tag{4.10.1}$$

Here we will assume that ε can be both a positive and a negative small parameter and a solution is sought in the domain $a \ll x \ll 1$, where $a = a(\varepsilon)$ is sought as a "smallest" value for which the perturbation theory fails. In general, though, we will pay attention to the case $\varepsilon > 0$, $a \leq 0$ and the computation of $y(0)$.

Passing from (4.10.1) to the system

$$\frac{dx}{dt} = x + \varepsilon y, \qquad \frac{dy}{dt} = -(2 + x)y,$$

we get the operator

$$X = x\frac{\partial}{\partial x} - (2 + x)y\frac{\partial}{\partial y} + \varepsilon y\frac{\partial}{\partial x}. \tag{4.10.2}$$

The leading operator has the eigenfunction x corresponding to the eigenvalue 1 and the invariant $g = yx^2e^x$. In these variables

$$X = x\frac{\partial}{\partial x} + \varepsilon\left(\frac{ge^{-x}}{x^2}\frac{\partial}{\partial x} + \frac{g^2e^{-x}(2 + x)}{x^3}\frac{\partial}{\partial g}\right).$$

From here it is immediately clear that the reconstruction takes place for $x \sim |\varepsilon|^{1/3}$. Then, since $y = gx^{-2}e^{-x}$, we have $y \sim 1/\Delta^2$ for $x \sim \Delta$ [when Δ is small; we start from $g \sim 1$; see (4.10.1)], and the reconstruction in old variables is $x = |\varepsilon|^{1/3}\tau$, $y = |\varepsilon|^{-2/3}z$ for $\tau \sim 1$, $z \sim 1$.

But we have no intention of making this transformation, noting only that $x \sim \varepsilon y$ for the reconstructed problem.

This and (4.10.2) make it possible to see (it is also possible to undertake temporarily the change $x = |\varepsilon|^{1/3}\tau$, $y = |\varepsilon|^{-2/3}z$) that the leading operator becomes

$$(x + \varepsilon y)\frac{\partial}{\partial x} - 2y\frac{\partial}{\partial y} \tag{4.10.3}$$

[and $-xy(\partial/\partial y)$ is "small" for a small x]. The operator (4.10.3) has eigenfunctions y corresponding to -2 and

$$x_1 = x + \tfrac{1}{3}\varepsilon y \tag{4.10.4}$$

corresponding to 1.

In variables y and x_1 [note that the change (4.10.4) is not singular] the operator (4.10.2) takes the form

$$X = x_1\frac{\partial}{\partial x_1} - (2 + x_1)y\frac{\partial}{\partial y} + \frac{\varepsilon}{3}y\left(y\frac{\partial}{\partial y} - x_1\frac{\partial}{\partial x_1}\right) + \frac{\varepsilon^2}{9}y^2\frac{\partial}{\partial x_1}. \tag{4.10.5}$$

Here the operator (4.10.3) is $x_1(\partial/\partial x_1) - 2y(\partial/\partial y)$ and the summand $-x_1y(\partial/\partial y)$ would have been part of the perturbation had we not performed the dilatation (see above).

Now we can again consider X in the same manner taking $x_1(\partial/\partial x_1) - (2 + x_1)y(\partial/\partial y)$ as a leading operator in (4.10.5).

This operator has an eigenfunction x_1 corresponding to 1 and an invariant $g_1 = yx_1^2e^{x_1}$. In these variables, (4.10.5) is of the form

$$\begin{aligned} X = {} & x_1\frac{\partial}{\partial x_1} - \varepsilon\frac{g_1e^{-x_1}}{3x_1}\left(\frac{\partial}{\partial x_1} + \frac{1 + x_1}{x_1}g_1\frac{\partial}{\partial g_1}\right) \\ & + \frac{\varepsilon^2 g_1^2 e^{-2x_1}}{9x_1^4}\left(\frac{\partial}{\partial x_1} + \frac{2 + x_1}{x_1}g_1\frac{\partial}{\partial g_1}\right). \end{aligned} \tag{4.10.5'}$$

From here we deduce that the reconstruction is now performed at $x_1 \sim |\varepsilon|^{2/5}$; i.e., we can make use of the leading operator

$$x_1\frac{\partial}{\partial x_1} - (2 + x_1)y\frac{\partial}{\partial y}$$

in the domain $x_1 \gg |\varepsilon|^{2/5}$, in particular, for $x_1 \sim |\varepsilon|^{1/3}$. If $\varepsilon > 0$, then the latter means that we have entered the domain $-\varepsilon^{1/3} \leq x \leq 1$ [see (4.10.4)]; for $\varepsilon > 0$, the singular point of Equation (4.10.1) $\xi + \varepsilon y(\xi) = 0$ is situated to the left of 0, which is clear from the consideration of the principal term of the asymptotics

$$g_1 = y\left(x + \frac{1}{3}\varepsilon y\right)^2 e^x = 1 \qquad [yx_1^2 e^{x_1} \approx 1,\ e^{x_1} \approx e^x,\ y(1) = \frac{1}{e}]$$

In the domain $0 \leq x \leq 1$, we can (see below) find the asymptotics and $y(0)$ [if $\varepsilon < 0$, the singular point is situated to the right of 0, and for $x_1 \sim |\varepsilon|^{1/3}$ we get $x \sim (-\varepsilon)^{1/3}$].

Remark 4.10.1 Making use of (4.10.5) we reach the domain $x_1 \sim |\varepsilon|^{2/5}$, $y \sim |\varepsilon|^{-4/5}$ and the leading operator becomes

$$x_1\frac{\partial}{\partial x_1} - 2y\frac{\partial}{\partial y} + \frac{\varepsilon^2}{9}y^2\frac{\partial}{\partial x_1}.$$

It has eigenfunctions y and $x_2 = x_1 + (\varepsilon^2/45)y^2$. In these variables,

$$X = x_1\frac{\partial}{\partial x_2} - (2 + x_2)y\frac{\partial}{\partial y} + \frac{\varepsilon}{3}y\left(y\frac{\partial}{\partial y} - x_2\frac{\partial}{\partial x_2}\right) + \frac{\varepsilon^2 y^2}{45}\left(y\frac{\partial}{\partial y} - 2x_2\frac{\partial}{\partial x_2}\right) + \frac{\varepsilon^3 y^3}{45}\frac{\partial}{\partial x_2} + \frac{2\varepsilon^4 y^4}{2025}\frac{\partial}{\partial x_2},$$

and the leading operator $x_2(\partial/\partial x_2) - (2 + x_2)y(\partial/\partial y)$ can be used when $x_2 \gg |\varepsilon|^{3/7}$.

This reconstruction process can be continued. At the n-th step the reconstruction is performed at $x_n \sim |\varepsilon|^{n+1/2n+3}$, $y \sim |\varepsilon|^{-2n+2/2n+3}$ $(x_0 \equiv x)$. In the limit as $n \to \infty$ we get $a(\varepsilon) \sim -\sqrt{\varepsilon}$ $(\varepsilon > 0)$ and $a(\varepsilon) \sim \sqrt{-\varepsilon}$ $(\varepsilon > 0)$.

Let us also discuss the computations for $\varepsilon > 0$ and $-\varepsilon^{1/3} \leq x \leq 1$. The principal term of the asymptotics, as has been already mentioned, is

$$g_1 = y_0\left(x + \frac{1}{3}\varepsilon y_0\right)^2 e^x = 1 \qquad (e^{x_1} \approx e^x),$$

implying $y_0(0) \approx (3/\varepsilon)^{2/3}$.

The construction of higher terms is achieved by the development $G = g_1 + G_1 + G_2 + \cdots$ of the desired invariant of X, where

$$X_0 G_1 + X_1 g_1 = 0, \qquad X_0 G_2 + X_1 G_1 = 0, \qquad \ldots,$$

and we have put [see (4.10.5′)] $X_0 = x_1(\partial/\partial x_1)$, $X_1 = X - X_0$, and X_1 is considered to be of "order" $\varepsilon^{1/3}$ for $y \sim \varepsilon^{-2/3}$ and $x_1 \sim \varepsilon^{1/3}$.

Depending on the assumed accuracy of computations, not all equations appearing for G_j should be solved exactly. For instance, instead of the exact solution

$$G_1 = \frac{g_1^2}{6}\left(\int_{x_1}^{\infty}\frac{e^{-t}}{t}\,dt - \frac{e^{x_1}(1 + x_1)}{x_1^2}\right)\varepsilon + \frac{g_1^3 e^{-2x_1}(24 + 3x_1 - 2x_1^2)}{540x_1^5}\varepsilon^2,$$

we can take

$$G_1 = \bar{G}_1 \equiv -\frac{g_1^2 e^{-x_1}}{6x_1^2}\varepsilon + \frac{2g_1^3 e^{-2x_1}}{45x_1^5}\varepsilon^2$$

if we ignore in G terms $\sim 0(\varepsilon^{2/3})$. Then, putting $G = h + \bar{G}_1 = C$, where $C = 1 + 0(\varepsilon)$ is a constant,

$$x_1 = x + \tfrac{1}{3}\varepsilon y, \qquad e^{x_1} = e^x\left(1 + \tfrac{1}{3}\varepsilon y + \cdots\right), \qquad y = y_0 + y_1 + \cdots,$$

where y_0 is a principal term and $y_1/y_0 \sim \varepsilon^{1/3}$, we get y with a relative error $0(\varepsilon^{2/3})$ on the whole segment $-\varepsilon^{1/3} \le x \le 1$:

$$y = y_0 - \frac{\varepsilon y_0^2(5x + 3\varepsilon y_0)}{30(x + \varepsilon y_0)}.$$

In particular,

$$y(0) = \left(\frac{3}{\varepsilon}\right)^{2/3} - \frac{3}{10}\left(\frac{3}{\varepsilon}\right)^{1/3} + 0(1).$$

Remark 4.10.2 As is clear (for example) from the exact form of G_1, in computing further terms, $\varepsilon \ln(x + \frac{1}{3}\varepsilon y_0)$ will appear.

The next example is a simple illustration of problems in which coefficients of X have singularities. In these problems it is often possible to make use of an appropriate change of variables to reduce the question to a reconstruction of the first type (or, in the case of resonance i-points, of the second type).

4.11 Example: Singularity of Coefficients of an Operator

$$\frac{du}{dx} = -\frac{\varepsilon u}{(1 - xu)^2}, \qquad u(0) = 1,\ 0 \le x \ll a(\varepsilon) \tag{4.11.1}$$

[as in the previous problem we are also interested in the "largest possible" value of $a(\varepsilon)$].

The operator corresponding to (4.11.1) is

$$X = \frac{\partial}{\partial x} - \varepsilon\frac{u}{(1 - xu)^2}\frac{\partial}{\partial u}.$$

The singular surface is $1 - xu = 0$. A change of variables

$$\frac{1}{1 - xu} = \tau \tag{4.11.2}$$

suggests itself. In variables τ and u $(u > 0)$ we get $X = \tau^2 Y$, where

$$Y = u\frac{\partial}{\partial\tau} - \varepsilon\left\{u\frac{\partial}{\partial u} + (\tau^2 - \tau)\frac{\partial}{\partial\tau}\right\}.$$

The leading operator $u(\partial/\partial\tau)$ is nilpotent and Y is normal with respect to $u(\partial/\partial\tau)$: the heights of the operators $u(\partial/\partial u)$, $\tau^2(\partial/\partial\tau)$, and $\tau(\partial/\partial\tau)$ are 2, 3, and 2 respectively.

Since the height of $\tau^2(\partial/\partial\tau)$ is maximal, we can take $u - \varepsilon\tau^2(\partial/\partial\tau)$ as a new leading operator. Its eigenfunction and invariant are

$$\varphi = \frac{\sqrt{u} - \sqrt{\varepsilon}\,\tau}{\sqrt{u} + \sqrt{\varepsilon}\,\tau}$$

and u. Putting

$$\delta = \sqrt{\varepsilon}, \qquad u = v^2, \qquad \varphi = \frac{v - \delta\tau}{v + \delta\tau}, \tag{4.11.3}$$

we get $Y = -2\delta v Z$, where

$$Z = Z_0 + \delta Z_1 = \varphi\frac{\partial}{\partial\varphi} + \delta\left\{\frac{3}{8v}(1-\varphi^2)\frac{\partial}{\partial\varphi} + \frac{1}{4}\frac{\partial}{\partial v}\right\}. \tag{4.11.4}$$

Now asymptotics are obtained in a standard fashion.

In particular, from (4.11.4) it is easy to deduce that the principal term of the asymptotics is

$$\varphi = \varphi_0 \exp\frac{4(v-1)}{\delta}, \quad \varphi_0 = \text{const}, \qquad (\text{here } M_1 = \tfrac{1}{4}\tfrac{\partial}{\partial v}),$$

and, according to (4.11.3) and (4.11.2) we have

$$x \simeq \frac{1}{u}\left(1 - \sqrt{\varepsilon/u}\, cth\left(\frac{2(1-\sqrt{u})}{\sqrt{\varepsilon}} + \sqrt{\varepsilon}\right)\right) \qquad \text{for } u(0) = 1.$$

Remark 4.11.1 As is clear from (4.11.4), asymptotic theory fails for $v \sim \delta$; i.e., for $u \sim \varepsilon$. Hence $a(\varepsilon) \sim 1/\varepsilon$.

Now it is not difficult to compute higher terms of the asymptotics.

In the literature the equation $\varepsilon y'' + p(x,\varepsilon)y' + q(x,\varepsilon)y = 0$, where $p(0,0) = 0$ and, perhaps, $q(0,0) = 0$, is often encountered. The asymptotics of its solutions have a quite diverse character. The change of the desired function

$$y \to y \exp\left(-\frac{1}{2\varepsilon}\int_0^x p(\xi,\varepsilon)\,d\xi\right)$$

reduces this equation to the WKB type: $\varepsilon^2 y'' = P(x;\varepsilon)y$. We turn in the next example to the special case $\varepsilon y'' + p(x)y' + q(x)y = 0$; more exactly, to a comparatively simple model equation $\varepsilon y'' + \mu x^m y' + \nu x^n y = 0$, where $m > 0$ and $n \geq 0$, and μ and ν are constants, considering real values of x: $0 \leq x \leq x_0 \sim 1$ (the case $x < 0$ is considered similarly; asymptotics are "matched" by continuity as in the Langer problem 3.20). We will not study concrete problems for this equation, noting only these main features: reconstructions and the possibility of reducing operators (systems) to the normal form. For simplicity's sake we will also restrict ourselves to the case $\mu = +1$ (an appropriate choice of scales for x and ε gives $|\mu| = 1$; the case $\mu = -1$ is discussed like $\mu = +1$).

4.12 Example: A Second Order Linear Equation

$$\varepsilon y'' + x^m y' + \nu x^n y = 0$$
$$(0 \leq x \leq x_0 \sim 1,\ m > 0,\ n \geq 0,\ \nu = \text{const}). \tag{4.12.1}$$

Passing from (4.12.1) to the system

$$\varepsilon \frac{dx}{dt} = \varepsilon, \qquad \varepsilon \frac{dy}{dt} = z, \qquad \varepsilon \frac{dz}{dt} = -x^m z - \varepsilon \nu x^n y,$$

we get the operator

$$X = X_0 + \varepsilon X_1 = z\frac{\partial}{\partial y} - x^m z \frac{\partial}{\partial z} + \varepsilon\left(\frac{\partial}{\partial x} - \nu x^n y \frac{\partial}{\partial z}\right).$$

As had been mentioned in Section 4.1, $x = 0$ is a singular surface. Putting $x = \Delta\tau$ we have

$$X = z\frac{\partial}{\partial y} - \Delta^m \tau^m z \frac{\partial}{\partial z} + \frac{\varepsilon}{\Delta}\frac{\partial}{\partial \tau} - \nu\varepsilon\Delta^n \tau^n y \frac{\partial}{\partial z},$$

and for a small Δ (but $\Delta \gg \varepsilon$) the nilpotent operator $\bar{X}_{00} = z(\partial/\partial y)$ becomes a leading one [see (4.3.2)] with respect to which X is normal.

Making use of the algebraic reconstruction method from Section 4.3, note that $X_0 = \bar{X}_{00} + \Delta^m \bar{X}_{01}$, where $\bar{X}_{01} = -\tau^m z(\partial/\partial z)$, is in a non-canonical normal form with respect to $\bar{X}_{00}$ in the sense of minimal height of perturbation: the height of $\bar{X}_{01}$ is 2 and canonical operators of minimal height are $y(\partial/\partial z)$ (height 3) and $y(\partial/\partial y) + z(\partial/\partial z)$ (height 1). Clearly, we can reduce X_0 to the canonical normal form

$$\bar{X}_{00} + \Delta^m c\tau^m \left(y\frac{\partial}{\partial y} + z\frac{\partial}{\partial z}\right) + 0(\Delta^{2m}),$$

but this is not necessary: it is clear that the "principal part" of X_0 (for a small Δ) is X_0 itself. We can at once apply a shearing transformation:

$y = \Delta^{-m/2}u$, $z = \Delta^{m/2}v$, "leveling" $\bar{X}_{00}$ and $\Delta^m \bar{X}_{01}$ (to $\bar{X}_{01}$, the height 1 is formally assigned). We get

$$X = \Delta^m Y = \Delta^m \left(v\frac{\partial}{\partial u} - \tau^m v \frac{\partial}{\partial v} + \frac{\varepsilon}{\Delta^{m+1}} \frac{\partial}{\partial \tau} - \frac{\varepsilon}{\Delta^{2m-n}} \nu \tau^n u \frac{\partial}{\partial v} \right),$$

and find Δ_{cr} as the maximal Δ for which in Y terms ~ 1 different from $v(\partial/\partial u) - \tau^m v(\partial/\partial v)$ ("the principal part" of $\Delta^{-m} X_0$) arise.

Depending on values of m and n we get three variants of reconstruction:

(a) $n > m - 1$. Then $\Delta_{cr} = \varepsilon^{1/m+1}$,

$$Y = Y_0 + \delta Y_1 = v\frac{\partial}{\partial u} - \tau^m v \frac{\partial}{\partial v} + \frac{\partial}{\partial \tau} - \delta \nu \tau^n u \frac{\partial}{\partial v},$$
$$\delta = \varepsilon^{n-m+1/m+1}. \tag{4.12.2}$$

(b) $n = m - 1$. Then $\Delta_{cr} = \varepsilon^{1/m+1}$; the problem loses a small parameter:

$$Y = v\frac{\partial}{\partial u} - (\tau^m v + \nu \tau^n u)\frac{\partial}{\partial v} + \frac{\partial}{\partial \tau}. \tag{4.12.3}$$

(c) $n < m - 1$. Then $\Delta_{cr} = \varepsilon^{1/2m-n}$,

$$Y = Y_0 + \delta Y_1 = v\frac{\partial}{\partial u} - (\tau^m v + \nu \tau^n u)\frac{\partial}{\partial v} + \delta \frac{\partial}{\partial \tau},$$
$$\delta = \varepsilon^{m-1-n/2m-n}. \tag{4.12.4}$$

The new leading operator fits in the whole domain since it "includes" the old operator.

The Case $n > m - 1$

This case is the simplest one. The leading operator Y_0 has the invariant $g = u + c(\tau)v$, where $c' - \tau^m c = -1$. Put

$$c(\tau) = e^{\tau^{m+1}/m+1} \int_\tau^{+\infty} e^{-t^{m+1}/m+1}\, dt. \tag{4.12.5}$$

In variables g, v, and τ [see (4.12.2)] we have

$$Y_0 = -\tau^m v \frac{\partial}{\partial v} + \frac{\partial}{\partial \tau},$$
$$Y_1 = \nu \tau^n c(\tau) \left(v\frac{\partial}{\partial v} - g\frac{\partial}{\partial g} \right) - \nu \tau^n \left(g \frac{\partial}{\partial v} - c^2(\tau) v \frac{\partial}{\partial y} \right). \tag{4.12.6}$$

The operator $Y = Y_0 + \delta Y_1$ can be reduced to the following (generalized) normal form M with respect to Y_0:

$$M = Y_0 + \delta \tau^n x(\tau; \delta) H, \qquad H = v\frac{\partial}{\partial v} - g\frac{\partial}{\partial g} \tag{4.12.7}$$

where $x(\tau,\delta) = \nu c(\tau) + \delta c_1(\tau) + \cdots$ and H commutes with Y_0. For this, it suffices to seek S_j in the form

$$S_j = A_j(\tau) g \frac{\partial}{\partial v} + B_j(\tau) v \frac{\partial}{\partial g}$$

enforcing, naturally, on solutions of equations for A_j and B_j [where $A_j' + \tau^m A_j = -\tau^n \bar{A}_{j-1}$, $B_j' - \tau^m B_j = -\tau^n \bar{B}_{j-1}$ and $\bar{A}_{j-1}, \bar{B}_{j-1}$ are recursively known, $\bar{A}_0 = -\nu$, $\bar{B}_0 = \nu c^2(\tau)$], conditions of weakest growth as $\tau \to +\infty$.

It is not difficult to show by induction, taking (4.12.5) and (4.12.6) into account, that A_j, B_j, and $c_j(\tau)$ grow at a rate no faster than $0(\tau^{(n-m-1)j})$; hence, the normal form (4.12.7) can be used for $\delta\tau^{n-m-1} \ll 1$ (i.e., for $x^{n-m-1}\varepsilon^{2/m+1} \ll 1$, where

$$x = \Delta_{\text{cr}}\tau = \varepsilon^{1/(m+1)}\tau, \qquad \delta = \varepsilon^{(n-m+1)/(m+1)}),$$

in particular, for $x \sim 1$.

The averaged system corresponding to M,

$$\frac{dv}{d\tau} = \tau^n[-1 + \delta x(\tau,\delta)]v, \qquad \frac{dg}{d\tau} = -\delta\tau^n x(\tau,\delta) g,$$

has, in all orders, a first integral

$$gv \exp\frac{\tau^{n+1}}{n+1} = \text{const}$$

due to commutability of Y_0 and H, and is clearly easy to solve.

The Case $n = m - 1$

In this case we have to solve exactly the equation [24]

$$\frac{d^2u}{d\tau^2} + \tau^m \frac{du}{d\tau} + \nu\tau^{m-1}u = 0 \tag{4.12.8}$$

corresponding to (4.12.3). The change $\zeta = \tau^{m+1}/(m+1)$ puts it into the form

$$\zeta\frac{d^2u}{d\zeta^2} + (\alpha + \beta + \zeta)\frac{du}{d\zeta} + \alpha u = 0, \qquad \alpha = \frac{\nu}{m+1}, \qquad \beta = \frac{m-\nu}{m+1},$$

which is integrated by the Laplace method [14, 43]. For example, functions

$$u_1 = \frac{1}{\Gamma(\alpha)\Gamma(\beta)}\int_0^1 t^{\alpha-1}(1-t)^{\beta-1}e^{-\zeta t}\,dt,$$

$$u_2 = \frac{1}{\Gamma(\beta)}\int_1^\infty t^{\alpha-1}(t-1)^{\beta-1}e^{-\zeta t}\,dt$$

where $\Gamma(\gamma)$ is the Euler gamma function and integrals (extended analytically) are considered as meromorphic functions of α and β, constitute a fundamental system of solutions. Then u_2, as a solution satisfying the condition of fastest decrease as $\zeta \to +\infty$, is unique (up to a constant multiple).

Remark 4.12.1 It we put $u = \bar{u}\exp(-\tau^{m+1}/(m+1))$, and $\bar{\nu} = \nu - m$ in (4.12.8), then we get for $\bar{u}$ the equation

$$\frac{d^2\bar{u}}{d\tau^2} - \tau^m \frac{d\bar{u}}{d\tau} + \bar{\nu}\tau^{m-1}\bar{u} = 0$$

corresponding to the case $\mu = -1$ (see the introductory paragraph).

The Case $n < m - 1$

The leading operator

$$Y_0 = v\frac{\partial}{\partial u} - (\tau^m v + \nu\tau^n u)\frac{\partial}{\partial v}$$

has two eigenfunctions $\varphi_{1,2} = \tau^n u - \nu\lambda_{1,2}(\tau)v$; i.e., $Y_0\varphi_i = \lambda_i(\tau)\varphi_i$, where eigenvalues $\lambda_{1,2}(\tau)$ are roots of the equation

$$\lambda^2 + \tau^m\lambda + \nu\tau^n = 0. \tag{4.12.9}$$

In the domain $\lambda_1 \neq \lambda_2$, the procedure of the reduction of Y [(4.12.4)] to the normal form with respect to Y_0 is a standard one.

But in the domain considered, new singular surfaces ($\lambda_1 = \lambda_2$) can arise:

$$\tau^{2m} - 4\nu\tau^n = 0 \tag{4.12.10}$$

[see (4.12.9)], and new reconstructions will be needed.

Equation (4.12.10) can have 0 as a root (if $n > 0$) and roots $\tau_0 \neq 0$. First, consider the reconstruction of the operator (4.12.4) in a neighborhood of the singular surface $\tau = \tau_0 = (4\nu)^{1/2m-n} \neq 0$.

Set

$$u = \tilde{u}\exp\left(-\frac{\tau_0^m}{2\delta}\tau\right), \qquad v = \left(\tilde{v} - \frac{\tau_0^m}{2}\tilde{u}\right)\exp\left(-\frac{\tau_0^m}{2\delta}\tau\right), \tag{4.12.11}$$

where $\tilde{u}$ and $\tilde{v}$ are new variables. Then, since $\tau_0^{2m} - 4\nu\tau_0^n = 0$ [see (4.12.10) and (4.12.11)] the operator (4.12.4) takes the form

$$Y = \tilde{v}\frac{\partial}{\partial\tilde{u}} + \left\{(\tau_0^m - \tau^m)\tilde{v} + \left(\frac{1}{2}\tau_0^m(\tau^m - \tau_0^m) + \nu(\tau_0^n - \tau^n)\right)\tilde{u}\right\}\frac{\partial}{\partial\tilde{v}} + \delta\frac{\partial}{\partial\tau}, \tag{4.12.12}$$

so that for a small $|\tau_0 - \tau|$ the canonical nilpotent operator $\tilde{v}(\partial/\partial\tilde{u})$ becomes the leading one in (4.12.12).

Now, putting $\tau = \tau_0 + \Delta\sigma$, we get that the "principal part" of Y_0 [the height of $\tilde{v}(\partial/\partial\tilde{v})$ is 2 and that of $\tilde{u}(\partial/\partial\tilde{v})$ is 3] is

$$\tilde{v}\frac{\partial}{\partial u} + \left(\frac{1}{2}m\tau_0^{2m-1} - \nu n\tau_0^{n-1}\right)\Delta\sigma\tilde{u}\frac{\partial}{\partial v} = \tilde{v}\frac{\partial}{\partial\tilde{u}} + \nu(2m-n)\tau_0^{n-1}\Delta\sigma\tilde{u}\frac{\partial}{\partial\tilde{v}} = \tilde{Y}_0$$

and the new leading operator is $\tilde{Y}_0 + (\delta/\Delta)(\partial/\partial\sigma)$, $\Delta = \Delta_{cr} = \delta^{2/3}$, whereas the shearing transformation is $\tilde{v} = \hat{v}\Delta^{1/4}$ and $\tilde{u} = \hat{u}/\Delta^{1/4}$.

To the new leading operator the Airy equation will evidently correspond. The problem of constructing the asymptotics in the domain containing the surface $\tau = \tau_0$ can be solved by matching the exterior asymptotics, obtained by the standard relation of Y to the normal form, with the inner ones, obtained by a power series development as has been done in Example 4.6. (In this problem the matching procedure is easier in principle since the equation is a linear one.)

The reconstruction in a neighborhood of $\tau = 0$ is similarly performed: putting $\tau = \Delta\rho$, we have

$$Y = v\frac{\partial}{\partial u} - (\Delta^m \rho^m v + \nu \Delta^n \rho^n u)\frac{\partial}{\partial v} + \frac{\delta}{\Delta}\frac{\partial}{\partial \rho}$$

and here the "principal part," Y_0, is

$$v\frac{\partial}{\partial u} - \nu \Delta^n \rho^n u \frac{\partial}{\partial v},$$

where $\Delta = \Delta_{cr} = \delta^{2/(n+2)}$. The shearing transformation is

$$v = \delta^{n/2(n+2)} q, \qquad u = \delta^{-n/2(n+2)} p.$$

The operator is

$$Y = \delta^{n/(n+2)} \left\{ q\frac{\partial}{\partial p} - \nu \rho^n p \frac{\partial}{\partial q} + \frac{\partial}{\partial \rho} - \delta^{(2m-n)/(n+2)} \rho^m q \frac{\partial}{\partial q} \right\}.$$

The problem arising is similar to Section 4.9, but here it is not possible, it seems, to obtain the asymptotics without matching (due to ρ^m in the perturbation). New reconstructions do not arise and the complete construction of the asymptotics can be carried out.

The next example—the problem of relaxation oscillations in a Van der Pol oscillator—illustrates once again applications of some of the above-mentioned arguments. This problem had been first solved in the well-known paper of A. A. Dorodnitsyn [15] with the use of some ad hoc tricks. The presentation below pursues only methodical goals and for that reason is perhaps too detailed (we skipped only some simple, easy to verify, computations).

4.13 Example: Van der Pol Oscillator (Relaxation Oscillations)

Here we consider the system

$$\varepsilon\frac{dx}{dt} = f(x,y) \equiv y + x - \frac{1}{3}x^3, \qquad \frac{dy}{dt} = -x \qquad (\varepsilon > 0), \qquad (4.13.1)$$

equivalent to the equation $\varepsilon\ddot{y} + (\dot{y} + \frac{1}{3}\dot{y}^3 + y) = 0$ and also to $\varepsilon\ddot{x} - (1 - x^2)x + x = 0$, which are called *Van der Pol equations* [44].

Preliminaries

In Figure 4.1 a typical phase trajectory of the point $P(x, y)$ starting from a "general position" is qualitatively depicted. The nature of the trajectory is intuitively clear from (4.13.1) if we take into account that "far" from

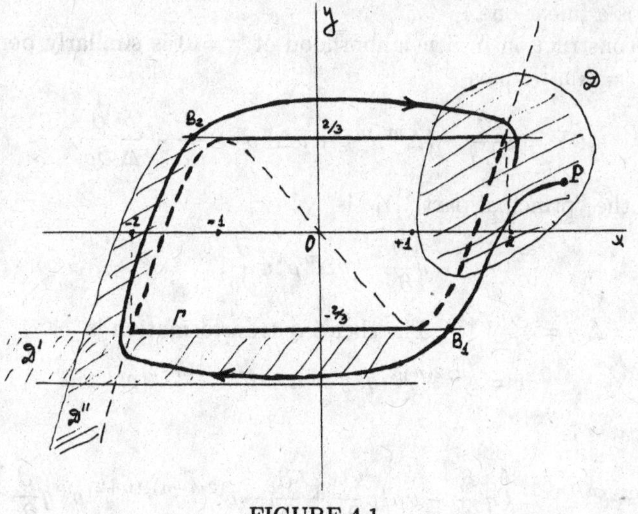

FIGURE 4.1.

the curve $f(x, y) = 0$ depicted by dots on the picture the velocity of P is great and is almost parallel to the x-axis. For small ε the trajectory is almost closed (as will be clear a little later, up to exponentially small terms) and the point encircles the origin clockwise, remaining near the contour Γ formed by parts of the curve $f(x, y) = 0$, where $1 \leq x \leq 2$, $-1 \leq x \leq 2$, and $-2 \leq x \leq 1$, respectively. The situation is called *relaxation oscillations*, which represent a special type of almost-discontinuous self-exciting oscillations [44].

The computation of the "period" $T(\varepsilon)$ of these oscillations is of main interest in this problem. By $T(\varepsilon)$ we understand henceforth the period of motion along the asymptotic relaxation trajectory which, as had been already said, will turn out to be closed in approximation when exponentially small terms are ignored.

Remark 4.13.1 For any ε there exists a limit cycle (with respect to t) ([44]) corresponding to some strictly periodic solution near the limit Γ as $\varepsilon \to 0$ of limit cycles. Some authors call the period of revolution along the limit (in t) cycle the *period of relaxation oscillations* [35].

We will consider only relaxation oscillations, being not interested in transition processes.

For the future it is worth mentioning that asymptotics of the transition process with initial values in "general position" might be constructed as in Example 3.9.

It can be done firstly in the following (canonical) way. The operator corresponding to (4.13.1) is

$$X = X_0 + \varepsilon X_1 = f(x,y)\frac{\partial}{\partial x} - \varepsilon x\frac{\partial}{\partial y}. \tag{4.13.2}$$

In the domain $\mathcal{D}$ (an example of such a domain is shown in Figure 4.1), where the equation $f(x,y) = 0$ with respect to x possesses the unique solution

$$x = \beta(y), \qquad |\beta'(y)| < c = \text{const}$$

for any y, we can construct an eigenfunction of the leading operator $X_0 = f(x,y)(\partial/\partial x)$:

$$\varphi(x,y) = (x - \beta(y))G(x,y), \tag{4.13.3}$$

$$G(x,y) = \exp\int_{\beta(y)}^{x}\left(\frac{1-\beta^2(y)}{f(\eta,y)} - \frac{1}{\eta-\beta(y)}\right)d\eta$$

corresponding to the eigenvalue

$$\lambda(y) = \left.\frac{\partial f}{\partial x}\right|_{x=\beta(y)} = 1 - \beta^2(y) \tag{4.13.4}$$

(y is an invariant of X_0) and reduce $X(M)$ to the canonical normal form

$$My = g(y,\varepsilon), M\varphi = (\lambda(y) + \varepsilon\omega_1(y) + \cdots)\varphi.$$

These computations are comparatively difficult because $x = x(\varphi, y)$ is implicit but do not contain any principal difficulties.

Secondly, we can construct asymptotics in the form of developments of the type (3.9.5) indicated in Example 3.9. It is equivalent to the canonical method but brings us to the goal somewhat faster.

Now note that the indicated methods allow one to seek asymptotics on segments of the relaxation trajectory situated "far" from points $(\pm 1, \mp 2/3)$, these points being the reconstruction points [in these points the eigenvalue (4.13.4) vanishes]. Formulas become cumbersome since they describe simultaneously asymptotics near straight and near curvilinear parts of Γ.

A way for simplification is quite evident; consider two domains of type $\mathcal{D}$; i.e., a narrow domain $\mathcal{D}'$ which fringes a straight segment of Γ, and a narrow domain $\mathcal{D}''$ which fringes a curvilinear part of Γ.

A construction of asymptotics in $\mathcal{D}'$ will be simplified considerably since here $\lambda = -3$ is constant $[y \approx \pm 2/3,\ \beta(y) \approx \pm 2]$ and the eigenfunction is expressed much more simply than in (4.13.3) since the operator in $\mathcal{D}'$ will be reconstructed due to the narrowness of $\mathcal{D}'$. It is true, though, that we should understand what the width of $\mathcal{D}'$ should be, but this will be clear after the trajectory will "exit" the neighborhood of the reconstruction point.

The construction of the asymptotics in $\mathcal{D}''$, i.e., near a curvilinear part of Γ, will be exceedingly simple. It is clear from (4.13.3) and (4.13.4) that the asymptotics in the domain of type $\mathcal{D}$ are of a boundary-value nature, and, since $\lambda < 0$, then near a curvilinear part of Γ the function φ (in new variables!) will be already exponentially small. This, incidentally, shows the closedness of the asymptotic trajectory mentioned above. Therefore, in $\mathcal{D}''$ the asymptotics coincide with the what is usually called the exterior solution; i.e., can be obtained via development in powers of ε.

But this simplification necessitates matching of asymptotics in a neighborhood of points $(\pm 2, \pm 2/3)$, which were artificially turned into the reconstruction points. This procedure is, in general, laborious.

However, we will take this way instead of constructing M in $\mathcal{D}$ making use of (4.13.3) and (4.13.4) since in this problem it is connected with simpler calculations.

Remark 4.13.2 An artificial reconstruction of this kind is often encountered in the so-called method of matching of asymptotic developments, where, as it seems to us, it often amounts to guessing.

Let us emphasize once more that until now we were speaking only about asymptotics "far" from "real" reconstruction points $(\pm 1, \pm 2/3)$, the main difficulty being in passing "through" them.

Let us mention also a symmetry of the problem, making use of which we will only calculate a part of the relaxation trajectory B_1B_2 (see Figure 4.1) from the moment when it crosses the line $y = -2/3$ near the point $(+1, -2/3)$ (putting here $t = 0$) until the moment when it crosses the line $y = +2/3$ near the point $(-1, +2/3)$, putting here $t = T(\varepsilon)/2$. (The choice of the starting point at B_1 is of no principal importance, nor is the symmetry.)

Elucidation of Computations to Follow

We begin computations from the reconstruction of the operators near the point $(+1, -2/3)$—the *first reconstruction*—and construction of asymptotics of the relaxation trajectory in a neighborhood of this point. At the same time we should also find the initial value $x(0)$, where the coordinates of B_1 are $(x(0), -2/3)$. This value is defined from the possibility of matching the asymptotics obtained with those "entering" the neighborhood of the point $(1, -2/3)$ and corresponding to $t < 0$. This matching will not be performed explicitly since the asymptotics of the "entering" part of the trajectory are not computed as yet; in a neighborhood of $(-1, 2/3)$ we have to perform such a matching, (see the third matching below) but will make use only of the fact that the trajectory "enters" from the part $x > 1$, which is sufficient to find $x(0)$.

Doing this we will pass only a small part of the trajectory from B_1 since the obtained asymptotics do not yet allow us to "go out" of a neighborhood of $(1, -2/3)$; a new (second) reconstruction is needed.

As a result of the second reconstruction we get an operator reconstructed in $\mathcal{D}'$ (see Figure 4.1) fringing a straight part of Γ, $y = -2/3$. Making use of it we will find (by the method of Example 3.9) the asymptotics applicable along the whole segment [a neighborhood of $(+1, -2/3)$ excluded] and depending on two unknown constants.

The latter will be defined matching the obtained asymptotics of the "artificial" point of reconstruction $(-2, -2/3)$.

Furthermore, we will find asymptotics near (in $\mathcal{D}''$) the curvilinear part of Γ, where $-2 < x < -1$, making use, as had been already said, of a development in powers of ε depending on one unknown constant.

This constant is found by matching the obtained asymptotics with the already known ones in a neighborhood of $(-2, -2/3)$ (the second matching).

Finally we should perform the third matching of asymptotics near the curvilinear part of Γ where $-2 < x < -1$, with the asymptotics in a neighborhood of $(-1, 2/3)$. The latter will have already been known from the symmetry of the problem: it is obtained by the change of x, y to $-x$, $-y$ and t to $t - T(\varepsilon)/2$ from asymptotics in a neighborhood of $(+1, -2/3)$ and will depend on only one constant $T(\varepsilon)$ (a shift in time). The third matching gives us $T(\varepsilon)$.

The First Reconstruction

Asymptotics near $(+1, -2/3)$. The reconstruction of X [(4.13.2)] at the "entrance" of the trajectory in a neighborhood of $(1, -2/3)$ is most easily performed by the "trajectory" method consideration (see Section 4.4).

Since P is moved along the line $f(x, y) = 0$, then for a small $x - 1$ we have (in the principal order) $y + 2/3 = 0((x-1)^2)$. Therefore, put $x = 1 + \Delta z$ and $y = -2/3 + \Delta^2 q$. Then

$$X = \Delta\left\{\left(q - z^2 - \frac{1}{3}\Delta z^3\right)\frac{\partial}{\partial z} - \frac{\varepsilon}{\Delta^3}(1 + \Delta z)\frac{\partial}{\partial q}\right\}. \qquad (4.13.5)$$

Let us emphasize that we are moving from the part where $z > 0$, $q > 0$, and $\Delta > 0$. It is immediately clear from (4.13.5) that in $(1/\Delta)X$ the summand

$$\bar{X}_0 = \left(q - z^2 - \frac{1}{3}\Delta z^3\right)\frac{\partial}{\partial z},$$

i.e., $(1/\Delta)X_0$, remains the leading one until $\varepsilon/\Delta^3 \ll 1$ [an eigenfunction of $\bar{X}_0$ differs only a little from an eigenfunction of $(q - z^2)(\partial/\partial z)$, namely from $(\sqrt{q}-z)/(\sqrt{q}+z)$, corresponding to $-2\sqrt{q}\neq 0$]. The reconstruction is performed at $\Delta=\varepsilon^{1/3}$. Putting

$$\delta=\varepsilon^{1/3},\qquad x=1+\delta z,\qquad y=-\tfrac{2}{3}+\delta^2 q, \tag{4.13.6}$$

we get a reconstructed operator $X=\delta Y$, where

$$Y=Y_0+\delta Y_1=(q-z^2)\frac{\partial}{\partial z}-\frac{\partial}{\partial q}+\delta\left(-\frac{z^3}{3}\frac{\partial}{\partial z}-z\frac{\partial}{\partial q}\right). \tag{4.13.7}$$

The operator Y_0 produces a differentiation along the field $\{q-z^2,-1\}$ without singular points so that in the domain $|z|\leq 1$, $|q|\leq 1$, the operator Y is reduced to the simplest normal form Y_0. It means that the solution of the system corresponding to Y, i.e.,

$$\frac{dz}{d\tau}=z^2-q+\delta\frac{z^3}{3},\qquad \frac{dq}{d\tau}=1+\delta z\qquad (\tau=\tfrac{t}{\delta^2}) \tag{4.13.8}$$

can be sought in the form of a development in powers of δ

$$z=z_0(\tau)+\delta z_1(\tau)+\cdots,\qquad q=q_0(\tau)+\delta q_1(\tau)+\cdots \tag{4.13.9}$$

[here $q_0(0)=q_1(0)=\cdots=0$ are known and $z_0(0)$, $z_1(0)$, ... are unknown].

Inserting (4.13.9) in (4.13.8) and collecting terms with the same powers of δ, consider consecutively the equations which arise.

For z_0 and q_0 we obtain the system $dz_0/d\tau=z_0^2-q$ and $dq_0/d\tau=1$. From here we immediately get

$$q_0=\tau, \tag{4.13.10}$$

and the equation for z_0 takes the form $dz_0/d\tau=z_0^2-\tau$ (the *Riccati equation*). The substitution

$$z_0=-\frac{A'}{A}\qquad (A'=\tfrac{dA}{d\tau}) \tag{4.13.11}$$

reduces it to the Airy equation

$$A''=\tau A. \tag{4.13.12}$$

Now note that matching with "entering" trajectory occurs at any rate for negative t, i.e., for large positive τ, and z_0 must be *positive* since $x>1$ and $x=1+\delta z$.

Therefore, taking (4.13.11) into account, we may take only the decreasing (as $\tau\to+\infty$) solution of (4.13.12) which, as is known, [4] is unique up to a constant factor. It is the so-called *Airy function of the first type* $Ai(\tau)$ defined by the initial values $Ai(0)=1/\sqrt[3]{9}\Gamma(2/3)$, $A'i(0)=-1/\sqrt[3]{3}\Gamma(1/3)$.

Thus $z_0(\tau)$ is actually uniquely defined and $z_0(0) = \sqrt[3]{3}\,\Gamma(2/3)/\Gamma(1/3)$. For the sake of convenience put

$$A = \sqrt[3]{9}\,\Gamma(2/3)Ai(\tau) \tag{4.13.13}$$

so that

$$A(0) = 1. \tag{4.13.14}$$

Notice also that the function z_0 is of order $\sqrt{\tau}$ as $\tau \to +\infty$, which is clear from the equation and also from the asymptotics of

$$Ai(\tau) \simeq \frac{1}{2\sqrt{\pi}\sqrt[4]{\tau}} \exp\left(-\frac{2}{3}\tau^{3/2}\right).$$

Now considering equations for z_1 and q_1 :

$$\frac{dz_1}{d\tau} = 2z_0z_1 - q_1 + \frac{1}{3}z_0^3, \qquad \frac{dq_1}{d\tau} = z_0.$$

From the second equation and (4.13.11) and (4.13.14) we find

$$q_1 = -\ln A(\tau) \tag{4.13.15}$$

$[q_1(0) = 0]$, and as a solution of a (linear) equation in z_1 we may only take

$$z_1 = \frac{1}{A^2(\tau)} \int_\tau^\infty \left(q_1(\eta) - \frac{1}{3}z_0^3(\eta)\right) A^2(\eta)\, d\eta, \tag{4.13.16}$$

since any other solution grows asymptotically, whereas "entering" asymptotics cannot have exponentially large terms. Notice also that z_1 grows like τ. The next terms of asymptotics are similarly found.

Remark 4.13.3 It is not difficult to demonstrate (here and in what follows we omit proofs) that z_n and q_n grow, as $\tau \to +\infty$, not faster than $\tau^{(n+1)/2}$ and $\tau^{(n+2)/2}$ respectively, so that the asymptotics (4.13.9) are applicable when $\delta\sqrt{\tau} \ll 1$; i.e., when $\tau \ll 1/\delta^2$. Putting $\tau \sim \Delta^2/\delta^2$, where $\Delta \to 0$, we will obtain

$$z \sim z_0 \sim \sqrt{\tau} \sim \frac{\Delta}{\delta}, \qquad q \sim q_0 \sim \tau \sim \frac{\Delta^2}{\delta^2}$$

and $x - 1 \sim \Delta$ and $y + 2/3 \sim \Delta^2$, yielding that the domain of matching with "entering" asymptotics is maximally "broad" ($\Delta = \delta^k$ and the critical value of k is 0).

The Second Reconstruction. Asymptotics Near the Straight Part of Γ $(y = -2/3)$

The asymptotics (4.13.9) do not yet allow us to "go out" from a neighborhood of the point $(1, -2/3)$ since $Ai(\tau)$ has zeros; hence they can be applied only when $\tau > \alpha$, where

$$\alpha \approx -2{,}338 \tag{4.13.17}$$

is the maximal zero of $Ai(\tau)$. [Since $\alpha < 0$, we enter the domain $z < 0$ via (4.13.9).]

The new reconstruction is quite evident:

$$\delta z = p \qquad (x = 1 + p,\ y = -\tfrac{2}{3} + \delta^2 q) \tag{4.13.18}$$

[(4.13.6) already hints that the trajectory will enter the domain $\mathcal{D}'$ of width $\sim \delta^2$ fringing Γ when the order of z will be $\sim 1/\delta$].

Remark 4.13.4 It is not difficult to show that functions $z_n(\tau)$ and $q_n(\tau)$ do not grow faster than $\ln^n \rho/\rho^{n+1}$, where $n \geq 0$, and $\ln^{n-1} \rho/\rho^{n-1}$, where $n \geq 1$, $q_1 \sim \ln \rho$, and $q_0 \sim \alpha$, respectively, as $\rho = \tau - \alpha \to +0$. Therefore, the asymptotics (4.13.9) are applicable for $\rho/|\ln \rho| \gg \delta$, which can be understood as $\rho \gg \delta$ if we take that $\rho \gtrsim \delta^{1-k}$, where $k > 0$ (the critical value of k is zero). Putting $\rho \sim \delta^{1-k}$, we have $\delta z \sim \delta z_0 = 0(\delta/\rho) = 0(\delta^k)$, $q \sim q_0 \sim \alpha \sim 1$, and, since $k_{\text{cr}} = 0$, we come to (4.13.18) due to "trajectory" considerations. As a result, the operator $X = \delta Y$ [see (4.13.7)] takes the form

$$X = Z_0 + \delta Z_1 + \delta^2 Z_2 = -R(p)\frac{\partial}{\partial p} - \delta(p+1)\frac{\partial}{\partial q} + \delta^2 q \frac{\partial}{\partial p}, \qquad R(p) = p^2 + \frac{p^3}{3}. \tag{4.13.18$'$}$$

The leading operator Z_0 has an invariant q and an eigenfunction $\varphi = ((3+p)/p)\exp(-(3+p)/p)$ corresponding to the eigenvalue $\lambda = -3$ (see Preliminaries), and X can be reduced to the normal form in the domain $q \sim 1$, $p < 0$.

As had been already said, the trick of constructing asymptotics from Section 3.9 leads more quickly to the goal. We now make use of it.

To the operator (4.13.18) the system

$$\delta \frac{dp}{d\rho} = R(p) - \delta^2 q; \qquad \delta \frac{dq}{d\rho} = \delta(p+1)$$

where

$$\rho \equiv \tau - \alpha = -(t/\delta^2) - \alpha \tag{4.13.19}$$

corresponds [clearly it is convenient to take ρ as an independent variable since (4.13.9) "breaks" as $\rho \to 0$].

A solution of (4.13.19) will be sought in the form [see (3.9.5)]

$$p = p_0(\rho,\zeta) + \delta p_1(\rho,\zeta) + \cdots, \qquad q = Q_0(\rho,\zeta) + \delta Q_1(\rho,\zeta) + \cdots, \tag{4.13.20}$$

where

$$\zeta = \exp \frac{3\rho}{\delta} \tag{4.13.21}$$

and p_j and Q_j are subject to regularity conditions of the form (3.9.6) as $\zeta \to 0$.

Initial values (at $\rho = 0$) will be

$$p_0(0,1) = -\frac{3}{1+a}, \quad p_1(0,1) = 0, \ \ldots; \quad Q_0(0,1) = h, \ Q_1(0,1) = 0, \tag{4.13.22}$$

where a, h $(0 < a < \infty)$ are constants, which may (and certainly will) depend on δ and will be found from matching with (4.13.9).

Inserting (4.13.20) into (4.13.19) and differentiating [taking (4.13.21) into account] according to the rule (3.9.7), let us collect summands with the same powers of δ and consider consecutively pairs of equations which arise.

The first pair is of the form

$$3\zeta\frac{\partial p_0}{\partial \zeta} = R(p_0), \quad 3\zeta\frac{\partial Q_0}{\partial \zeta} = 0 \qquad (R(p) = p^2 + \tfrac{p^3}{3}),$$

whence

$$m_0(\rho)\zeta = -\frac{3(3+p_0)}{p_0}e^{-(3+p_0)/p_0}, \qquad Q_0 = n_0(\rho),$$

where $m_0(0) = 3ae^a$, $n_0(0) = h$, due to (4.13.22).

Furthermore, the next pair is

$$\frac{\partial p_0}{\partial \rho} + 3\zeta\frac{\partial p_1}{\partial \zeta} = R'(p_0)p_1, \qquad \frac{\partial Q_0}{\partial \rho} + 3\zeta\frac{\partial Q_1}{\partial \zeta} = p_0 + 1.$$

From here, taking into account that $p_0 = -3 + m_0(\rho)\zeta + 0(\zeta^2)$,

$$p_1 = \bar{p}_1(\rho) + \hat{p}_1(\rho)\zeta + 0(\zeta^2), \qquad Q_1 = \bar{Q}_1(\rho) + \hat{Q}_1(\rho)\zeta + 0(\zeta^2)$$

for small ζ (regularity conditions), we get

$$3\bar{p}_1(\rho) = 0, \quad m_0'(\rho) = -4m_0(\rho)\bar{p}_1(\rho), \quad n_0'(\rho) = -2, \quad 3\hat{Q}_1(\rho) = m_0(\rho),$$

implying

$$\bar{p}_1(\rho) = 0, \quad m_0(\rho) = \text{const} = 3ae^a, \quad n_0(\rho) = h - 2\rho, \quad \hat{Q}_1(\rho) = ae^a.$$

Thus we finally get the principal term of the asymptotics

$$ae^a\zeta = -\frac{3+p_0}{p_0} - \exp\left(-\frac{3+p_0}{p_0}\right), \qquad Q_0 = h - 2\rho, \tag{4.13.23}$$

and also find, integrating equations for p_1 and Q_1, and taking

$$3\zeta\frac{\partial p_0}{\partial \zeta} = R(p_0), \qquad \frac{\partial p_0}{\partial \rho} = 0, \qquad \frac{\partial Q_0}{\partial \rho} = -2$$

into account [see also (4.13.22)], that

$$p_1 = m_1(\rho)R(p_0), \quad m_1(0) = 0; \quad Q_1 = n_1(\rho) - \frac{3+p_0}{p_0}, \quad n_1(0) = -a,$$

where $m_1(\rho)$ and $n_1(\rho)$ are to be found from the regularity conditions from the next pair of equations, as m_0 and n_0 had been defined.

Computing p_k and Q_k of any order is elementary. Omitting these calculations, let us give formulas which will be used in the concrete example of matching:

$$p_1 = -\frac{4}{9}\rho(h-\rho)R(p_0), \qquad Q_1 = -a - \frac{3+p_0}{p_0} \tag{4.13.24}$$

[from regularity conditions we have

$$m_1'(\rho) = -\frac{4}{9}n_0(\rho), \qquad n_1'(\rho) = 0,$$

implying $m_1(\rho) = -\frac{4}{9}\rho(h-\rho)$, $n_1(\rho) \equiv -a$], and

$$\begin{aligned} p_2 = R(p_0)\Big\{ & m_2(\rho) + \frac{m_1^2(\rho)}{2}\left(R'(p_0) - R'\left(-\frac{3}{1+a}\right)\right) \\ & + m_1'(\rho)\int_{-3/1+a}^{p_0} \frac{9/4 - R(\eta)}{R^2(\eta)}\,d\eta \Big\} \end{aligned} \tag{4.13.25}$$

$$Q_2 = n_2(\rho) + m_1(\rho)(p_0+3), \qquad m_2(0) = n_2(0) = 0,$$

$$\bar{p}_2(\rho) = -\frac{3}{4}m_1'(\rho)$$

[we will only need the term $m_1'(\rho)(p_0/2 + 1/2 - 3/4p_0)$ of p_2].

Remark 4.13.5 It is not difficult to establish the following properties of p_k and Q_k. Functions $m_k(\rho)$ and $n_k(\rho)$ are polynomial. For a small p_0 a singularity in p_k and Q_k is no greater than $|p_0|^{1-k}$. Therefore the asymptotics (4.13.20) are applicable for $\delta \ll -p_0$, which corresponds to $\rho \ll 1$, since, due to (4.13.23), the smallness of $-p_0$ implies $-p_0 \sim 3/\ln\zeta = \delta/\rho$ [see (4.13.21)]. On the other hand, the asymptotics (4.13.9) are applicable when $\rho \gg \delta$ (see Remark 4.13.4). Therefore both (4.13.9) and (4.13.20) are applicable in the domain $\delta \ll \rho \ll 1$ (the maximally "wide" matching domain).

Let us give also formulas for the "exterior" solution:

$$\begin{aligned} \bar{p} &= p(\rho,0) = p_0(\rho,0) + \delta p_1(\rho,0) + \cdots = \bar{p}_0(\rho) + \delta\bar{p}_1(\rho) + \cdots, \\ \bar{Q} &= q(\rho,0) = Q_0(\rho,0) + \delta Q_1(\rho,0) + \cdots = \bar{Q}_0(\rho) + \delta\bar{Q}_1(\rho) + \cdots \\ &= n_0(\rho) + \delta n_1(\rho) + \cdots \end{aligned} \tag{4.13.26}$$

$$\begin{aligned} \bar{p} &= -3 + \delta^2\frac{1}{3}(h-2p) + \delta^3\left(-\frac{a}{3} - \frac{2}{9}\right) + \cdots, \\ \bar{Q} &= h - 2p + \delta(-a) \end{aligned}$$

(we have also found $\bar{p}_3 = -a/3 - 2/9$), which can be used in a neighborhood of the "artificial" reconstruction point $(-2, -2/3)$. These asymptotics will be matched with the ones near the curvilinear part of Γ.

The First Matching

Thus we have shown above how to find asymptotics

$$x = 1 + \delta z_0 + \delta^2 z_1 + \cdots, \qquad y = -\frac{2}{3} + \delta^2 q_0 + \delta^3 q_1 + \cdots,$$

$$z_j = z_j(\tau), \qquad q_j = q_j(\tau), \qquad \tau = -\frac{t}{\delta^2}$$

in a neighborhood of the reconstruction point $(1, -2/3)$ and have computed z_0, z_1, q_0, and q_1 [formulas (4.13.10), (4.13.11), (4.13.13), (4.13.15), and (4.13.16)], which allows us to find x and y with error $0(\delta^2)$, $0(\delta^3)$, respectively, ($\delta = \varepsilon^{1/3}$) when $|\tau| \lesssim 1, \tau > \alpha$ [(4.13.17)].

With lesser accuracy, depending on $\rho = \tau - \alpha$, these asymptotics are applicable when $\rho \gg \delta$ (see Remark 4.13.4).

We have also shown how to find asymptotics

$$x = 1 + p_0 + \delta p_1 + \delta^2 p_2 + \cdots, \qquad y = -\tfrac{2}{3} + \delta^2 Q_0 + \delta^3 Q_1 + \cdots,$$

$$p_j = p_j(\rho, \zeta), \qquad Q_j = Q_j(\rho, \zeta), \qquad \zeta = e^{3\rho/\delta}$$

near the straight part of $\Gamma(y = -2/3)$ and have given formulas [(4.13.23), (4.13.24), (4.13.25)] for p_0, p_1, and p_2 [the function $m_2(\rho)$ in p_2 had not been found but we will not need it in our computations] and Q_0 and Q_1, which allows us to find x and y with error $0(\delta^2)$, $0(\delta^3)$, respectively, for $-\rho \lesssim 1$ if constants a and h are known. With lesser accuracy, depending on ρ, these asymptotics are applicable for positive $\rho \ll 1$ (see Remark 4.13.5).

Besides, we have given formulas (4.13.26) for computing $x = 1 + \bar{p}$, $y = -(2/3) + \delta^2 \bar{Q}$ which follow from these asymptotics at $\zeta = 0$, are valid for $-\rho \sim 1$ and enable one to find x, y with accuracy $0(\delta^3) = 0(\varepsilon)$ near the point $(-2, -2/3)$.

Now our aim is to compute a and h; i.e., to match the obtained asymptotics in the domain $\delta \ll \rho \ll 1$ where we may employ both asymptotics.

For a small ρ we have

$$\begin{aligned} q_0 &= \alpha + \rho, \\ z_0 &= -\frac{1}{\rho} - \frac{\alpha}{3}\rho + 0(\rho^2) \\ q_1 &= -\ln \rho - \ln A'(\alpha) + 0(\rho^2), \\ z_1 &= -\frac{\ln \rho}{3\rho^2} - \frac{k}{\rho^2} + \frac{\alpha}{9}\ln \rho + \frac{\alpha}{3}\left(k - \frac{2}{3}\right) + 0(\rho \ln \rho), \end{aligned}$$

where

$$k = \frac{1}{6} + \frac{1}{3}\ln A'(\alpha) + \frac{2}{3A'^2(\alpha)} \int_\alpha^\infty A^2(\eta) \ln A(\eta)\, d\eta. \tag{4.13.27}$$

Remark 4.13.6 It is convenient to seek asymptotic developments of q_j and z_j for a small ρ directly from equations for q_j and z_j but constants that enter these equations can only be found by making use of exact formulas for q_j and z_j. We can obtain a formula for

$$k = -\lim_{\rho\to 0}\left(\rho^2 z_1 + \tfrac{1}{3}\ln\rho\right)$$

expressing z_1 [see (4.13.16)] in the form

$$\begin{aligned} z_1 = \frac{1}{A^2(\tau)}\Big[& -\frac{1}{3}A'^2(\tau)\ln A(\tau) + \frac{1}{3}\tau A^2(\tau)\ln A(\tau) \\ & - \frac{1}{6}A'^2(\tau) - \frac{2}{3}\int_\tau^\infty A^2(\eta)\ln A(\eta)\,d\eta\Big], \end{aligned}$$

taking $A'' = \tau A$ into account and integrating by parts.

Terms of developments for a small ρ which are not written explicitly can be ignored due to the matching principle from Section 4.5, since the value of their contribution to $x = 1 + \delta z_0 + \delta^2 z_1$, $y \sim -(2/3) + \delta^2 q_0 + \delta^3 q_1$ are $\ll \delta^2$ and $\ll \delta^3$, respectively, in the domain of matching, $\delta \ll \rho \ll 1$.

Introducing a more convenient (in what follows) variable $\nu = 3\rho/\delta$ instead of ρ, we get

$$\begin{aligned} x \simeq 1 & - \frac{3}{\nu} - \frac{\delta^2\alpha}{9}\nu - \frac{3(\ln\nu + \ln\delta - \ln 3)}{\nu^2} \\ & - \frac{9k}{\nu^2} + \frac{\delta^2\alpha}{9}(\ln\nu + \ln\delta - \ln 3) + \frac{\delta^2\alpha}{3}\left(k - \frac{2}{3}\right), \qquad (4.13.28) \\ y = & -\frac{2}{3} + \delta^2\alpha + \frac{\delta^3\nu}{3} - \delta^3(\ln\nu + \ln\delta - \ln 3) - \delta^3\ln A'(\alpha). \end{aligned}$$

Furthermore, for a large $\nu = \ln\zeta = 3\rho/\gamma$ (in the domain of matching $1 \ll \nu \ll 1/\delta$) we have

$$\begin{aligned} p_0 &= -\frac{3}{\nu}\frac{3\ln\nu}{\nu^2} + \frac{3(1 + a + \ln a)}{\nu^2} + 0\left(\frac{\ln\nu^2}{\nu^3}\right), \\ p_1 &= 0\left(\frac{\delta}{\nu}\right), \\ p_2 &= -\frac{h}{9}\nu + \frac{h}{9}\ln\nu - \frac{3 + a + \ln a}{9}h + 0(\delta\nu^2), \\ Q_0 &= h - \frac{2}{3}\delta\nu, \\ Q_1 &= \nu - \ln a + \ln\nu + 0\left(\frac{\ln^2\nu}{\nu}\right), \end{aligned}$$

as follows from (4.13.23)–(4.13.25) $[R(p_0) = 0(1/\nu^2)]$.

Inserting these expressions into $x \simeq 1+p_0+\delta p_1+\delta^2 p_2$, $y \simeq -\frac{2}{3}+\delta^2 Q_0+\delta^3 Q_1$, we can ignore terms not written explicitly thanks to the matching principle. Setting also $a = a_0 + \delta a_1 + \cdots$, $h = h_0 + \delta h_1 + \cdots$ we get (also ignoring terms of values $\ll \delta^2$ and $\ll \delta^3$, for x and y, respectively)

$$\begin{aligned} x &\simeq 1 - \frac{3}{\nu} - \frac{3\ln\nu}{\nu^2} + \frac{3(1+a_0+\ln a_0)}{\nu^2} - \delta^2\frac{h_0}{9}\nu \\ &\quad + \delta^2\frac{h_0}{9}\ln\nu - \delta^2\frac{3+a_0+\ln a_0}{9}h_0, \qquad (4.13.29) \\ y &\simeq -\frac{2}{3} + \delta^2 h_0 + \delta^3 h_1 + \frac{1}{3}\delta^3\nu - \delta^3\ln\nu + \delta^3\ln a_0. \end{aligned}$$

Now due to the matching principle there should exist a_0, h_0, and h_1 for which formulas (4.13.28) and (4.13.29) *identically* coincide (in ν). In fact, this is so if

$$\begin{aligned} a_0 + \ln a_0 &= \ln 3 - 3k - 1 - \ln\delta, \\ h_0 &= \alpha, \\ h_1 &= 1 + a_0 + 3k - \ln A'(\alpha). \end{aligned} \qquad (4.13.30)$$

Remark 4.13.7 The system of equations arising for a and h is highly overdefined (especially for computations with high accuracy). We have not dealt with the question of an "economic" mode of computation (excluding a remark in 4.5 on a possible shortening of calculations). Note, however, that compatibility of the arising system is a nice check for the correctness of computations.

Remark 4.13.8 The quantity $a_0 = a_0(\ln\delta)$ is defined by the first (implicit) formula in (4.13.30). From a computational point of view it seems that it is unwise to seek an asymptotic representation of a_0. In general it is convenient to ignore a weak logarithmic dependence on δ (considering $\ln\delta$ formally as a finite parameter). Then all developments are performed in powers of δ.

Now (a and h being already found) we find the asymptotics of $x = 1+\bar{p}$ and $y = -\frac{2}{3} + \delta^2\bar{Q}$ making use of (4.13.30), (4.13.26), and (4.13.27) near the point $(-2, -2/3)$:

$$\begin{aligned} x &\simeq -2 + \delta^2\left(\alpha - \frac{2}{3}\tau\right) + \delta^3\left(\frac{5}{18} + \frac{2}{3}\ell\right) + \cdots, \\ y &= -\frac{2}{3} + \delta^2(3\alpha - 2\tau) + \delta^3\left(\frac{3}{2} + 2\ell\right) + \cdots, \end{aligned} \qquad (4.13.31)$$

where $\tau = -t/\delta^2$ and

$$\ell = \frac{1}{A'^2(\alpha)}\int_\alpha^\infty A^2(\eta)\ln A(\eta)\,d\eta, \qquad (4.13.32)$$

and we "go out" for good from the neighborhood of the reconstruction point $(1, -2/3)$.

It is permissible to use (4.13.31) for $\tau \sim 1$.

Asymptotics Near the Curvilinear Part of $\Gamma(-2 < x < -1)$

We seek these asymptotics in the form

$$x = x_0 + \varepsilon x_1 + \cdots, \qquad y = y_0 + \varepsilon y_1 + \cdots$$
$$[x_j = x_j(t),\ y_j = y_j(t)]. \tag{4.13.33}$$

Inserting (4.13.33) in (4.13.1) we have $0 = y_0 + x_0 - \frac{1}{3}x_0^3$ and $dy/dt = -x_0$ in the principal order.

Denoting the (as yet unknown) moment when the trajectory intersects the line

$$y = -\tfrac{2}{3} \qquad [y_0(\delta^2\xi) = -\tfrac{2}{3},\ y_1(\delta^2\xi) = 0, \ldots]$$

near the point $(-2, -2/3)$ by $\delta^2\xi <$ we get

$$y_0 = -x_0 + \frac{1}{3}x_0^3 \qquad [x_0(\delta^2\xi) = -2],$$
$$\ln\left(-\frac{x_0}{2}\right) + \frac{1}{2}(4 - x_0^2) = t - \delta^2\xi. \tag{4.13.34}$$

In the next order we obtain equations $dx_0/dt = y_1 + (1 - x_0^2)x_1$ and $dy_1/dt = -x_1$, and making use of the above we find

$$y_1 = x_0 \ln\left(-\frac{2\sqrt{x_0^2 - 1}}{x_0\sqrt{3}}\right), \qquad x_1 = \frac{x_0}{x_0^2 - 1}\frac{dy_1}{dx_0}, \tag{4.13.35}$$

etc.

Remark 4.13.9 It is easy to show that the asymptotics obtained are applicable for $-1 - x_0 \gg \delta$ [since x_n, y_n do not grow faster than $(1 - x_0)^{1-3n}$ and $(-1 - x_0)^{3-3n}$, where $n \geq 2$, respectively, as $x_0 \to -1$].

The Second Matching

Now, matching (4.13.33) and (4.13.31), we find ξ.

Putting $\xi = \xi_0 + \delta\xi_1 + \cdots$ and substituting $t = -\delta^2\tau$ we find from (4.13.33)–(4.13.35) developments when $\tau \sim 1$:

$$x \simeq -2 + \delta^2\left(-\frac{2}{3}\xi_0 - \frac{2}{3}\tau\right) + \delta^3\left(-\frac{2}{9} - \frac{2}{3}\xi_1\right) + \cdots,$$
$$y = -\frac{2}{3} + \delta^2(-2\xi_0 - 2\tau) + \delta^3(-2\xi_1) + \cdots.$$

These formulas must coincide identically with (4.13.31), which actually takes place for

$$\xi = -\frac{3}{2}\alpha - \delta\left(\frac{3}{4} + \ell\right) + \cdots \tag{4.13.36}$$

$[\xi > 0$ since $\alpha < 0$, *see*(4.13.17)$]$.

The Third Matching

At last we should match the asymptotics (4.13.33), where ξ is already known, with the asymptotics in a neighborhood of $(-1, 2/3)$. These last asymptotics, as had been already mentioned, can be obtained due to the symmetry of the problem changing in (4.13.9) $\tau = -t/\delta^2$ to

$$\hat{\tau} = -\frac{t - T(\varepsilon)/2}{\delta^2}$$

and inserting $z(\hat{\tau})$ and $q(\hat{\tau})$ in $x = -1 - \delta z$ and $y = 2/3 - \delta^2 g$.

Let us take $\hat{\tau}$ for the variable of matching so that the domain of matching is $1 \ll \hat{\tau} \ll 1/\delta^2$ (maximally "wide"; see Remarks 4.13.3 and 4.13.9).

Let us perform the matching with the same accuracy as the first one. Let us insert $t = T(\varepsilon)/2 - \delta^2\hat{\tau}$ in (4.13.33) and, setting $T(\varepsilon) = T_0 + \delta T_1 + \delta^2 T_2 + \delta^3 T_3 + \cdots$, compute x up to terms of value $\ll \delta^2$ in the domain of matching.

First, setting $x_0 = -1 - \delta s_1 + \delta^2 s_2 + \cdots$, we get from (4.13.34)

$$\frac{3}{2} - \ln 2 - \delta^2 s_1^2 + \delta^3\left(2s_1 s_2 + \frac{1}{3}s_1^3\right) + \cdots$$
$$= \frac{T_0}{2} + \delta\frac{T_1}{2} + \delta^2\left(\frac{T_2}{2} - \hat{\tau} + \frac{3}{2}\alpha\right) + \delta^3\left(\frac{T_3}{2} + \frac{3}{4} + \ell\right)$$

[since $\xi = (-3/2)\alpha - \delta(3/4 + \ell)$, see (4.13.36)], which, incidentally, implies

$$T_0 = 3 - 2\ln 2, \qquad T_1 = 0, \tag{4.13.37}$$

and also

$$x_0 \simeq 1 - \delta\sqrt{\hat{\tau} - \frac{3}{2}\alpha - \frac{T_2}{2}}$$
$$+ \delta^2\left(-\frac{1}{6}\left(\hat{\tau} - \frac{3}{2}\alpha - \frac{T_2}{2}\right) + \frac{3/8 + \ell/2 + T_3/4}{\sqrt{\hat{\tau} - (3/2)\alpha - T_3/2}}\right) + \cdots.$$

With the same accuracy, we have

$$x \simeq x_0$$
$$\simeq -1 - \delta\sqrt{\hat{\tau} - \frac{3}{2}\alpha - \frac{T_2}{2}} - \frac{\delta^2}{6}\left(\hat{\tau} - \frac{3}{2}\alpha - \frac{T_2}{2}\right). \tag{4.13.38}$$

Furthermore, let us compute y from (4.13.33) up to terms of value $\ll \delta^3$ in the domain of matching. With this accuracy we may assume

$$y_1 \simeq -\frac{1}{2}\ln\frac{8\delta s_1}{3},$$

and, making use of (4.13.34) and the formula for x_0 found earlier, we get

$$y \simeq \frac{2}{3} - \delta^2\left(\hat{\tau} - \frac{3}{2}\alpha - \frac{T_2}{2}\right) + \delta^3\left[\frac{3}{4} + \ell + \frac{T_3}{2} - \frac{1}{2}\ln\frac{8\delta\sqrt{\hat{\tau} - (3/2)\alpha - T_2/2}}{3} - \frac{2}{3}\left(\hat{\tau} - \frac{3}{2}\alpha - \frac{T_2}{2}\right)^{3/2}\right]. \tag{4.13.39}$$

Now let us turn to similar computations for $x = -1 - \delta z(\hat{\tau})$, $y = (2/3) - \delta^2 q(\hat{\tau})$. For a large τ we have

$$Ai(\tau) \simeq \frac{1}{2\sqrt{\pi}\sqrt[4]{\tau}} e^{-(2/3)\tau^{2/3}}\left(1 + 0\left(\frac{1}{\tau^{3/2}}\right)\right),$$
$$q_0 = \tau,$$
$$z_0(\tau) \simeq \sqrt{\tau} + 0\left(\frac{1}{\tau}\right),$$
$$q_1(\tau) \simeq \ln\frac{2\sqrt{\pi}}{\sqrt[3]{9}\Gamma(2/3)} + \frac{2}{3}\tau^{3/2} + \frac{1}{4}\ln\tau + 0\left(\frac{1}{\tau^{3/2}}\right),$$
$$z_1(\tau) \simeq \frac{1}{6}\tau + 0\left(\frac{\ln\tau}{\sqrt{\tau}}\right)$$

(see formulas for q_j and z_j).

Thus, up to terms reaching values $\ll \delta^2$ and $\ll \delta^3$ for x and y, respectively, in the matching domain $1 \ll \hat{\tau} \ll 1/\delta^2$, we get

$$x \simeq -1 - \delta\sqrt{\hat{\tau}} - \frac{\delta^2}{6}\hat{\tau}, \tag{4.13.40}$$

$$y \simeq \frac{2}{3} - \delta^2\hat{\tau} - \delta^3\left(\frac{2}{3}\hat{\tau}^{3/2} + \frac{1}{4}\ln\hat{\tau} + \ln\frac{2\sqrt{\pi}}{\sqrt[3]{9}\Gamma(2/3)}\right). \tag{4.13.41}$$

Formulas (4.13.38) and (4.13.40) and formulas (4.13.39) and (4.13.41), respectively, coincide, identically at

$$T_2 = -3\alpha, \qquad T_3 = \ln\delta - 2\ell - \frac{3}{2} + \ln\left(\frac{2\sqrt[3]{3}}{\pi}\Gamma^2\left(\frac{2}{3}\right)\right). \tag{4.13.42}$$

The formula for T_3 can be somewhat simplified due to (4.13.32) and (4.13.13) the constant ℓ can be written in the form

$$\ell = \frac{1}{Ai'^2(\alpha)}\int_\alpha^\infty Ai^2(\eta)\ln Ai(\eta)\,d\eta + \ln(\sqrt[3]{9}\Gamma(2/3)),$$

making use of the fact that $(\tau Ai^2 - Ai'^2)' = Ai^2$. Now setting

$$m = \frac{1}{Ai'^2(\alpha)}\int_\alpha^\infty Ai^2(\eta)\ln Ai(\eta)\,d\eta, \tag{4.13.43}$$

we derive from (4.13.42) that $T_3 = \ln \delta - 2m - (3/2) + \ln(2/3\pi)$. Thus [see (4.13.37)],

$$T(\varepsilon) = 3 - 2\ln 2 - 3\alpha\varepsilon^{2/3} + \frac{1}{3}\varepsilon \ln \varepsilon - \left(2m + \frac{3}{2} + \ln \frac{3\pi}{2}\right)\varepsilon + \cdots,$$

where m is defined by (4.13.43) and $\alpha \approx -2,338\ldots$ [see (4.13.17)] is the maximal zero of $Ai(\tau)$.

5

Equations in Partial Derivatives

Below we generalize the above formalism to make it applicable to equations in partial derivatives.

5.1 Functional Derivatives

Earlier we have assigned to the system $du_k/dt = a_k(u)$, where $k = 1, \ldots, m$ the operator

$$AF = a_k(u)\frac{\partial F}{\partial u_k} = \lim_{\delta \to 0} \frac{1}{\delta}[F(u + \delta a) - F(u)]$$

acting in the space of functions in m variables $u_1, \ldots, u_m$. In problems with partial derivatives we deal with functions or vector functions $u(x)$ and equations of the type $\partial u/\partial t = \ell(u)$, where $\ell(u)$ is a differential operator. These problems are continual analogues of those considered earlier. The discrete parameter k which indexes indeterminates u_k turns into continual parameter x which runs an n-dimensional domain. The role of functions $F(u_1, \ldots, u_m)$ is now played by functionals $F(u, x)$ depending on x as on a parameter. We may surely speak about operators $F(u; x)$ that transform a vector function $u(x)$ into the function $F(u; x)$.

Further the following notations are used: U is the linear space of vector functions $(u_1(x), \ldots, u_m(x))$, ϕ is the linear space of vector functionals $\varphi = (\varphi_1(u; x), \ldots, \varphi_m(u; x))$, where $\varphi_i(u, x)$ is a functional in U depending on x as on a parameter. As always, we do not specify the domains of definition of U, ϕ.

Let $v, \varphi \in \Phi$. Put

$$v[\varphi] = \frac{d}{d\varepsilon}\varphi(u + \varepsilon v)\Big|_{\varepsilon=0} . \tag{5.1.1}$$

For a fixed v the expression $v[\varphi]$ is a linear operator which assigns $v[\varphi] \in \phi$ to any $\varphi \in \phi$. (This expression is a linear operator that maps Φ into itself if φ is fixed and v is varied.)

Definition 5.1.1 $v[\varphi]$ *is called the* functional derivative *of* φ *along* v.

Set $\varphi \cdot \psi = \sum_{1 \leq k \leq m} \varphi_k(u,x)\psi_k(u,x)$ and define $v[\varphi]$ by the formula (5.1.1). Then

$$v[\varphi \cdot \psi] = (v[\varphi]) \cdot \psi + \varphi(v[\psi]). \tag{5.1.2}$$

Clearly we deal here with the direct generalization of the notion of a first order differential operator.

Often instead of $v[\varphi]$ we will write $\hat{v}\varphi = v[\varphi]$ to stress that $\hat{v}$ is a linear operator in Φ. We will call v the density of the operator $\hat{v}$, and put $\hat{\Phi}$ for the space of these densities.

Lemma 5.1.2 *Let $\hat{a}, \hat{b} \subset \hat{\phi}$. Then*

$$\hat{b}(\hat{a}\varphi) - \hat{a}(\hat{b}\varphi) = (b[a])[\varphi] - (a[b])[\varphi] = (b[a] - a[b])[\varphi]. \tag{5.1.3}$$

This important formula generalizes the corresponding fact of the first order operators theory. Its meaning is the statement that if $\hat{a}, \hat{b} \subset \hat{\phi}$, then $\hat{b}\hat{a} - \hat{a}\hat{b} \subset \hat{\phi}$ and the density of this operator is $b[a] - a[b]$, i.e.,

$$\hat{b}\hat{a} - \hat{a}\hat{b} = \widehat{b[a] - a[b]}.$$

To prove this let us consider

$$\begin{aligned}\varphi(u + \varepsilon a + \delta b; x) = \varphi(u; x) + \varepsilon\hat{a}\varphi + \delta\hat{b}\varphi + \varepsilon^2\Gamma(a,\varphi) \\ + \delta^2\Gamma(b,\varphi) + \varepsilon\delta\langle a,\varphi,b\rangle + O(\varepsilon^2 + \delta^2).\end{aligned} \tag{5.1.4}$$

Here the right-hand side is simply the development of the left-hand side in the power of ε and δ. Clearly

$$\langle a,\varphi,b\rangle = \langle b,\varphi,a\rangle. \tag{5.1.5}$$

Set

$$\begin{aligned}\mathcal{J} &= \varphi(u + \delta b + \varepsilon a(u + \delta b)) \\ &= \varphi(u + \varepsilon a + \delta(b + \varepsilon b[a])) + O(\delta^2 + \varepsilon^2).\end{aligned} \tag{5.1.6}$$

The expression for $\mathcal{J}$ with accuracy up to $O(\delta^2 + \varepsilon^2)$ can be computed in two ways: first putting $\delta = 0$ in (5.1.4) we get $\varphi(u+\varepsilon a) \simeq \varphi(u) + \varepsilon\hat{a}\varphi + \varepsilon^2\Gamma(a,\varphi)$ yielding

$$\begin{aligned}\mathcal{J} &\simeq \varphi(u) + \varepsilon\hat{a}\varphi + \varepsilon^2\Gamma(a,\varphi)\Big|_{u+\delta b} \\ &\simeq \varphi(u) + \delta\hat{b}\varphi + \delta^2\Gamma(b,\varphi) + \varepsilon(\hat{a}\varphi + \delta\hat{b}\hat{a}\varphi) + \varepsilon^2\Gamma(a,\varphi)\end{aligned}$$

after substituting $u + \delta b$ instead of a. On the other hand making use of the right-hand side of (5.1.6) and (5.1.4) we get

$$\mathcal{J} \simeq \varphi(u,x) + \varepsilon\hat{a}\varphi + \delta((b + \varepsilon b[a])[\varphi]) + \varepsilon^2\Gamma(a,\varphi) + \delta^2\Gamma(b,\varphi) + \varepsilon\delta\langle a,\varphi,b\rangle.$$

The comparison of coefficients of $\varepsilon\delta$ in these two expressions gives

$$\hat{b}\hat{a}\varphi = (b[a])[\varphi] + \langle a, \varphi, b\rangle. \tag{5.1.7}$$

Interchanging a and b in (5.1.7) and using (5.1.5) one gets the proof of (5.1.3).

Remark 5.1.3 For two operators $\hat{a}, \hat{b} \subset \hat{\phi}$ put $[\hat{a}, \hat{b}] = \hat{a}\hat{b} - \hat{b}\hat{a}$. We have shown that $\hat{\phi}$ is closed with respect to this operation, i.e., $\hat{a}, \hat{b} \in \hat{\phi}$ implies $[\hat{a}, \hat{b}] \in \hat{\phi}$.

We can say that $\hat{\phi}$ is a *Lie algebra* with respect to $[\cdot, \cdot]$. Clearly the Jacobi identity

$$[[\hat{a}, \hat{b}], \hat{c}] + [[\hat{c}, \hat{a}], \hat{b}] + [[\hat{b}, \hat{c}], \hat{a}] = 0 \tag{5.1.8}$$

holds. Now for any $a, b \subset \phi$ put $(a, b) = (a[b] - b[a])$. Then (5.1.8) and (5.1.3) imply that the Jacobi identity

$$((a, b), c) + ((c, a), b) + ((b, c), a) = 0 \tag{5.1.9}$$

holds also. Unlike the trivial formula (5.1.8), formula (5.1.9) is not trivial. We have shown that ϕ is also a Lie algebra with respect to ( ) isomorphic to $\hat{\phi}$.

Algebras ϕ_∂ and $\hat{\Phi}_\partial$

In what follows to simplify notations we will only consider functions $u(x)$ but not vector functions and Φ will be understood as a linear space of functionals $\varphi(u, x)$. Put $\mathcal{D}^\alpha u$ for the differential operator

$$\frac{\partial^{\bar{\alpha}_1 + \cdots + \bar{\alpha}_m}}{\partial x_1^{\bar{\alpha}_1}, \ldots, \partial x_m^{\bar{\alpha}_m}} u(x_1, \ldots, x_m),$$

where $\alpha = (\bar{\alpha}_1, \ldots, \bar{\alpha}_m)$, and consider the subspace ϕ_∂ of Φ consisting of functionals of the form $\varphi = \varphi(D^{\alpha_1}u, \ldots, D^{\alpha_p}u; x)$ where φ is any function in arbitrary but finite number of some $D^\alpha u$.

If $\varphi \in \Phi_\partial$, then, clearly,

$$\hat{v}\varphi = v[\varphi] = \sum_\alpha D^\alpha(v)\frac{\partial\varphi}{\partial D^\alpha u}. \tag{5.1.10}$$

Formula (5.1.10) implies that if $a \in \Phi_\partial$ and $\varphi \in \Phi_\partial$, then

$$\hat{a}\varphi \in \Phi_\partial. \tag{5.1.11}$$

In particular, this implies that $a \in \Phi_\partial$ and $b \in \Phi_\partial$ implies (since $[\hat{a}\hat{b}]\varphi = (a[b] - b[a])[\varphi]$) that the operator $[\hat{a}, \hat{b}]$ has a density generated by the element $a[b] - b[a]$ which belongs to ϕ_∂, i.e., operators $\hat{a}$ generated by elements from Φ_∂ constitute a Lie subalgebra $\hat{\Phi}_\partial$ of the Lie algebra $\hat{\Phi}$ with respect to [,]. Similarly, Φ_∂ is a Lie subalgebra with respect to (,); this algebra is a direct generalization of the Lie algebra with the Poisson brackets. Finally, note the formula which will be repeatedly used and is subject to a straightforward verification: if $f \in \Phi_\partial$, then

$$f\left[\frac{\partial u}{\partial x_\nu}\right] = \frac{\partial f}{\partial x_\nu} + \frac{\partial u}{\partial x_\nu}[f] \qquad [f = f(u,x),\ \nu = 1,\ldots,n]. \tag{5.1.12}$$

Liouville's equation

Now consider two problems: to solve the equations

$$\frac{\partial v(x,t)}{\partial t} = a(v,x), \qquad v(x,0) = u(x); \tag{5.1.13}$$

$$\frac{\partial \varphi(u,x,t)}{\partial t} = a(u;x)[\varphi], \qquad \varphi(u,x;0) = u(x). \tag{5.1.14}$$

Lemma 5.1.4 *If problems (5.1.13) and (5.1.14) have the unique solution for an arbitrary initial value u, then* $\varphi(u,x;t) = v(x,t)$.

A similar lemma is well known in the theory of continuous groups and follows from the group property of solutions of (5.1.13) and (5.1.14) which in turn follows from the uniqueness theorem. The lemma means that if we consider a solution $v(x,t)$ of (5.1.13) as a functional depending on the initial value u, i.e., $v(x,t) = \varphi(u;x;t) \in \Phi$, then it satisfies the *linear* equation (5.1.14).

PROOF

Write a solution of (5.1.13) as $v(x,t) = v(u;x,t)$ indicating explicitly the dependence on the initial condition u. Then $v(u;x,t+\tau) = v(v(u;x,\tau);x,t)$. Formula (5.1.13) implies that $v(u;x,\tau) = u(x) + \tau a(u;x) + o(\tau)$, yielding

$$\begin{aligned} v(u;x,t+\tau) &= v(u(x) + \tau a(u;x);x,t) + o(\tau) \\ &= v(u(x);x,t) + \tau a[v] + o(\tau). \end{aligned}$$

But then $\partial v(u;x,t)/\partial t = a[v]$ and $v(u;x,0) = u$. Q.E.D.

It is useful to consider a more general problem

$$\frac{d\psi}{dt} = a(u,x)[\psi] = \hat{a}\psi, \tag{5.1.15}$$

where $\psi(u;x,0) = \psi_0$ and ψ_0 is an arbitrary functional from Φ.

Lemma 5.1.5 (Liouville's equation) *If (5.1.15) has a unique solution, then*

$$\psi(u;x,t) = \psi_0(v(u;x,t);x) \tag{5.1.16}$$

where v is a solution of (5.1.13).

To prove the lemma it suffices to show that the right-hand side of the (5.1.16) satisfies (5.1.15).

First let us mention an identity which follows directly from definition:

$$a(u)[\psi(v(u))] = (a(u)[v(u)])[\psi(v)]. \tag{5.1.17}$$

[Dependence of the functional on x is not indicated; whereas in the right-hand side an independent variable is $v = v(u)$. This identity is just the "Chain Rule."] Now from (5.1.16) we obtain

$$\begin{aligned}\frac{\partial\psi_0(v(u;x,t),x)}{\partial t} &= \left(\frac{\partial v(u;x,t)}{\partial t}\right)[\psi_0(v;x)] = a[v][\psi_0(v,x)] \\ &= a(u)[\psi_0[v(u,x,t);x]]. \qquad \text{Q.E.D.}\end{aligned}$$

Lemmas 5.1.4 and 5.1.5 which reduce the nonlinear problem (5.1.13) to the linear one (5.1.14) or to (5.1.15) and the existence of $\hat{\Phi}_\partial$ make it possible to generalize the formalism developed in Chapters 1 and 2 almost literally onto the case of equations with partial derivatives. Let us discuss this in more detail.

A solution of (5.1.15) can be formally written in the form

$$\psi(u;x,t) = e^{t\hat{a}}\psi(u;x) \tag{5.1.18}$$

or taking (5.1.16) into account in the form

$$\psi(u;x,t) = e^{t\hat{a}}\psi_0(u;x) = \psi(e^{t\hat{a}}u;x). \tag{5.1.18$'$}$$

Formula (5.1.18) shows that $e^{\hat{a}}$ can be considered as a change of variables operator $e^{\hat{a}}\psi(u;x) = \psi(e^{\hat{a}}u;x)$. Now as in Chapter 2 we can pass from the equation $d\psi/dt = \hat{a}\psi$ to the equation

$$e^{-\hat{s}}\frac{d\psi}{dt} = \frac{d}{dt}e^{-\hat{s}}\psi = e^{-\hat{s}}\hat{a}e^{\hat{s}}e^{-\hat{s}}\psi$$

or putting $e^{-\hat{s}}\psi = \eta$, $e^{-\hat{s}}\hat{a}e^{\hat{s}} = \hat{b}$ to the equation

$$\frac{d\eta}{dt} = \hat{b}\eta, \tag{5.1.19}$$

where $\hat{b} = e^{-\hat{s}}\hat{a}e^{\hat{s}}$. Note that $\hat{b}$ is well defined since

$$e^{-\hat{s}}\hat{a}e^{\hat{s}} = \hat{a} + [\hat{a}\hat{s}] + \frac{1}{2!}[[\hat{a},\hat{s}]\hat{s}] + \cdots$$

and $\hat{\Phi}$ is a Lie algebra, i.e., all summands in this sum belong to $\hat{\Phi}$. (It goes without saying that it is possible to verify directly that $e^{-\hat{s}}\hat{a}e^{\hat{s}} \in \hat{\Phi}$ without using this series.) Great difficulties which can arise in the construction of a rigorous theory in concrete cases are manifest. For instance the existence of $e^{\hat{s}}$ does not necessary imply the existence of $e^{-\hat{s}}$. So if $s = \partial^2 u/\partial x^2$ then $e^{t\hat{s}}u$ is a solution of $\partial\varphi/\partial t = \partial^2\varphi/\partial x^2$, where $\varphi(x,0) = u(x)$. Clearly, $e^{-t\hat{s}}$ is not, generally speaking, well defined. However, since in formal procedures we make use of only finite number of terms of the series $e^{\hat{s}}$, formal theory is still possible.

Following the reasonings of Chapter 2 we need to define a notion of a canonical form of the operator $\hat{a}_0 + \varepsilon\hat{a}_1$. However, even in Chapter 2 we have seen that this notion is only effective if the leading operator possesses some specific properties. Therefore in what follows we will restrict ourselves to systems of a specific form.

5.2 Equations with Partial Derivatives Whose Principal Part Is an Ordinary Differential Equation

We will consider equations

$$\varepsilon^p \frac{du}{dt} = a_0(u;x) + \varepsilon a_1(u;x), \tag{5.2.1}$$

where $u = (u_1, \ldots, u_m)$ and $x = (x_1, \ldots, x_n)$. Here a_0 is a vector function depending on x as on a parameter and $a_1(u;x) \in \Phi_\partial$. (Now we again assume that Φ, Φ_∂ are the spaces of vector functionals.) If we put $\varepsilon = 0$ on the right-hand side of (5.2.1), then the system turns into a system of ordinary differential equations depending on x as on a parameter.

First consider the operator $\hat{a}_0$ related with the equation of the first approximation. Put $\tilde{F}$ for the linear subspace of Φ_∂ consisting of functions $F(u,x)$, i.e. the set of elements of ϕ_∂ that do not depend on derivatives of u. Then (5.1.10) yields

$$\hat{a}_0 F = a_0[F] = a_{0,k}(u;x)\frac{\partial F(u,x)}{\partial u_k} = A_0 F. \tag{5.2.2}$$

Formula (5.2.2) means that on $\tilde{F}$ the operator $\hat{a}_0$ acts as a usual first order differential operator which depends on x as on a parameter.

In what follows we will not make the most general assumptions. The general scheme is already presented in Chapter 2 and here we will restrict ourselves with the following assumptions concerning A_0.

Suppose that as x ranges a domain $\mathcal{D}_1$, $u(x)$ ranges a domain $\mathcal{D}_2$ in the space $\bar{u}$ of variables (u). Let $\varphi_1(u;x), \ldots, \varphi_p(u;x)$; $e_1(u;x), \ldots, e_q(u;x)$, where $p + q = m$ and $x \subset \mathcal{D}_1$ is fixed, be a basic system of functions in $\mathcal{D}_2$ so that e are invariants and φ are eigenfunctions of A_0, i.e.,

$$A_0 e_i = 0, \qquad A_0(\varphi_k) = \lambda_k(e;x)\varphi_k. \tag{5.2.3}$$

Conceding $\lambda_j(e;x) \neq \lambda_k(e;x)$ for $j \neq k$ in the domain $\mathcal{D} = \mathcal{D}_1 \times \mathcal{D}_2$ we will construct in what follows a formal analogue of the notion of the normal form of the operator $\hat{a}_0 + \varepsilon\hat{a}_1$ introduced in Chapter 1. Put

$$\begin{aligned}\psi_k &= \varphi_1^{\nu_{1k}}\varphi_2^{\nu_{2k}}\dots\varphi_p^{\nu_{pk}} \qquad (k = 1,\dots,r),\\ Y &= \mathcal{D}^{\alpha_1}e_{i_1}\dots\mathcal{D}^{\alpha_h}e_{i_h}\mathcal{D}^{\beta_1}\psi_1\dots\mathcal{D}^{\beta_r}\psi_r \quad (\beta_j = \beta_{j1}\dots\beta_{jr_j}), \qquad (5.2.4)\\ \tilde{\lambda}_k &= \nu_{1,k}\lambda_1 + \dots + \nu_{p,k}\lambda_p,\\ \|\beta_j\| &= \beta_{j1} + \dots + \beta_{jr_j}.\end{aligned}$$

Call $\|\beta\| = \sum \|\beta_j\|$ the weight of Y.

Lemma 5.2.1 $\hat{a}_0 Y - (\tilde{\lambda}_1 + \dots + \tilde{\lambda}_r)Y = \sum'$, *where* $\sum'$ *is a finite sum of terms of the same form as* Y*, the weight of each of them being less than* $\|\beta\|$.

PROOF

In fact, (5.1.2), (5.1.12), and (5.2.2) implies

$$\begin{aligned}\hat{a}_0(\mathcal{D}^{\alpha_1}e_{i_1}\dots\mathcal{D}^{\alpha_h}e_{i_h}) &= \mathcal{D}^{\alpha_1}(\hat{a}_0 e_{i_1})\mathcal{D}^{\alpha_2}e_{i_2}\dots\mathcal{D}^{\alpha_h}e_{i_h}\\ &\quad + \mathcal{D}^{\alpha_1}e_{i_1}\mathcal{D}^{\alpha_2}(\hat{a}_0 e_{i_2})\mathcal{D}^{\alpha_3}e_{i_3}\dots\mathcal{D}^{\alpha_h}e_{i_h} + \cdots\\ &= 0. \qquad (5.2.5)\end{aligned}$$

Analogously,

$$\hat{a}_0\mathcal{D}^{q_i}(\psi_i) = \mathcal{D}^{q_i}\hat{a}_0\psi_i = \mathcal{D}^{q_i}(\tilde{\lambda}_i(e,x)\psi_i) = \tilde{\lambda}_i(e,x)\mathcal{D}^{q_i}\psi_i + \sum{}' . \qquad (5.2.6)$$

where $\sum'$ is a finite sum of terms of the same form as Y but whose weights are less than $|q_i|$. Now making use of (5.1.2), (5.2.5), and (5.2.6) we immediately get the statement of the lemma.

Lemma 5.2.1 implies that

$$(\hat{a}_0 - (\tilde{\lambda}_1 + \dots + \tilde{\lambda}_r))^{\|\beta\|+1}Y = 0. \qquad (5.2.7)$$

Formula (5.2.7) is a justification for the following definition. The functional $\varphi \in \phi_\partial$ corresponds to the eigenvalue λ, where $\hat{a}_0(\lambda) = 0$ if for some N

$$(\hat{a}_0 - \lambda)^N\varphi = 0. \qquad (5.2.8)$$

It follows from (5.2.7) that the application of $\mathcal{D}^\beta$ to φ gives the result corresponding to λ but with, generally speaking, higher N. As in Chapter 2, if φ corresponds to λ and ψ corresponds to μ, then $\varphi\psi$ corresponds to $\lambda + \nu$.

Following the scheme considered in Chapter 2, suppose now that all expressions encountered are decomposable into series constructed from expressions of the form (5.2.4). In this case the following statement holds.

Theorem 5.2.2 *There exists an operator* $\hat{S} = \varepsilon \hat{S}_1 + \varepsilon^2 \hat{S}_2 + \cdots$ *such that*

$$e^{-\hat{S}}(\hat{a}_0 + \varepsilon \hat{a}_1)e^{\hat{S}} = \hat{a}_0 + \varepsilon \hat{m}_1 + \cdots = \hat{M}, \tag{5.2.9}$$

where $\hat{M}e_i$ *corresponds to* 0 *and* $M\varphi_k$ *corresponds to* λ_k.

It goes without saying that this statement is purely formal and should be understood in the same sense as the corresponding statements of Chapter 2.

To prove it, it is necessary as in Chapter 2 to be able to define operators $\hat{s}$ and $\hat{m}$ on functionals e and φ only so that $\hat{m}e_i$ corresponds to 0 and $\hat{m}\varphi_k$ to λ_k and

$$[\hat{a}_0, \hat{S}] + \hat{a}_1 = \hat{m}. \tag{5.2.10}$$

Under these assumptions and due to the linearity of the problem we may assume that $\hat{a}_1$ is of the form (5.2.4), i.e. $\hat{a}_1 e_i = Y_i$ and $\hat{a}_1 \varphi_k = \tilde{Y}_k$. First let us define $\hat{S}e_i$ so that $[\hat{a}_0, \hat{S}]e_i + Y_i = \hat{m}e_i$ corresponds to 0, i.e. so does $\hat{a}_0\hat{S}e_i - \hat{S}\hat{a}_0 e_i + Y_i = \hat{a}_0\hat{S}e_i + Y_i$. Put $\hat{S}e_i = -(1/(\tilde{\lambda}_1 + \cdots + \tilde{\lambda}_r))Y_i + \hat{g}e_i$, where Y_i is defined by (5.2.4). Then $\hat{a}_0 Se_i = -(1/(\tilde{\lambda}_1 + \cdots + \tilde{\lambda}_2))\hat{a}_0 Y_i + \hat{a}_0\hat{g}e_i$. But due to Lemma 5.2.1, $\hat{a}_0 Y_i = (\tilde{\lambda}_1 + \cdots + \tilde{\lambda}_r)Y_i + \sum'$ where the weight of terms in $\sum'$ is less than that of Y_i. Therefore $\hat{a}_0\hat{S}e_i + Y_i = \hat{a}_0\hat{g}e_i + \sum'$, where the weight of summand in $\sum'$ is less than that of Y_i. Clearly this process solves the problem: $\hat{m}e_i$ is the sum of terms of the development of $\hat{a}_1 e_i$ that correspond to 0. We will not consider how to search for $\hat{S}\varphi_k$. It is done according to the already described method.

Remark 5.2.3 Formula (5.1.12) often enables one to simplify computations. In particular it implies that a solution of the problem $\hat{a}_0 b = \mathcal{D}^\alpha \psi$ is $b = \mathcal{D}^\alpha C$, where $\hat{a}_0 C = \psi$.

A Connection with N. N. Bogolyubov's Ideas

Now we can literally repeat the said in Section 2.6. From

$$\varepsilon^\alpha \frac{du}{dt} = (\hat{a}_0 + \varepsilon \hat{a}_1)u \tag{5.2.11}$$

it follows that

$$\varepsilon^\alpha \frac{de^{-\hat{S}}u}{dt} = e^{-\hat{S}}(\hat{a}_0 + \varepsilon \hat{a}_1)e^{\hat{S}}e^{-\hat{S}}u$$

or putting $e^{-\hat{S}}u = u^*$ that

$$\varepsilon^\alpha \frac{du^*}{dt} = Mu^*. \tag{5.2.12}$$

Passing to coordinates φ^* and e^* we get

$$\varepsilon^\alpha \frac{de_i^*}{dt} = \varepsilon \hat{m}_1 e_i^* + \cdots, \qquad \varepsilon^\alpha \frac{d\varphi_k^*}{dt} = \lambda_k(e^*, x)\varphi_k^* + \varepsilon \hat{m}_1 \varphi_k^* + \cdots, \tag{5.2.13}$$

where $\hat{m}e_i^*$ corresponds to 0 and $\hat{m}\varphi_k^*$ to λ_k. Since

$$e_i = e^{\hat{s}} e_i^*, \qquad \varphi_j = e^{\hat{s}} \varphi_j^*, \tag{5.2.14}$$

then we again arrive to a formulation typical for Bogolyubov's ideology: find a change of variables (5.2.14) such that if e and φ satisfy equations derived from (5.2.11), then e^* and φ^* satisfy (5.2.1).

It only remains to note that since at each step more and more differentiations in $\hat{S}$ arises, it is as a rule more convenient to employ in the search of $\hat{S}$, $\hat{m}$ the indicated here "operator" method.

5.3 Partial Derivatives. On Whitham Method

(1) Preliminaries

Consider a linear equation with constant coefficients of the form

$$\frac{\varepsilon}{i}\frac{\partial u}{\partial t} + \sum_{0\le \ell \le n} a_\ell \left(\frac{\varepsilon}{i}\right)^\ell \frac{\partial^\ell u}{\partial x^\ell} = 0. \tag{5.3.1}$$

A solution of (5.3.1) can be presented as a Fourier integral

$$u(x,t) = \int e^{\frac{i}{\varepsilon}(kx - \omega(k)t)} f(k)\, dk, \tag{5.3.2}$$

where

$$\omega(k) = \sum_{0\le \ell \le u} a_\ell k^\ell. \tag{5.3.3}$$

Formula (5.3.3) is called a *dispersion relation.* It expresses the fact that for a harmonic wave $\exp i(kx - \omega t)$ which is a solution of (5.3.1) for $\varepsilon = 1$ the frequency ω is a function of wavenumber k.

We will suppose that the time (and the period of oscillation) and the spatial coordinate (and the wavelength) are measured in units characteristic for the given problem.

Put the following question. Let $\varepsilon = 1$ and suppose $u(x,0)$ is known. How will $u(x,t)$ behave at $x, t \gg 1$ and $c = x/t = \text{const}$, i.e., what shall we observe moving along x axis with a speed c after a sufficiently long time after the signal's emittance?

We have

$$u(x,t) = \int e^{i(ck - \omega(k))t} f(k)\, dk.$$

Making use of the classical formulas of the stationary phase method we get

$$u(x,t) \simeq f(\bar{k}) \left(\frac{2\pi}{th''(\bar{k})}\right)^{1/2} \exp\left(ih(\bar{k})t + \frac{i\pi}{4}\operatorname{sgn} h''(\bar{k})\right), \tag{5.3.4}$$

where $h(k) = ck - \omega(k)$ and $\omega'(\bar{k}) = c = x/t$. [To avoid irrelevant details we consider the simplest situation. When several stationary points are present in (5.3.4) one has to write the corresponding sum.] Formula (5.3.4) shows that for $t \gg 1$ all solutions have a very simple form

$$u \simeq \frac{1}{\sqrt{t}} A(x/t) \exp \theta(x,t), \qquad \theta(x,t) = \omega(\bar{k})t - \bar{k}x \tag{5.3.5}$$

where $\bar{k}$ is defined from (5.3.4) and $\bar{k} = \bar{k}(x/t)$. Consider (5.3.5) and let us find out what speed we should move with to remain on the curve of a constant phase or of a constant amplitude. Differentiating the equation $\theta(x,t) = \text{const}$ we get

$$\omega(\bar{k}) - \bar{k}\frac{dx}{dt} + (\omega'(\bar{k})t - x)\frac{d\bar{k}}{dt} = 0$$

whence, see (5.3.4), $dx/dt = \omega(\bar{k})/\bar{k}$. We have obtained an expression which coincides with the usual expression for the phase velocity. Concerning $A(x/t) = \text{const}$, we see that (5.3.4) implies that A is a function of $\bar{k}$ which is constant if so is $\bar{k}$. But since $t\omega'(\bar{k}) - x = 0$, then $dx/dt = \omega'(\bar{k})$ for a constant $\bar{k}$. This is the so-called *group velocity*, i.e., the velocity needed to observe all the time a constant module of the amplitude. It is the group velocity which is connected with the speed of transmission of a signal's energy (far from the source).

The problem can be solved somewhat differently, putting $t = \tau/\varepsilon$ and $x = \xi/\varepsilon$ in the formula $u = \int \exp(i(kx - \omega(k)t)) f(k)\, dk$, i.e., in other notations to return to (5.3.2) and find the asymptotics $u(\xi,\tau)$ as $\varepsilon \to 0$. This formulation makes the problem more general. Namely considering ξ and τ as initial variables we now simply seek an asymptotics of $u(\xi,\tau)$ assuming that (ξ,τ) is contained in a bounded domain. Besides we can now assume that $f(k)$ depends on ε. For instance, putting $F(k) = f_0(k)\exp(i/\varepsilon)\varphi(k)$ we obtain an answer in the form

$$u(\xi,\tau) = F(\bar{k}) \left(\frac{2\pi\varepsilon}{|h''(\bar{k})|} \right)^{1/2} \exp \frac{i}{\varepsilon} h(\bar{k},\xi,\tau),$$

where

$$h = k\xi - \omega(k)\tau + \varphi(k), \qquad h'(\bar{k}) = \xi - \omega'(k)\tau + \varphi'(\bar{k}) = 0.$$

Note that in this formula it is not necessary to assume that $|\tau|/\varepsilon \to \infty$. Since $\theta(\xi,\tau) = h(\bar{k},\xi,\tau)$, then $\theta_\xi(d\xi/d\tau) + \theta_\tau = 0$ and $\theta_\xi = \bar{k} + h(\bar{k})(d\bar{k}/d\xi) = \bar{k}$, $\theta_\tau = -\omega(\bar{k})$ imply that the formula for the phase velocity $d\xi/d\tau = \omega(\bar{k})/\bar{k}$ did not change and lines of constant amplitude are $\bar{k} = \text{const}$, i.e., $\xi - \omega'(c)\tau + \varphi'(c) = 0$, and the group velocity is $d\xi/d\tau = \omega'(\bar{k})$. We have also established the following important fact: if $\theta(\xi,\tau)$ is an asymptotics of $u(\xi,\tau)$, then $\theta_\tau = -\omega(\theta_\xi)$ (*dispersion relation* between θ_ξ and θ_τ). This "experimental" fact shows that the phase θ satisfies a first order equation in partial derivatives which is defined not by a concrete solution $u(\xi,\tau)$ but only by the *initial differential equation.*

Now let us discuss another problem. Consider

$$u(x,t)=\frac{1}{2\delta}\int e^{i(kx-\omega(k)t)}f(k)\,dk$$

where $f(k)$ is nonzero only in the interval $(k_0-\delta,k_0+\delta)$ and δ is small; $u(x,t)$ is a so-called *wave packet.* Then assuming $\delta^2|t|\ll 1$, we easily get

$$u(x,t)\simeq f(k_0)e^{i(k_0x-\omega(k_0)t)}\frac{\sin[\delta(x-z'(k_0)t]}{[\delta(x-\omega'(k_0)t)]}.$$

A phase that varies quickly as compared with amplitude again appeared and phase and group velocities are again given by old formulas.

Considered examples show that if an equation of type (5.3.1) contains a small parameter ε or if we seek a solution at $|t|\gg 1$, i.e., far from the source as in the first formulation, then a solution is of the form $A(x,t,\varepsilon)\exp(i/\varepsilon)\times\theta(x,t)$ or is a sum of summands of this form. Therefore it is clear that if from the very beginning we will seek its asymptotics in the form $A(x,t)\exp(i/\varepsilon)\theta$, then we can get a considerable amount of information without exactly solving the initial equations. Below we will discuss in more details shortcomings of considerations of this kind; presently we will distinguish a sufficiently broad class of equations in the solution of which these considerations are applicable.

Introduce the standard notations putting

$$\mathcal{D}_x=\left(\frac{\partial}{\partial x_1},\dots,\frac{\partial}{\partial x_n}\right),\qquad \left(\frac{1}{i}\mathcal{D}_x\right)^\alpha=\left(\frac{1}{i}\frac{\partial}{\partial x_1}\right)^{\alpha_1}\cdots\left(\frac{1}{i}\frac{\partial}{\partial x_n}\right)^{\alpha_n},$$

$$\alpha=(\alpha_1,\dots,\alpha_n),\qquad p^\alpha=p_1^{\alpha_1}\dots p_n^{\alpha_n},\qquad L(x,p)=\sum_{0\le|\alpha|\le m}a_\alpha(x)p^\alpha.$$

Let us consider the equation

$$L\left(x;\frac{\varepsilon\mathcal{D}_x}{i}\right)u=0. \tag{5.3.6}$$

It is linear and we can try to find the asymptotics of its partial solution in the form

$$u=e^{i\theta(x)/\varepsilon}(A_0(x)+\varepsilon A_1(x)+\cdots)=e^{i\theta/\varepsilon}A. \tag{5.3.7}$$

Since

$$\frac{\varepsilon}{i}\frac{\partial}{\partial x_s}u=e^{i\theta/\varepsilon}\left(\theta_{x_s}A+\frac{\varepsilon}{i}\frac{\partial A}{\partial x_s}\right), \tag{5.3.8}$$

then (5.3.8) and (5.3.6) yield

$$\left\{L(x,p)+\frac{\varepsilon}{i}\left(\left(\frac{\partial A_0}{\partial x},\frac{\partial L}{\partial p}\right)+\frac{1}{2}\sum_{1\le j,k\le m}\frac{\partial^2 L}{\partial p_j\partial p_k}\frac{\partial^2\theta}{\partial x_j\partial x_k}A_0\right)\right.$$
$$\left.+\left(\frac{\varepsilon}{i}\right)^2[\ldots]+\cdots\right\}_{p=\mathrm{grad}\theta}=0,$$

whence after equating to 0 coefficients of powers of ε we have

$$L\left(x,\frac{\partial\theta}{\partial x}\right)=0 \tag{5.3.9}$$

and a linear equation with first order partial derivatives for A_0. (In what follows, this equation will be considered in more detail. Now we will only mention that for A_1, A_2, ... similar recursive equations also arise.) Equation (5.3.9) is completely analogous to the above constructed "dispersive" equation and is called the eikonal equation.

Now let $L(x,u,p)=\sum_{0\le|\alpha|\le m}a_\alpha(x,u)p^\alpha$. Let us consider a quasilinear equation

$$L(x,u;\varepsilon\mathcal{D}_x)u=0. \tag{5.3.10}$$

[Now we write $\varepsilon\mathcal{D}_x$ instead of $(\varepsilon/i)\mathcal{D}_x$ which is more convenient in the general case. Note also that quasilinearity is not necessary and is assumed only to simplify notations.] It is no more possible to seek a solution of (5.3.10) in the form (5.3.7). However, it is possible to seek it in the form

$$u=w(\theta,x)=v(z,x),\qquad z=\theta/\varepsilon \tag{5.3.11}$$

or in the form $y(\xi,x),\xi=\exp\theta/\varepsilon$, where the function u is subject to the following:

Condition I: $$\frac{\partial^k w}{\partial x^m\partial\theta^{k-m}}=\varepsilon^{m-k}F(\theta,x).$$

Condition I means that "large" terms in the computation of $\partial^\alpha u(\theta,x)/\partial x^\alpha$ arise only from differentiating with respect to θ. For instance, $\exp(\theta\lambda(x)/\varepsilon)$ does not satisfy this condition if λ is not a constant.

Put $\varepsilon(\partial w/\partial\theta)=w_1$, then $\varepsilon(\partial u/\partial x_r)=w_1\theta_r+\varepsilon(\partial w/\partial x_r)$, where $\theta_r=\partial\theta/\partial x_r$ and putting $\varepsilon(\partial w_1/d\theta)=w_2$ we get

$$\varepsilon^2\frac{\partial^2 u}{\partial x_2\partial x_e}=w_2\theta_r\theta_\ell+\varepsilon\frac{\partial w_1}{\partial x_\ell}\theta_r+\varepsilon\frac{\partial w_1}{\partial x_r}\theta_\ell+\varepsilon w_1\frac{\partial^2\theta}{\partial x_r\partial x_\ell}+\varepsilon^2\frac{\partial^2 w}{\partial x_r\partial x_\ell}.$$

In these formulas w, w_1, ... are considered as functions in independent variables θ, x_1, ..., x_n.

Continuing this process further we obtain equations

$$\varepsilon\frac{dw}{d\theta}=w_1,\ \ldots,\ \varepsilon\frac{\partial w_{m-2}}{\partial\theta}=w_{m-1},$$

$$\sum_{|\alpha|=0}^{m} a_\alpha(x,w)\theta_1^{\alpha_1}\ldots\theta_n^{\alpha_n}\varepsilon^{|\alpha|}\frac{\partial^{|\alpha|}w}{\partial\theta^{|\alpha|}} = \varepsilon a_1 + \varepsilon^2 a_2 + \cdots.$$

This system clearly belongs to the type considered in Section 5.2 if θ is considered as the time. [This method of reducing the problem to an equation of the type (5.2.1) is clearly not unique. Different methods of reduction not only lead to different computational procedures but may introduce superfluous variables.] In what follows we consider corollaries of our initial assumptions.

(2) The Whitham Method

Let us write our equations in the form

$$\varepsilon\frac{dw_i}{d\theta} = a_0[w_i] + \varepsilon a_1[w_i] + \varepsilon^2 a_2[w_i] + \cdots, \tag{5.3.12}$$

where $a_0[w_i] = a_0(x,\bar{w})$ is a function in $\bar{w} = (w,\ldots,w_{m-1})$ and x, while a_1 and a_2 are some operators in partial derivatives. Let A_0 be a first order operator corresponding to a_0. First consider the case when A_0 possesses only two purely imaginary eigenvalues $\lambda = \pm i\sigma(e,x)$, i.e., $A_0\varphi = \lambda\varphi$, $A_0\bar{\varphi} = -\lambda i$ and $(m-2)$ invariants $(e_1,\ldots,e_{m-2}) = e(x)$. Note that all operators in (5.3.12) contain partial derivatives of an unknown function θ. After reducing (5.3.12) to the normal form the equations will take the form

$$\left.\begin{aligned} \varepsilon\frac{d\varphi^*}{d\theta} &= i\sigma\varphi^* + \varepsilon m_1[\varphi^*] + \varepsilon^2 m_2[\varphi^*] + \cdots, \\ \varepsilon\frac{d\bar{\varphi}^*}{d\theta} &= -i\sigma\bar{\varphi}^* + \varepsilon\overline{m_1[\varphi^*]} + \varepsilon^2\overline{m_2[\varphi^*]} + \cdots, \end{aligned}\right\} \tag{5.3.13}$$

$$\varepsilon\frac{de_k^*}{d\theta} = \varepsilon m_1[e_k^*] + \varepsilon^2 m_2[e_k^*] + \cdots, \tag{5.3.13$'$}$$

where $m_j[e_k^*]$ corresponds to 0 and $m_j[\varphi^*]$ corresponds to λ. If we assume for simplicity's sake that the system (5.3.13), (5.3.13′) does not require an extension, i.e., $m_j[e_k^*]$ only contains e^*, hence (5.3.13′) constitute a closed system, then from (5.3.13′) we deduce that e_k^* as a function in θ and x does not contain "large" derivatives.

But then since θ is also a function in x, we may assume that e^* does *not* depend on θ, i.e., e is a function of $x = (x_1,\ldots,x_n)$ only and the equations for e^* are

$$m_1[e_k^*] + \varepsilon m_2[e_k^*] + \cdots = 0. \tag{5.3.14}$$

But it is still impossible to solve either (5.3.14) or (5.3.13) since they contain partial derivatives of θ. However an approximate solution of (5.3.13) is $\varphi^* = \psi^*\exp(i/\varepsilon)\sigma\theta$ which shows that in order to fulfill Condition I it is necessary and sufficient to satisfy the following generalized eikonal equation:

$$\sigma(e;x) = 1. \tag{5.3.14$'$}$$

[$\sigma(e, x)$ depends on x also via $\nabla\theta$ and we may assume that either $\sigma = \text{const}$ or $\sigma = \text{const} + \varepsilon\sigma_1(x) + \varepsilon^2\sigma_2(x) + \cdots$ which is evidently unimportant. The condition (5.3.14$'$) should be imposed *after* the system is reduced to the normal form.]

Now putting $\varphi^* = e^{i\theta/\varepsilon}\psi^*$ we get an equation for ψ^*

$$\frac{d\psi^*}{d\theta} = \left(m_1[e^{i\theta/\varepsilon}\psi^*] + \varepsilon m_2[e^{i\theta/\varepsilon}\psi^*] + \cdots\right) e^{-i\theta/\varepsilon}$$

where the right-hand side *corresponds to* 0. Therefore we again assume that ψ^* does not depend on θ and get an equation for ψ^*:

$$\left(m_1[e^{i\theta/\varepsilon}\psi^*] + \varepsilon m_2[e^{i\theta/\varepsilon}\psi^*] + \cdots\right) e^{-i\theta/\varepsilon} = 0. \tag{5.3.14$''$}$$

As a result a system of $m + 1$ equations, (5.3.14), (5.3.14$'$), and (5.3.14$''$) arises to define m functions e and ψ, and the function θ. Much more complicated is the case of several essentially different eigenfunctions. It suffices to consider for definiteness sake the situation when eigenfunctions are φ_1, $\bar{\varphi}_1$, φ_2 and $\bar{\varphi}_2$ and $A_0\varphi_1 = i\sigma_1\varphi_1$, $A_0\varphi_2 = i\sigma_2\varphi_2$, where σ_2 is *not* a function in σ_1. The reduction of the system to the normal form is carried out without alterations and we get an analogue of (5.3.13) and (5.3.13$'$). However there are now two essentially different eikonal equations and they cannot be satisfied simultaneously with equations for φ^*, e^*. Therefore we have either to seek an initial solution as a function in two functions θ or to restrict ourselves with a special case considered below.

Let the equation for φ_1^* and φ_2^* be

$$\varepsilon\frac{d\varphi_1^*}{d\theta} = i\sigma_1\varphi_1^* + \varepsilon m_1[\varphi_1^*] + \cdots, \quad \varepsilon\frac{d\varphi_2^*}{d\theta} = i\sigma_2\varphi_2^* + \varepsilon m_1[\varphi_2^*] + \cdots.$$

Suppose that for the second one $\varphi_2^* = 0$ is a solution in a domain of initial variables interesting to us. Then putting $\varphi_2^* = 0$ and taking the eikonal equation in the form $\sigma_1 = 1$ we can repeat all the above considerations and obtain a "1-frequency" solution of our problem.

Remark 5.3.1 From the very beginning it was evident that we were seeking partial solutions. The just considered procedure of obtaining a "1-frequency" solution diminishes their quantity still more. The principal difference from the linear problems is the impossibility in the nonlinear case to recover from two partial solutions a new one.

Remark 5.3.2 Consider the linear case more attentively. In this case the popular method is to seek a solution in the form $u = e^{i\theta/\varepsilon}a$, where $a = a_0 + \varepsilon a_1 + \cdots$ (a geometric optic). If we were seeking a solution using the just considered scheme the problem of the form $\varepsilon(dw/d\theta) = Aw + \varepsilon a_1[w] + \cdots$ would arise. Then we should expect that some of the eigenvalues of the matrix A are real since the eikonal equation should be real. Further, assuming for simplicity that the corresponding first order linear operator has no zero eigenvalues, we will after reducing to the normal form deal with equations of the type (5.3.13). Due to linearity each of them has zero solution and therefore it is possible to construct a 1-frequency solution.

(Eikonal equations are clearly the same under arbitrary methods of construction.) The above-mentioned transform $\varphi^* = \psi^* e^{i\theta/\varepsilon}$ corresponds to the formula $u = e^{i(\theta/\varepsilon)}a$. However later an essential distinction arises: in the first case we get a differential equation for ψ^* of the form $m_0[\psi^*] + \varepsilon m_1[\psi^*] + \cdots = 0$ and when we make use of the standard method the function a is searched for in the form $a_0 + \varepsilon a_1 + \cdots$ and a recursive system of equations for $a_0, a_1, \ldots$ arises. It is not difficult to understand that if we seek ψ^* in the form $\psi^* = \psi_0^* + \varepsilon\psi_1^* + \cdots$, then systems for a_j and ψ_j^* are equivalent. This is the same distinction which takes place between the considered in Section 2.7 methods (a) and (b). The domain where we can make use of differential equations for ψ^* should be broader than the domain where the decomposition of the type $\psi^* = \psi_0^* + \varepsilon\psi_1^* + \cdots$ is valid.

The trivial example which illustrates the difference of the two methods is the following one: $\varepsilon(dx/dt) = i(1+\varepsilon+\varepsilon^2)x$. The equation is already in the normal form. However, if we will seek a solution in the form $x = e^{it/\varepsilon}y$, where $y = y_0 + \varepsilon y_1 + \cdots$, then for y the equation $dy/dt = i(1+\varepsilon)y$ arises. If now we will seek its solution in the form $y_0 + \varepsilon y_1 + \cdots$, then $y_0 = c_0 e^{it}$. Further, secular terms will arise and this solution will fit only for $t\varepsilon \ll 1$.

Remark 5.3.3 (Grad's Example) This example also demonstrates the difference of both methods. It shows that unsolvable equations can arise.

Consider the system

$$\varepsilon\left(\frac{\partial\sigma_1}{\partial t} + \frac{\partial\sigma_2}{\partial x}\right) = -\lambda_1\sigma_1, \qquad \varepsilon\left(\frac{\partial\sigma_2}{\partial t} + \frac{\partial\sigma_1}{\partial x}\right) = -\lambda_2\sigma_2. \tag{5.3.15}$$

Putting $\sigma_1 = \sigma_1(\theta,x,t)$, $\sigma_2 = \sigma_2(\theta,x,t)$, $a = \partial\sigma_1/\partial t + \partial\sigma_2/\partial x, b = \partial\sigma_2/\partial t + \partial\sigma_1/\partial x$ and $\theta_1 = \partial\theta/dt, \theta_2 = \partial\theta/dx$, $\rho = \theta_1^2 - \theta_2^2$ we will reduce the system to the form

$$\varepsilon\frac{d\sigma_1}{d\theta} = a_0^{(1)}(\sigma_1,\sigma_2) + \varepsilon a_1^{(1)}(\sigma_1,\sigma_2),$$

$$\varepsilon\frac{d\sigma_2}{d\theta} = a_0^{(2)}(\sigma_1,\sigma_2) + \varepsilon a_1^{(2)}(\sigma_1,\sigma_2)$$

where

$$a_0^{(1)} = -\frac{\lambda_1\theta_1}{\rho}\sigma_1 + \frac{\lambda_2\theta_2}{\rho}\sigma_2, \qquad a_0^{(2)} = \frac{\lambda_1\theta_2}{\rho}\sigma_1 - \frac{\lambda_2\theta_1}{\rho}\sigma_2,$$

$$a_1^{(1)} = -\frac{\theta_1}{\rho}a + \frac{\theta_2}{\rho}b, \qquad a_1^{(2)} = \frac{\theta_2}{\rho}a - \frac{\theta_1}{\rho}b$$

or $\varepsilon(d\sigma/d\theta) = a_0[\sigma] + \varepsilon a_1[\sigma]$, where $a_k[\sigma_i] = a_k^{(i)}$ $(k = 0, 1,\ i = 1, 2)$. The equation for eigenfunctions φ of the operator a_0, where $\varphi = p\sigma_1 + q\sigma_2$, is $pa_0^{(1)} + qa_0^{(2)} = \lambda(p\sigma_1 + q\sigma_2)$. A partial solution of the eikonal equation is $\theta = t$. When θ is so chosen, eigenfunctions of a_0 are σ_1 and σ_2, i.e., $a_0[\sigma_1] = -\lambda_1\sigma_1$, $a_0[\sigma_2] = -\lambda_2\sigma_2$.

We can make use of the Bogolyubov method and seek a solution in the form

$$\begin{aligned} \sigma_1 &= \sigma_1^* + \varepsilon p_1(\sigma_1^*, \sigma_2^*) + \varepsilon^2 p_2(\sigma_1^*, \sigma_2^*) + \cdots, \\ \sigma_2 &= \sigma_2^* + \varepsilon q_1(\sigma_1^*, \sigma_2^*) + \varepsilon^2 q_2(\sigma_1^*, \sigma_2^*) + \cdots, \end{aligned} \tag{5.3.16}$$

assuming that equations

$$\begin{aligned} \varepsilon\frac{d\sigma_1^*}{d\theta} &= -\lambda_1\sigma_1^* + \varepsilon m_1^{(1)}(\sigma_1^*, \sigma_2^*) + \varepsilon^2 m_2^{(1)}(\sigma_1^*, \sigma_2^*) + \cdots, \\ \varepsilon\frac{d\sigma^{*2}}{d\theta} &= -\lambda_2\sigma_2^* + \varepsilon m_1^{(2)}(\sigma_1^*, \sigma_2^*) + \varepsilon^2 m_2^{(2)}(\sigma_1^*, \sigma_2^*) + \cdots, \end{aligned} \tag{5.3.16$'$}$$

where $m_j^{(1)}, \ldots, m_j^{(2)}$ correspond to $-\lambda_1$, $-\lambda_2$ respectively, are satisfied (clearly p, m are not functions, they are operators).

Rewriting (5.3.16′) in the form $\varepsilon(d\sigma^*/d\theta) = \lambda\sigma^* + \varepsilon m_1[\sigma_1^*] + \varepsilon^2 m_2[\sigma^*]$ and inserting (5.3.16) and (5.3.16′) into the equation which under our choice of θ is

$$\begin{aligned} \varepsilon\frac{d\sigma_1}{d\theta} &= -\lambda_1\sigma_1 - \varepsilon\left(\frac{\partial\sigma_1}{\partial t} + \frac{\partial\sigma_2}{\partial x}\right), \\ \varepsilon\frac{d\sigma_2}{d\theta} &= -\lambda_2\sigma_2 - \varepsilon\left(\frac{\partial\sigma_2}{\partial t} + \frac{\partial\sigma_1}{\partial x}\right), \end{aligned}$$

we get

$$\begin{aligned} &(\lambda\sigma^* + \varepsilon m_1 + \varepsilon^2 m_2)[\sigma_1^* + \varepsilon p_1 + \varepsilon^2 p_2] \\ &\qquad = -\lambda_1(\sigma_1^* + \varepsilon p_1^* + \varepsilon^2 p_2^*) \\ &\qquad\quad - \varepsilon\left(\frac{\partial}{\partial t}(\sigma_1^* + \varepsilon p_1 + \varepsilon^2 p_2) + \frac{\partial}{\partial x}(\sigma_2^* + \varepsilon q_1 + \varepsilon^2 q_2)\right) \\ &(\lambda\sigma^* + \varepsilon m_1 + \varepsilon^2 m_2)[\sigma_2^* + \varepsilon q_1 + \varepsilon^2 q_2] \\ &\qquad = -\lambda_2(\sigma_2^* + \varepsilon q_1^* + \varepsilon^2 q_2^*) \\ &\qquad\quad - \varepsilon\left(\frac{\partial}{\partial t}(\sigma_2^* + \varepsilon q_1 + \varepsilon^2 q_2) + \frac{\partial}{\partial x}(\sigma_1^* + \varepsilon p_1 + \varepsilon^2 p_2)\right). \end{aligned}$$

From here equating coefficients of the same powers of ε we get

$$(\lambda\sigma^*)[p_1] + m_1[\sigma_1^*] = -\lambda_1 p_1^* - \frac{\partial\sigma_1^*}{\partial t} - \frac{\partial\sigma_2^*}{\partial x}, \tag{5.3.17}$$

$$(\lambda\sigma^*)[q_1] + m_1[\sigma_2^*] = -\lambda_2 q_1^* - \frac{\partial\sigma_2^*}{\partial t} - \frac{\partial\sigma_1^*}{\partial x}, \tag{5.3.17$'$}$$

$$(\lambda\sigma^*)[p_2] + m_1[p_1] + m_2[\sigma_1^*] = -\lambda_1 p_2 - \frac{\partial p_1}{\partial t} - \frac{\partial q_1}{\partial x}, \qquad (5.3.18)$$

$$(\lambda\sigma^*)[q_2] + m_1[q_1] + m_2[\sigma_2^*] = -\lambda_2 q_2 - \frac{\partial q_1}{\partial t} - \frac{\partial p_1}{\partial x}. \qquad (5.3.18')$$

Since $m_1[\sigma_1^*] = m_1^{(1)}$ should correspond to $-\lambda_1$, then (5.3.17) gives $m_1^{(1)} = -(\partial/\partial t)\sigma_1^*$ and $[\lambda\sigma^*][p_1] + \lambda_1 p_1^* = -(\partial/\partial x)\sigma_2^*$. Putting $p_1 = \alpha\partial\sigma_2^*/\partial x$ we get $-\lambda_2\alpha\partial\sigma_2^*/\partial x + \alpha\lambda_1\partial\sigma_2^*/\partial x = -\partial\sigma_2^*/\partial x$, whence $p_1 = (\partial\sigma_2^*/\partial x)/\lambda_2 - \lambda_1$. Analogously (5.3.17′) gives $m_1^{(2)} = -\partial\sigma_2^*/\partial t$, $q_1 = -(\partial\sigma_1^*/\partial x)/\lambda_2 - \lambda_1$. Inserting these results in (5.3.18) and (5.3.18′), we get

$$m_2[\sigma_1^*] = m_2^{(1)} = \frac{1}{\lambda_2 - \lambda_1}\frac{\partial^2\sigma_1^*}{\partial x^2},$$
$$m_2[\sigma_2^*] = m_2^{(2)} = -\frac{1}{\lambda_2 - \lambda_1}\frac{\partial^2\sigma_2^*}{\partial x^2}.$$

Now the equations are

$$\varepsilon\frac{\partial\sigma_1^*}{\partial\theta} = -\lambda_1\sigma_1^* - \varepsilon\frac{\partial\sigma_1^*}{\partial t} + \frac{1}{\lambda_2 - \lambda_1}\frac{\partial^2\sigma_1^*}{\partial x^2},$$
$$\varepsilon\frac{\partial\sigma_2^*}{\partial\theta} = -\lambda_2\sigma_2^* - \varepsilon\frac{\partial\sigma_2^*}{\partial t} - \frac{1}{\lambda_2 - \lambda_1}\frac{\partial^2\sigma_2^*}{\partial x^2}.$$

Putting $\sigma_j^* = \tau_j^* \exp(-\lambda\theta/\varepsilon)$, where $j = 1, 2$, and assuming that τ_j^* does not depend on θ we get

$$\frac{\partial\tau_1^*}{\partial t} = \frac{1}{\lambda_2 - \lambda_1}\frac{\partial^2\tau_1^*}{\partial x^2}, \qquad \frac{\partial\tau_2^*}{\partial t} = -\frac{1}{\lambda_2 - \lambda_1}\frac{\partial^2\tau_2^*}{\partial x^2}. \qquad (5.3.19)$$

If $\lambda_2 > \lambda_1$ then the second equation in (5.3.19) is unsolvable. Intuitively the reason is evident: the solution corresponding to λ_2 is clearly unstable and the unsolvability of the equation mirrors this fact.

Note that if both λ are purely imaginary, then both equations (5.3.19) are solvable.

This example shows that if the exponential decreasing plays a role in the problem then it is necessary to restrict ourselves with the consideration of only the minimal eigenvalue. However, when the domain varies the minimal eigenvalues may be interchanged making it impossible, generally speaking, to restrict ourselves with a fixed branch of a solution.

5.4 Geometric Optics and the Maslov Method

(A) Geometric Optics

Under problems of the geometric optics and the ray method we will understand problems of formal (asymptotic) solutions of equations

$$L\left(x, \frac{\varepsilon}{i}d; \varepsilon\right) u = 0, \tag{5.4.1}$$

where

$$x = (x_1, \ldots, x_n), \qquad d = \left(\frac{\partial}{\partial x_1}, \ldots, \frac{\partial}{\partial x_n}\right), \qquad d^\alpha = \frac{\partial^{\alpha_1 + \cdots + \alpha_n}}{\partial x_1^{\alpha_1}, \ldots, \partial x_n^{\alpha_n}},$$
$$(\alpha = \alpha_1, \ldots, \alpha_n), \qquad |\alpha| = \textstyle\sum_i \alpha_i, p^\alpha = p_1^{\alpha_1} \ldots p_n^{\alpha_n},$$
$$L(x, p, \varepsilon) = L_0(x, p) + \varepsilon L_1(x, p) + \cdots$$

and $L_i(x, p)$ are polynomials in $p = (p_1, \ldots, p_n)$ whose coefficients are smooth functions in x provided a solution is searched in the form

$$u = e^{i\theta/\varepsilon}(A_0(x) + \varepsilon A_1(x) + \cdots). \tag{5.4.2}$$

It is often possible to find a solution of the form (5.4.2) only in a "small" domain. Obstructions to its continuation are singular manifolds in the sense of Chapter 4 (caustics) since when we reach them it is no longer possible to seek a solution in the form (5.4.2). In this section we recite the classical method of a solution of the problem in small and give several formulas needed in what follows.

Remark 5.4.1 Instead of (5.4.2) we may seek u in the form $u = \exp(i/\varepsilon) \times (\theta_0 + \varepsilon\theta_1 + \cdots)$. This equivalent form is sometimes more convenient. We may seek u in the form $u = u(\theta/\varepsilon, x)$ and use method (b) from Section 5.3. We will get equations for 1-frequency solutions. If we seek their solution in the form (5.4.2) we get the usual formulas but near caustics these equations can be used for a reconstruction.

Keeping in mind what will follow let us make use of the following formal trick: let us present L in the form

$$L = L(x, p, \varepsilon) = \int \tilde{L}(x, \nu; \varepsilon) e^{i(p,\nu)}\, d\nu. \tag{5.4.3}$$

[Formula (5.4.3) is understood in the sense of generalized functions theory; i.e., the right-hand side of (5.4.3) is defined if (5.4.3), being multiplied by an arbitrary smooth function with a compact support, yields an identity.]

Equation (5.4.1) will turn into

$$L\left(x, \frac{\varepsilon d}{i}; \varepsilon\right) u = \int \tilde{L}(x, \nu, \varepsilon) e^{\varepsilon(d,\nu)} u\, d\nu$$
$$= \int \tilde{L}(x, \nu, \varepsilon) u(x + \varepsilon\nu)\, d\nu = 0 \tag{5.4.4}$$

[we have made use of the formula $e^{\varepsilon(d,\nu)}u(x) = u(x+\varepsilon\nu)$].

Substituting (5.4.2) into (5.4.4) gives

$$\int \tilde{L}(x,\nu;\varepsilon)e^{i\theta(x+\varepsilon\nu)/\varepsilon}(A_0(x+\varepsilon\nu)+\varepsilon A_1(x+\varepsilon\nu)+\cdots)\,d\nu = 0. \quad (5.4.5)$$

But

$$e^{i\theta(x+\varepsilon\nu)/\varepsilon} = e^{i\theta(x)/\varepsilon}e^{i(\nu,\nabla\theta)}\left(1+\frac{i\varepsilon}{2}\frac{\partial^2\theta}{\partial x_\alpha \partial x_\beta}\nu_\alpha\nu_\beta+\cdots\right). \quad (5.4.6)$$

Substituting (5.4.6) and developments of $\tilde{L}$ and A_i into power series in ε into (5.4.5) we get, collecting terms of the same powers of ε, equations of which we give below the first two ones:

$$\int e^{i(\nu,\nabla\theta)}\tilde{L}(x,\nu)A_0(x)d\nu = 0,$$

$$\int e^{i(\nu,\nabla\theta)}\left(\tilde{L}_0(x,\nu)\left(A_1+\frac{i}{2}\frac{\partial^2\theta}{\partial x_\alpha\partial x_\beta}\nu_\alpha\nu_\beta A_0+\nu\nabla A_0\right)+i\tilde{L}_1A_0\right)d\nu = 0.$$

Taking (5.4.3) into account and putting $\nabla\theta = p$ we easily get

$$L_0(x,p) = 0 \quad \text{(eikonal equation)}, \quad (5.4.7)$$

$$\frac{\partial L_0(x,p)}{\partial p_k}\frac{\partial A_0}{\partial x_k}+\frac{1}{2}\frac{\partial^2 L_0(x,p)}{\partial p_\alpha \partial p_\beta}\frac{\partial^2\theta}{\partial x_\alpha\partial x_\beta}A_0+iL_1(x,p)A_0. \quad (5.4.7')$$

Equations for other A are easily written and investigated as (5.4.7′).

Before we proceed further let us make a digression which will be used later. The form (5.4.2) in which we seek u can be generalized: put

$$u = \int e^{i((\check{x},\check{p})/\varepsilon)}w(\bar{x},\check{p})\,d\check{p} \quad (5.4.8)$$

where $x = (\bar{x},\check{x}), \bar{x} = (x_1,\ldots,x_k), \check{x} = (x_{k+1},\ldots,x_n), w = e^{ig(\bar{x},\check{p})/\varepsilon}(v_0(\bar{x},\check{p})+\varepsilon v_1(\bar{x},\check{p})+\cdots)$.

As we have already seen, if in (5.4.8) the stationary phase method can be used, then computing the asymptotics of u we will come to the presentation (5.4.2) which makes it possible to assume (5.4.8) as a generalization of (5.4.2). Let us now present L in the form

$$L(\bar{x},\check{x},\bar{p},\check{p}) = \int \tilde{L}(\bar{x},\xi;\nu,\check{p})e^{i(\check{x},\xi)}e^{i(\bar{p},\nu)}\,d\nu\,d\xi. \quad (5.4.9)$$

The equation is

$$\begin{aligned} 0 = Lu &= \int \tilde{L}(\bar{x},\xi;\nu,\frac{\varepsilon}{i}\check{d})e^{\varepsilon(\check{x},\xi)}e^{\varepsilon(\bar{d},\nu)}e^{\varepsilon(\check{x},\check{p})/\varepsilon}w(\bar{x},\check{p})\,d\check{p}\,d\nu\,d\xi \\ &= \int \tilde{L}(\bar{x},\xi;\nu,\check{p})e^{i(\check{x},\check{p}+\varepsilon\xi)/\varepsilon}w(\bar{x}+\varepsilon\nu;\check{p})\,d\check{p}\,d\nu\,d\xi \\ &= \int e^{i(\check{x},\check{p})/\varepsilon}\,d\check{p}\int \tilde{L}(\bar{x},\xi;\nu,\check{p}-\varepsilon\xi)w(\bar{x}+\varepsilon\nu,\check{p}-\varepsilon\xi)\,d\nu\,d\xi \end{aligned}$$

or finally

$$\int \tilde{L}(\bar{x},\xi;\nu,\check{p}-\varepsilon\xi)w(\bar{x}+\varepsilon\nu,\check{p}-\varepsilon\xi)\,d\xi\,d\nu = 0. \tag{5.4.10}$$

We see that proceeding as above we can obtain equations similar to (5.4.7) and (5.4.7′). The matter of principle is the deduction from these formulas that the problems of seeking u either in the form (5.4.2) or in the form (5.4.8) are local.

Now let us search for u in the form (5.4.2), i.e., let us solve a (local) problem, assuming that u is defined on a $(n-1)$-dimensional surface

$$S : x_i = \bar{x}_i(\alpha_1,\dots,\alpha_{n-1}) \qquad \text{for } i = 1,\dots,n,$$

where $\theta|_S = \theta_0(\alpha)$ and $A_i|_S = A_i(\alpha)$.

Solving (5.4.7)

The Cauchy Method. Under certain conditions we may find

$$\left.\frac{\partial\theta}{\partial x_k}\right|_S = \bar{p}_k(\alpha)$$

on S. In fact, $\theta(\bar{x}(\alpha)) = \theta_0(\alpha)$ implies

$$\frac{\partial\theta}{\partial\bar{x}_k}\frac{\partial\bar{x}_k}{\partial\alpha_s} = \bar{p}_k(\alpha)\frac{\partial\bar{x}_k}{\partial\alpha_s} = \frac{\partial\theta_0(\alpha)}{\partial\alpha_s}, \qquad s = 1,\dots,n-1 \tag{5.4.11}$$

and besides

$$L_0(\bar{x}(\alpha),\bar{p}(\alpha)) = 0. \tag{5.4.12}$$

Let us suppose that the system (5.4.11), (5.4.12) is solvable and its Jacobian does not vanish

$$\frac{\partial(\bar{x},L_0)}{\partial\bar{p}} = \frac{\partial L_0}{\partial\bar{p}_1}N_1 + \cdots + \frac{\partial L_0}{\partial\bar{p}_n}N_n \neq 0. \tag{5.4.13}$$

The vector $N = (N_1,\dots,N_n)$ is a normal vector to S. Thus we demand that the vector $\partial L/\partial\bar{p}$ does not belong to the tangent plane to S.

Now consider a Hamiltonian system

$$\frac{dx_k}{dt} = \frac{\partial L_0}{\partial p_k}, \quad \frac{dp_k}{dt} = -\frac{\partial L_0}{\partial x_k} \qquad \text{for } k = 1,\dots,n \tag{5.4.14}$$

where $x_k(0) = \bar{x}_k(\alpha)$, $p_k(0) = \bar{p}_k(\alpha)$. Denote its solution by $x(\alpha,t)$ and $p(\alpha,t)$. We have obtained a manifold $\mathcal{L}$ of trajectories in $2n$-dimensional space $[H{:}(x,p)]$ depending on n parameters (α,t). In what follows we will assume that $\mathcal{L}$ is n-dimensional, i.e., $\operatorname{rank}(\partial(x,p)/\partial(\alpha t)) = n$.

The following statements hold:

1. If the required solution θ of (5.4.7) exists, then $(\partial\theta/\partial x_k)(x(\alpha,t)) = p_k(\alpha,t)$ for $k = 1,\dots,n$.

2. Additionally suppose $\mathcal{T} = \partial x(\alpha,t)/\partial(\alpha,t) \neq 0$. Since (1) implies that

$$\frac{d\theta(x(\alpha,t))}{dt} = \sum p_k \frac{\partial L_0(x(\alpha,t),p_k)}{\partial p_k}, \tag{5.4.15}$$

then

$$\theta(x(\alpha,t)) = \theta_0(\alpha) + \int_0^t p_k \frac{\partial L_0}{\partial p_k}\, d\tau \tag{5.4.15$'$}$$

and $\mathcal{T} \neq 0$ implies the solvability of equations $x(\alpha,t) = y$ with respect to (α,t); hence (5.4.15′) allows us to find $\theta(x,t)$ if it exists.

3. The problem $\mathcal{L}(x,p) = 0$, where $p_k = \partial\theta/\partial x_k$, can be clearly formulated in coordinates (α,t) as follows:

   Find $p(\alpha,t)$, $x(\alpha,t)$ satisfying $\mathcal{L}(x(\alpha,t),p(\alpha,t)) = 0$, where $\sum P_k(\alpha,t)\,d\alpha_k + P_0(\alpha,t)\,dt$ is the total differential of a function $\theta(\alpha,t)$. If

$$P_k(\alpha,t) = \sum p_\ell(\alpha,t)\frac{\partial x_\ell}{\partial \alpha_k} \qquad \text{and} \qquad P_0(\alpha,t) = \sum p_\ell(\alpha,t)\frac{\partial x_\ell}{\partial t}$$

then the following theorem holds:

Theorem 5.4.2 (The Existence Theorem) *If $x(\alpha,t)$, $p(\alpha,t)$ is a solution of (5.4.14), then $P_k(\alpha,t)\,d\alpha_k + P_0(\alpha,t)\,dt$ is the total differential of a function $\theta(\alpha,t)$ and $\theta(\alpha,0) = \theta_0(\alpha)$.*

We will verify only (1) [the validity of (3) is almost evident when (1) is established].

Proof of 1

Since $p(x) = \nabla\theta(x)$ is known as a function in x, then it is possible to solve the system

$$\frac{dx_k}{dt} = \frac{\partial L_0(x,p(x))}{\partial p_k}, \qquad x_k(0) = \bar{x}_k(\alpha). \tag{5.4.16}$$

But $L_0(x,p) = 0$ implies $0 = \partial L_0/\partial x_k + (\partial L_0/\partial p_s)(\partial p_k/\partial x_s)$, i.e., on the trajectory (5.4.16) we have $0 = \partial L/\partial x_k + \partial p_k/\partial t$. Then $x(t)$ and $\nabla_x\theta(x(t)) = p(t)$ is a solution of (5.4.14) proving (1).

The final result is as follows: integrating the system of ordinary differential equations (5.4.14) and (5.4.15′) we may define the function $\theta(\alpha,t)$. If $\mathcal{T} = \partial x(\alpha,t)/\partial(\alpha,t) \neq 0$, then we may find a desired function θ as a function in x. This is the main result of the local theory. A possibility to define $\theta(x)$ is based on the condition $\mathcal{T} = \partial x(\alpha,t)/\partial(\alpha,t) \neq 0$. This condition often fails in the large on the so-called caustics and also in boundary problems for several points of boundary surfaces, e.g., in the semishade in diffraction problems which will be discussed separately. In several important works by V. P. Maslov a method was proposed permitting in several problems to extend a solution "beyond the caustic." It will be considered below, but in a preliminary we will discuss a solution of equations of the type (5.4.7′) in the local theory.

Equation (5.4.7′) can be rewritten in the form

$$\frac{dA_0}{dt} - \sum_k \frac{1}{2}\frac{\partial^2 L_0}{\partial p_k \partial x_k} A_0 + \frac{1}{2}\frac{1}{\mathcal{T}}\frac{d\mathcal{T}}{dt} A_0 + iL_1(x,p)A_0 = 0, \tag{5.4.17}$$

where d/dt is understood as a differentiation along the trajectory (5.4.14). To show this it suffices to make use of the identity

$$\frac{1}{\mathcal{T}}\frac{d\mathcal{T}(\alpha,t)}{dt} = \frac{\partial^2 L_0}{\partial p_k \partial x_k} + \frac{\partial^2 L_0}{\partial p_k \partial p_\ell}\frac{\partial^2 \theta}{\partial x_k \partial x_\ell}$$

which is a direct corollary of the following

Theorem 5.4.3 (Liouville's Theorem) *For any autonomous system $dx_k/dt = f_k(x)$ for $k = 1, \ldots, n$, where $x_k = x_k(\alpha,t), x_k(\alpha,0) = \bar{x}_k(\alpha)$, we have*

$$\frac{1}{\mathcal{T}}\frac{d\mathcal{T}(\alpha,t)}{dt} = \sum_k \frac{\partial f_k(x)}{\partial x_k}.$$

Proof of the Liouville Theorem

From $dx_k/dt = f_k(x)$ it follows that $x_k(\alpha, t+\delta) \simeq x_k(\alpha,t) + \delta f_k$. The differentiation gives

$$\frac{\partial x_k(\alpha,t+\delta)}{\partial \alpha_s} \simeq \frac{\partial x_k(\alpha,t)}{\partial \alpha_s} + \delta \sum_k \frac{\partial f_k}{\partial x_\ell}\frac{\partial x_\ell}{\partial \alpha_s},$$

$$\frac{\partial x_k(\alpha,t+\delta)}{\partial t} \simeq \frac{\partial x_k(\alpha,t)}{\partial t} + \delta \sum_k \frac{\partial f_k}{\partial x_\ell}\frac{\partial x_\ell}{\partial t}.$$

Putting $A(t) = (\partial^2 x/\partial\alpha\partial t)$ we rewrite this in the form

$$A(t+\delta) \simeq A(t) + \delta \frac{\partial f}{\partial x} A$$

or

$$A(t+\delta)A^{-1} \simeq \left(E + \delta\frac{\partial f}{\partial x}\right).$$

But then

$$\det(A(t+\delta)A^{-1}(t)) = \frac{\mathcal{T}(t+\delta)}{\mathcal{T}(t)} \simeq \left(1 + \frac{\delta}{\mathcal{T}}\frac{d\mathcal{T}}{dt}\right) \simeq \det\left(E + \delta\frac{\partial f}{\partial x}\right)$$
$$\simeq 1 + \delta Sp\frac{\partial f}{\partial x} = 1 + \delta \sum_k \frac{\partial f_k}{\partial x_k}$$

implying the proof.

In our case $dx_k/dt = \partial L(x,p)/\partial p_k$, $p = \nabla_x \theta$. From (5.4.17) we deduce

$$A_0(x,t) = A_0(x(\alpha,0))(|\mathcal{T}(\alpha,0)|/|\mathcal{T}(\alpha,t))^{1/2} \cdot \exp \int_0^t \left(\frac{1}{2} \sum \frac{\partial^2 L_0}{\partial x_k \partial p_k} - iL_1(x,p) \right) dt. \tag{5.4.18}$$

Equations for A_i are similarly solved. From (5.4.18) it is clear that vanishing of $|\mathcal{T}(\alpha,t)|$ is an obstruction for a search of an asymptotics in the given form even if θ is known. The manifold $\mathcal{T}(\alpha,t) = 0$ is singular.

(B) The Maslov Method

The singular manifold of our problem is $\mathcal{T}(\alpha,t) = 0$. However note the following: by assumption the manifold $[\mathcal{L} : x(\alpha,t), p(\alpha,t)]$ is an n-dimensional submanifold in $2n$-dimensional space (x,p). It means that the rank of $\partial(x,r)/\partial(\alpha,t)$ equals n.

In what follows we will prove the following theorem which serves as a base of the method.

Theorem 5.4.4 *There is a numeration of variables x [and consequently of conjugate variables p] and a number k (both the numeration and the number k depend on the selected point (α,t)) such that at any point (α,t) of $\mathcal{L}$ we have*

$$\tilde{\mathcal{T}}(\alpha,t) = \frac{\partial(x_1,\ldots,x_k,p_{k+1},\ldots,p_n)}{\partial(\alpha,t)} \neq 0.$$

It is important that variables p in this formula are conjugate to the remaining variables $x_{k+1},\ldots,x_n$. Nonvanishing of $\tilde{\mathcal{T}}(\alpha,t)$ means that the function $\theta(\alpha,t)$ defined in (a) can be considered as a function in $(\bar{x},\check{p})$, where

$$\bar{x} = (x_1,\ldots,x_k), \qquad x = (x_{k+1},\ldots,x_n),$$

$$\bar{p} = (p_1,\ldots,p_k), \qquad p = (p_{k+1},\ldots,p_n).$$

The Maslov method consists in the search of a solution in the form (5.4.8) in the domain, where $\tilde{\mathcal{T}}(\alpha,t) \neq 0$. We show that $g(\bar{x},\check{p})$ is connected with $\theta(x)$ by the Legendre transformation if we can simultaneously make use of (5.4.2) and (5.4.8), and formulas are deduced which enable us to connect these representations. When the neighborhood of the point where $\mathcal{T}(\alpha,t) = 0$ is passed over by this method we may return to the problem of the search of a solution in the form of (5.4.2).

Now let us pass to detailed considerations. Suppose that while computing u and integrating (5.4.14) and (5.4.15) along the trajectory starting from $(\alpha_0,0)$ we arrive at the point $x_0(\alpha_0,t_0)$ where $\mathcal{T}(\alpha_0,t_0) = 0$. Let us draw a tube of trajectories from a small neighborhood of α_0 on S_0. Now suppose

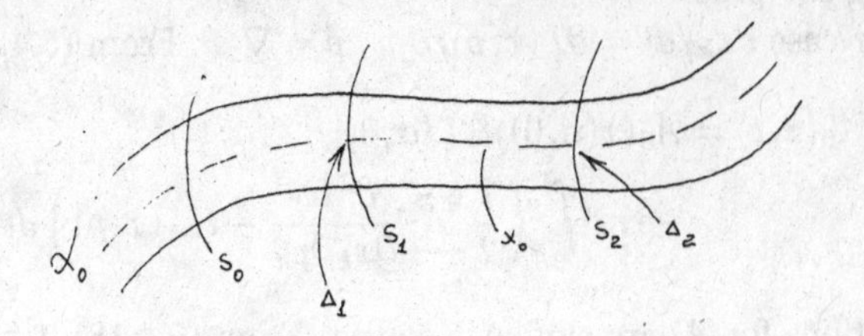

FIGURE 5.1.

that in the space $X = (x_1, \ldots, x_n)$ it is possible to draw in a vicinity of $x_0 = x(\alpha_0, t_0)$ two surfaces S_1 and S_2 with the property that if Δ_1 and Δ_2 are sets of intersecting points of the projections in X of curves of the tube with S_1 and S_2, then $|\partial(x)/\partial(\alpha, t)| = T(\alpha, t) > \gamma > 0$ if $x(\alpha, t) \in \Delta_1, \Delta_2$. For the set $L_{1,2}$ of points belonging to curves of the tube whose projection of X belongs to a domain between S_1 and S_2 we have

$$|\bar{T}(\alpha, t)| = \left|\frac{\partial(\bar{x}, p)}{\partial(\alpha, t)}\right| > \gamma > 0.$$

The last requirement may be considered to be satisfied due to Theorem 5.4.4.

Remark 5.4.5 A presence of a close to S_1 surface S_2 on which again $\bar{T}(\alpha, t) \neq 0$ is a quite serious requirement though often satisfied.

We seek a solution of (5.4.1) in the domain $L_{1,2}$ in the form (5.4.8). Making use of (5.4.10) we get

$$\int \tilde{L}(\bar{x}, \xi; \nu, \check{p} - \varepsilon\xi) e^{(i/\varepsilon) g(\bar{x}+\varepsilon\nu, \check{p}-\varepsilon\xi)} v(\bar{x} + \varepsilon\nu, \check{p} - \varepsilon\xi)\, d\nu\, d\xi = 0. \quad (5.4.19)$$

A procedure which has led to (5.4.7), (5.4.7′) is repeated word for word. Let us write out only results:

$$L_0^*\left(\bar{x}, \frac{\partial g}{\partial \bar{p}}, \frac{\partial g}{\partial \bar{x}}, \check{p}\right) = L_0\left(\bar{x}, -\frac{\partial g}{\partial \check{p}}; \frac{\partial g}{\partial \bar{x}}, \check{p}\right) = 0, \qquad (5.4.20)$$

$$\begin{aligned} &\left(\frac{\partial v_0}{\partial \bar{x}}, \frac{\partial L_0^*}{\partial \bar{p}}\right) - \left(\frac{\partial v}{\partial \bar{p}}, \frac{\partial L_0^*}{\partial \check{x}}\right) \\ &\quad + \frac{1}{2}\sum\left(\frac{\partial^2 g}{\partial \bar{x}_\alpha \partial \bar{x}_\beta}\frac{\partial^2 L_0^*}{\partial \bar{p}_\alpha \partial \bar{p}_\beta} - \frac{\partial^2 g}{\partial \bar{x}_\alpha \partial \check{p}_\beta}\frac{\partial^2 L_0^*}{\partial \bar{p}_\alpha \partial \check{y}_\beta} + \frac{\partial^2 g}{\partial \check{p}_\alpha \partial \check{p}_\beta}\frac{\partial^2 L_0^*}{\partial \check{x}_\alpha \partial \check{x}_\beta}\right) v_0 \\ &\quad + \sum_\alpha \frac{\partial^2 L_0^*}{\partial \check{p}_\alpha \partial \check{x}_\alpha} v_0 + iL_1^* v_0 = 0. \end{aligned} \qquad (5.4.21)$$

Let us also require that S_1 and S_2 be chosen so that $\det(\partial^2\theta(\bar{x},\check{x}))/(\partial\hat{x}_\alpha\partial\hat{x}_\beta) \neq 0$ on Δ_1 and Δ_2. It means the solvability of $\partial\theta(\bar{x},\check{x})/\partial\check{x} = \check{p}$ in a vicinity of corresponding points of the trajectory. Now assuming that u is defined on S_1 by (5.4.2) as well as by (5.4.8), where

$$w = \varepsilon^h \left(\exp\frac{ig(\bar{x},\check{p})}{\varepsilon}\right)(v_0(\bar{x},\check{p}) + \varepsilon v_1 + \cdots),$$

we will establish a connection between θ and g, and A and v. First of all it is clear that in the formula

$$u = \int e^{(i/\varepsilon)(\check{x},\check{p})+(i/\varepsilon)g(\bar{x},\check{p})}(v_0(\bar{x},\check{p}) + \varepsilon v_1 + \cdots)\,d\check{p}$$

we may make use of the usual stationary phase method. A stationary point in this formula is defined by the equation

$$\check{x} + \frac{\partial g(\bar{x},\check{p})}{\partial\check{p}} = 0 \tag{5.4.22}$$

and since by assumption $\partial\theta/\partial\check{x} = \check{p}$ is solvable, then θ and g are connected by the Legendre transformation

$$\theta(\bar{x},\check{x}) = (\check{x},\check{p}) + g(\bar{x},\check{p}). \tag{5.4.23}$$

We must also show that the form $(\partial^2 g/\partial\check{p}_\alpha\partial\check{p}_\beta)\xi_\alpha\xi_\beta = G$ is nondegenerate. But $\partial g/\partial\check{p}_\alpha = -\check{x}_\alpha(\bar{x},\check{p})$, hence $(\partial^2 g/\partial\check{p}_\alpha\partial\check{p}_\beta) = -\partial\check{x}_\alpha(\bar{x},\check{p})/\partial\check{p}_\beta$. However by assumption on S_1 we have $T = \partial(\bar{x},x)/\partial(\alpha,t) \neq 0$, $\tilde{T} = \partial(\bar{x},\check{p})/\partial(\alpha,t) \neq 0$ hence $\partial(\bar{x},\check{x})/\partial(\bar{x},\check{p}) = T/\tilde{T} \neq 0$ immediately implying our statement.

Thus, we know that $g(\bar{x},\check{p}) = (\check{x},\check{p}) - \theta(\bar{x},\check{x})$ on S_1 may be considered as an $(n-1)$-dimensional surface in the space $(\bar{x},\check{p})$ and we may apply the Cauchy method for solving (5.4.20). Putting $-\check{x} = \partial g/\partial\check{p}$, $\partial g/\partial\bar{x} = \bar{p}$ we get equations coinciding with (5.4.14). Moreover as follows from (5.4.21) and the equality $\partial\theta/\partial\bar{x} = \partial g/\partial\bar{x}$ [which follows after we differentiate (5.4.22) with respect to $\bar{x}$ and take (5.4.21) into account since $\check{p} = \check{p}(\bar{x},\check{x})$] the values of indeterminates coincide on S_1 for both systems. Thus we deal with the same trajectory ℓ. Values of g on this trajectory are computed making use of the auxiliary equation $\partial g(\bar{x},\check{p})/\partial t = \bar{p}(\partial L_0/\partial\bar{p}) - (\partial g/\partial\check{p})(\partial L_0/\partial\check{x})$ similar to (5.4.15). We obtain g as a function in (α,t) but since $\tilde{T}(\alpha,t) \neq 0$ in $L_{1,2}$ we may consider g as a function in $(\bar{x},\check{p})$.

Computation of $v_0, v_1, \ldots$ on S_1 is performed making use of the well-known formula of the stationary phase method whose first term is given below: from

$$u = \iint e^{i\theta(\check{p})/\varepsilon} v(\check{p})\,d\check{p}, \qquad \theta(\check{p}) = (\check{x},\check{p}) + g(\bar{x},\check{p})$$

it follows that

$$u \simeq (2\pi\varepsilon)^{(n-k)/2} \left| \det \left(\frac{\partial^2 \theta}{\partial \check{p}_\alpha \partial \check{p}_\beta} \right) \right|^{-1/2} \times v(\check{p}) \exp \left(\frac{i\theta(\check{p})}{\varepsilon} + i\frac{\pi}{4} \operatorname{sgn} \sum \left(\frac{\partial^2 \theta}{\partial \check{p}_\alpha \partial \check{p}_\beta} \right) \right) \quad \text{for } \varepsilon > 0. \tag{5.4.24}$$

where $\operatorname{sgn}(\partial^2\theta/\partial\check{p}_\alpha\partial\check{p}_\beta)$ is the difference between the number of positive and negative squares of the form $(\partial^2\theta/\partial\check{p}_\alpha\partial\check{p}_\beta)\xi_\alpha\xi_\beta$. (Above we have shown that this form is nondegenerate.) This formula (and its subsequent terms) allows us to compute v_0, v_1, ... on S_1 and later find them in $L_{1,2}$ by the known method. After we have reached S_2 we similarly perform an inverse recount to coordinates x and computations are continued according to the local theory.

Remark 5.4.6 We have presented only a "computative" part of the Maslov theory, i.e., a construction of a formal solution along a narrow tube of trajectories. The consideration of the problem in the large gives rise to a number of topological questions and respectively deep relations with topology. For instance two trajectories starting from different ("far") points of the surface may have points with the same projection in the space X. If the solution is unique the computation of the function by both methods should produce similar results thus clearly leading to topological problems. The complete theory is discussed in details in the book [33] by Maslov and Fedoryuk.

Proof of Theorem 5.4.4

When working with determinants it is convenient to make use of the so-called Grassmann algebra; introducing n symbols $e_1, \dots, e_n$ and defining their (associative) multiplication by the formula $e_\alpha e_\beta = -e_\beta e_\alpha$. We may constitute linear forms $\omega = \sum \alpha_k e_k$, where α_k are numbers and multiply them by the usual algebra rules. The product of k forms $\omega_1, \dots, \omega_k$ can be uniquely presented in the form

$$\sum_{\alpha_1 < \alpha_2 < \cdots < \alpha_k} A_{\alpha_1 \dots \alpha_k} e_1 \dots e_k$$

and the condition $\omega_1 \dots \omega_k = 0$ is equivalent to $A_{\alpha_1 \dots \alpha_k} = 0$. Put $\omega_i = \alpha_{ik} e_k$. The formula $\omega_1 \dots \omega_n = (\det(\alpha_{ik})) e_1 \dots e_n$ is true and easy to verify. It shows a deep connection of this algebra with the theory of determinants.

Theorem 5.4.4 follows from the next statement.

Let $P_1, \dots, P_k$, $Q_1, \dots, Q_k$ be linear forms depending on e and $P_1 Q_1 + \dots + P_k Q_k = 0$. Further, suppose that among forms P there are exactly ℓ linearly independent ones, say $P_1, \dots, P_\ell$.

Then forms $Q_1, \ldots, Q_\ell$ depend linearly on $P_1, \ldots, P_e$; $Q_{\ell+l}, \ldots, Q_k$.

Suppose that this statement is true. Put α_0 for the variable t and write (see the discussion of the Cauchy method) $d\theta = P_k\, d\alpha_k$, where $P_k = p_\nu(\partial x_\nu/\partial \alpha_k)$ and $\partial P_k/\partial \alpha_s = \partial P_s/\partial \alpha_k$. The last formula is equivalent to

$$\sum_\nu \frac{\partial p_\nu}{\partial \alpha_s}\frac{\partial x_\nu}{\partial \alpha_k} - \frac{\partial p_\nu}{\partial \alpha_k}\frac{\partial x_\nu}{\partial \alpha_s} = 0$$

which can be written in the form $\tilde{P}_1\tilde{Q}_1 + \cdots + \tilde{P}_n\tilde{Q}_n = 0$, where $\tilde{P}_i = \sum_s (\partial p_i/\partial \alpha_s)e_s$; $\tilde{Q}_i = \sum_s (\partial x_i/\partial \alpha_s)e_s$. Assuming that $P_1, \ldots, P_\ell$ are all linearly independent forms among P we will see that forms $\tilde{Q}_1, \ldots, \tilde{Q}_\ell$ and $\tilde{P}_{\ell+1}, \ldots, \tilde{P}_n$ depends linearly on $\tilde{P}_1, \ldots, \tilde{P}_\ell$ and $\tilde{Q}_{\ell+1}, \ldots, \tilde{Q}_n$.

On the other hand, $\mathcal{L}$ is an n-dimensional manifold meaning that among $2n$ linear forms P and Q there are exactly n linearly independent ones. But then forms $\tilde{P}_1, \ldots, \tilde{P}_\ell, \tilde{Q}_{\ell+1}, \ldots, \tilde{Q}_n$ should be linearly independent proving the validity of Theorem 5.4.4.

Now let us prove the initial statement. The assumption $P_s = \sum_{1 \le i \le \ell} c_{s,i}P_i$, where $s > \ell$ and $\sum_\mu P_\mu Q_\mu = 0$, implies

$$P_1\left(Q_1 + \sum_{s>\ell} c_{s,1}Q_s\right) + \cdots + P_\ell\left(Q_\ell + \sum_{s>\ell} c_s Q_{s\ell}\right) = 0.$$

Multiplying this equality by $P_2 \ldots P_\ell$ we get

$$P_1 \ldots P_\ell\left(Q_1 + \sum_{s>\ell} c_{s,1}Q_s\right) = 0.$$

But forms $P_1, \ldots, P_\ell$ are linearly independent. The reader will easily prove that in this case $P_1 \ldots P_\ell \omega = 0$ implies that ω is a linear combination of $P_1, \ldots, P_\ell$ yielding the validity of the statement for Q_1 and analogously for $Q_2, \ldots, Q_\ell$.

Further we consider several problems which are generalizations of problems of the geometric optics onto the nonlinear case. What will follow is mainly a discussion and an extension of results of Whitham.

5.5 Problem (Whitham)

The problem is to investigate a solution $u(\theta, x, t)$

$$\varepsilon^2\left(\frac{\partial^2 u}{\partial t^2} - \frac{\partial^2 u}{\partial x^2}\right) + T'(u) = 0, \tag{5.5.1}$$

where $\theta = \theta(x, t)$, under the assumption that derivatives of $u(\theta, x, t)$ considered as a function in three variables with respect to x, t, and θ are finite.

Set $\theta_x = \theta_1$, $\theta_t = \theta_2$, $\varepsilon u_\theta = v$. Then $\varepsilon(\partial u/\partial x) = v\theta_1 + \varepsilon u_x$, where u_x is a partial derivative of $u(\theta, x, t)$. Differentiating this relation and computing similarly derivatives with respect to t we get the system

$$\begin{aligned} u_\theta &= v, \\ \varepsilon v_\theta &= -\frac{T'(u)}{q} - \frac{2\varepsilon(v_t\theta_2 - v_x\theta_1)}{q} \\ &\quad - \frac{\varepsilon v(\theta_{tt} - \theta_{xx})}{q} - \frac{\varepsilon^2}{q}(u_{tt} - u_{xx}), \end{aligned} \tag{5.5.2}$$

where $q = \theta_2^2 - \theta_1^2$. The 0-th order approximation system is

$$\varepsilon u_\theta = v, \quad \varepsilon v_\theta = -\frac{T'(u)}{q},$$

i.e., the system of ordinary differential equations depending via q on parameters x, t.

To the system (5.5) the operator

$$A_0 = v\frac{\partial}{\partial u} - \frac{T'(u)}{q}\frac{\partial}{\partial v}$$

corresponds. An invariant of A_0 is $E = T(u) + (v^2/2)q$. For an eigenfunction φ we obtain in coordinates E, u the equation

$$\frac{d\varphi}{\varphi} = \lambda\sqrt{q/2}\frac{du}{\sqrt{E - T(u)}}, \tag{5.5.3}$$

where $\lambda = \lambda(E, x, t)$.

Now, assume that the equation $E - T(u) = 0$ has two simple roots $u_0(E) < u_1(E)$ and that $u_0 \le u \le u_1$ and $E - T(u) > 0$ in this interval. Since v is real, then $q > 0$. Let us make a cut in the u-plane joining points u_0 and u_1. Let us introduce the function

$$\nu(u, E) = \frac{1}{2\tau}\int_{u_0}^{u} \frac{du}{\sqrt{E - T(u)}}, \tag{5.5.4}$$

where

$$\tau = \tau(E) = \int_{u_0}^{u_1} \frac{du}{\sqrt{E - T(u)}}.$$

In what follows for convenience sake we will assume $T(u)$ to be analytical function in a sufficiently narrow domain with the boundary S containing the cut $[u_0, u_1]$. The function $\sqrt{E - T(u)}$ is a single-valued analytical function in $\mathcal{D}$ outside of the cut. We will assume that $\sqrt{E - T(u)} > 0$ on the upper part of the cut. Then

$$2\tau(E) = \int_c \frac{du}{\sqrt{E - T(u)}}.$$

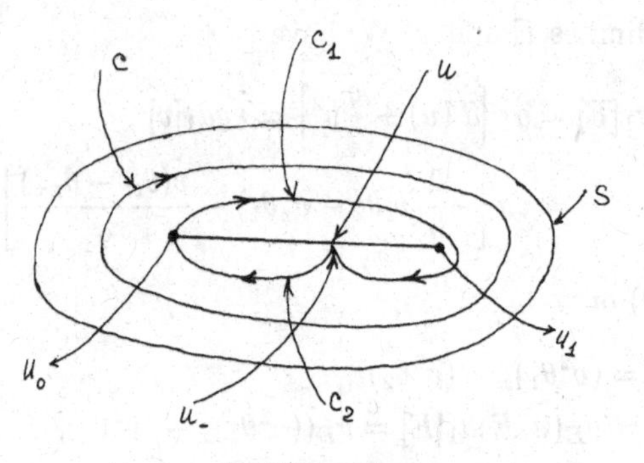

FIGURE 5.2.

where c is the contour described in Fig. 5.2.

It follows from (5.5.4) that u is a periodic function in ν with period 1 and

$$\frac{\partial u(\nu, E)}{\partial \nu} = 2\tau(E)\sqrt{E - T(u)}. \tag{5.5.5}$$

Passing to coordinates u and ν we get from (5.5.5)

$$A_0(\nu) = \frac{1}{2\tau\sqrt{E - T(u)}}$$

$$A_0(u) = \frac{v}{2\tau\sqrt{E - T(u)}}$$

$$= \frac{\sqrt{(2/q)(E - T(u))}}{2\tau\sqrt{E - T(u)}} = \frac{1}{\tau\sqrt{2q}}. \tag{5.5.6}$$

The formula (5.5.3) gives, taking (5.5.5) into account,

$$\frac{d\varphi}{\varphi} = \lambda\sqrt{q/2}\frac{du}{\sqrt{E - T(u)}} = \lambda\sqrt{q/2} \cdot 2\tau(E)\, d\nu.$$

Now it is necessary to choose λ so that u could be expressed in terms of E and φ. The evident choice is

$$\lambda = \frac{2\pi i}{\tau\sqrt{2q}} \tag{5.5.7}$$

since this choice implies

$$\varphi = e^{2\pi i\nu} \tag{5.5.8}$$

and a periodic function u is expandable into the Fourier series in $e^{2\pi i\kappa\nu}$.

All is ready to find the normal form of the operator $a = a_0 + \varepsilon a_1 + \varepsilon^2 a_2$, corresponding to (5.5.2). Below we define in the first approximation the normal form $\tilde{a}$ of a and the corresponding equations of the first approximation will be found.

Pass to coordinates E and φ. We have

$$a_1[E] = a_1\left[T(u) + \frac{q}{2}v^2\right] = vqa_1[v]$$
$$= vq\left[-\frac{2}{q}(v_t\theta_2 - v_x\theta_1) - \frac{v(\theta_{tt} - \theta_{xx})}{q}\right]$$

(since $a_1[u] = 0$) or

$$a_1[E] = (v^2\theta_1)_x - (v^2\theta_2)_t, \tag{5.5.9}$$
$$a_1[\nu] = \nu_E(u, E)a_1[E] = \nu_E((v^2\theta_1)_x - (v^2\theta_2)_t). \tag{5.5.9$'$}$$

To find the required normal form

$$\tilde{a} = e^{-\varepsilon S-\varepsilon^2 S_1-\cdots}(a_0 + \varepsilon a_1 + \varepsilon^2 a_2 + \cdots)e^{\varepsilon S+\varepsilon^2 S_1+\cdots} = a_0 + \varepsilon M_1 + \cdots$$

we should solve two equations

$$a_0[S[E]] + a_1[E] = M_1[E], \tag{5.5.10}$$
$$a_0[S[\nu]] - S[a_0[\nu]] + a_1[\nu] = M_1[\nu], \tag{5.5.10$'$}$$

where $M_1[E]$ corresponds to zero and so does $M_1(\nu)$ since ν does. Let us rewrite (5.5.10) in the form

$$a_0[S[E]] + (v^2\theta_1)_x - (v^2\theta_2)_t = M_1[E]$$

and put $S[E] = \tilde{S}_1 + \tilde{S}_2$, $\tilde{S}_1 = (S_1)_x$, $\tilde{S}_2 = (S_2)_t$, $M_1[E] = \tilde{m}_1 + \tilde{m}_2$, $\tilde{m}_1 = (m_1)_x$, $\tilde{m}_2 = (m_2)_t$. Consider the equations $a_0[\tilde{S}_1] + (v^2\theta_1)_x = \tilde{m}_1$ and $a_0[\tilde{S}_2] - (v^2\theta_2)_t = \tilde{m}_2$, where $\tilde{m}_1$ and $\tilde{m}_2$ should correspond to 0. Since $\hat{a}_0$ commutes with differentiation we deduce that these equations are equivalent to $(a_0[S_1])_x + (v^2\theta_1)_x = (m_1)_x$, $(a_0[S_2])_t - (v^2\theta_2)_t = (m_2)_t$. This immediately implies that it suffices to solve equations

$$a_0[S_1] + v^2\theta_1 = m_1, \qquad a_0[S_2] - v^2\theta_2 = m_2,$$

where m_1 and m_2 correspond to 0, and then put

$$S[E] = (S_1)_x + (S_2)_t, \qquad M_1[E] = (m_1)_x + (m_2)_t.$$

The problem is reduced to the usual problem for ordinary *differential equations.* We will not seek formulas for S. For m_i we clearly get

$$m_1 = \theta_1 \int_0^1 v^2\,d\nu, \qquad m_2 = -\theta_2 \int_0^1 v^2\,d\nu.$$

But for arbitrary F

$$\int_0^1 F(E, u(\nu, E))\,d\nu = \int_c F(E, u)\frac{\partial \nu}{\partial u}\,du = \int_c \frac{F(E, u)}{2\tau\sqrt{E - T(u)}}\,du, \tag{5.5.11}$$

where c is the contour depicted on Fig. 5.2. Making use of (5.5.11) we get

$$\begin{aligned} m_1 &= \theta_1 \int_0^1 v^2\, d\nu = \frac{\theta_1}{q\tau} \int_c \sqrt{E - T(u)}\, du, \\ m_2 &= -\frac{\theta_2}{q\tau} \int_c \sqrt{E - T(u)}\, du, \qquad (5.5.12) \\ M_1[E] &= \left(\frac{\theta_1}{q\tau} \int_c \sqrt{E - T(u)}\, du \right)_x - \left(\frac{\theta_2}{q\tau} \int_c \sqrt{E - T(u)}\, du \right)_t . \end{aligned}$$

Now let us deal with (5.5.10′). Passing to coordinates E, ν let us rewrite $\mathcal{T} = \nu_E(v^2\theta_1)_x$ in the form

$$\mathcal{T} = \frac{2\theta_1}{q}[E_x(\nu_E - T'(u)u_E\nu_E) - T'(u)\nu_E u_\nu \nu_x] + \left(\nu_E(E - T(u)) \left(\frac{2\theta_1}{q} \right)_x \right).$$

Since E_x and ν_x correspond to 0, then the same considerations as above yield

$$\begin{aligned} M_1[\nu] = \frac{2\theta_1}{q} E_x &\left(\int_0^1 (\nu_E - T'(u)u_E\nu_E)\, d\nu \right) \\ &- \left(\int_0^1 \frac{2\theta_1}{q} (T'(u)\nu_E u_\nu)\, d\nu \right) \nu_x \\ &+ \left(\frac{2\theta_1}{q} \right)_x \int_0^1 \nu_E(E - T(u))\, d\nu. \qquad (5.5.13) \end{aligned}$$

The problem is again reduced to the usual one. [Let us clarify what has been done. The expression for $\mathcal{T}$ contains a partial derivative with respect to an independent variable x. We have the problem *that does not contain partial derivatives with respect to independent variables* making use of the passage to variables that *correspond to zero* as E and $\nu = (\ln \varphi)/2\pi i$ do in this case.]

It only remains to compute integrals which enter (5.5.13). First note that if u and u_- denote points that belong to upper and lower sides of the cut respectively (Fig. 5.2) then

$$\nu(u_-) = \frac{1}{2\tau} \int_{c_1} \frac{du}{\sqrt{E - T(u)}}, \qquad \frac{1}{2\tau} \left(\int_{c_1} + \int_{c_2} \right) \frac{du}{\sqrt{E - T(u)}} = 1$$

imply that $\nu(u, E) + \nu(u_-, E) = 1$ or

$$\nu_E(u, E) = -\nu_E(u_-, E). \qquad (5.5.14)$$

Now

$$\int_0^1 \nu_E\, d\nu = \int_c \nu_E \frac{du}{2\tau\sqrt{E - T(u)}} = 0$$

since (5.5.14) implies that the integrand takes the same values on both sides of the cut. Further, from (5.5.4) we deduce that

$$\nu_E + \frac{1}{2\tau(E)} \frac{u_E}{\sqrt{E - T(u)}} = 0.$$

Therefore

$$\int_0^1 T'(u) u_E \nu_E \, d\nu = -\int_c (\nu_E)^2 2\tau(E) \sqrt{E - T(u)} \frac{T'(u)\, du}{2\tau(E)\sqrt{E - T(u)}} = 0$$

(by the same reason). Finally

$$\begin{aligned}\int_0^1 T'(u) \nu_E u_\nu \, d\nu &= \int_c T'(u) \nu_e u_\nu \frac{\partial \nu}{\partial u} \, du = \int_c \nu_E T'(u) \, du \\ &= \frac{\partial}{\partial E} \int_c \nu T'(u) \, du = -\frac{\partial}{\partial E} \int_c \nu d(E - T(u)) \\ &= \frac{\partial}{\partial E} \int_c (E - T(u)) \frac{\partial \nu}{\partial u} \, du = \frac{\partial}{\partial E} \int_c \frac{1}{2\tau} \sqrt{E - T(u)} \, du.\end{aligned}$$

Clearly

$$\int_0^1 \nu_E (E - T(u)) \, d\nu = 0.$$

All these formulas yield

$$M_1[\nu] = \left(\frac{\partial}{\partial E} \int_c \frac{1}{q\tau(E)} \sqrt{E - T(u)} \, du \right) (\theta_2 \nu_t - \theta_1 \nu_x). \qquad (5.5.15)$$

It remains to write out a system of equations in the normal form for E and ν which follows from the above and an equation for θ.

First in the 0-th approximation $\varphi = e^{\lambda\theta/\varepsilon}$. To avoid unbounded partial derivatives in x and t of the function φ of variables θ, x, and t, we should require $\lambda = \text{const}$. Returning to (5.5.7) we put

$$\tau(E)\sqrt{2q} = 1. \qquad (5.5.16)$$

Clearly if this condition holds, then φ has bounded derivatives in t and x in any approximation. Equation (5.5.16) is a generalized eikonal equation.

The equation $\varepsilon(d\varphi/d\theta) = \lambda\varphi + \varepsilon M_1[\varphi]$, (5.5.8) and (5.5.12), and (5.5.16) being taken into account, gives

$$\varepsilon \frac{d\eta}{d\theta} = \left(\frac{\lambda}{2\pi i} - 1 \right) + \varepsilon M_1[\eta] = \varepsilon M_1(\eta)$$

if we put $\nu = \theta/\varepsilon + \eta$ or $d\eta/d\theta = M_1[\eta]$ for $\lambda = 2\pi i$. Besides, $dE/d\theta = M_1[E]$. Since these equations do not contain a small parameter in the left-hand side, we may assume that η and E do not depend on θ at all. Therefore (5.5.12) and (5.5.15) imply equations of the first approximation of the form

$$\left(\theta_1\sqrt{2/q}\int_c \sqrt{E-T(u)}\,du\right)_x - \left(\theta_2\sqrt{2/q}\int_c \sqrt{E-T(u)}\,du\right)_t = 0, \quad (5.5.17)$$

$$\theta_2\eta_t - \theta_1\eta_x = 0. \quad (5.5.18)$$

[We have taken (5.5.16) into account] which together with (5.5.16) gives three equations to define three unknowns, E, θ, and η.

Remark 5.5.1 In the above computations we have repeatedly used formulas (5.1.12), which often enables us to reduce the problem to ordinary differential equations.

5.6 Problem. Diffraction of Short Waves on a Circle (Semishade)

The general diffraction problem of short waves on a convex body is well studied and we do not intend to describe it here. In this section we will consider without details a simple example of how to apply the "standard" approach to problems of this type in the semishade and the shade, where difficulties arise. These difficulties were overcome in their time by Fock and Keller, respectively, by physical considerations (for detailed references see [47]). Only principal terms of asymptotics were found at the time. The problem of computing all terms turned out to be nontrivial and had been solved mainly by V. M. Babich and his colleagues who developed the so-called *model equation method.* The method is based on an explicit solution of a certain model problem. With the help of this solution a general form of the asymptotics with undetermined coefficients is guessed and these coefficients found afterwards. Below we will show by the model example how to find asymptotics without explicitly solving the problem. The general case differs from the considered one only by the quantity of needed computations.

Suppose a planar wave $u_P = -e^{-ix/\varepsilon}$ is incident on the unit circle S with a boundary $\bar{S}$. We wish to find the asymptotics of the arising field $u = u_P + u_0$ satisfying $\varepsilon^2\Delta u + u = 0$ outside of S and the boundary value $u|_{\bar{S}} = 0$ as $\varepsilon \to 0$. (Values on infinity are discussed below.) Further, let us seek u_0. Denoting u_0 once again by u we get $u|_{\bar{S}} = e^{-ix/\varepsilon}$. According to the usual recipe set

$$u = e^{i\theta/\varepsilon}A, \qquad \theta|_{\bar{S}} = -x. \quad (5.6.1)$$

The eikonal equation is

$$\theta_x^2 + \theta_y^2 = 1 \quad (5.6.2)$$

and the equation for A is

$$2i\varepsilon(\theta_x A_x + \theta_y A_y) + i\varepsilon(\Delta\theta)A + \varepsilon^2 \Delta A = 0, \qquad A|_{\bar{S}} = 1. \tag{5.6.3}$$

For the circle being defined by $x = \cos\varphi$, $y = \sin\varphi$ we easily get, solving the Hamiltonian system corresponding to (5.6.2),

$$x = \cos\varphi + \tau\cos 2\varphi, \qquad y = \sin\varphi + \tau\sin 2\varphi,$$
$$\theta = -\cos\varphi + \tau. \tag{5.6.4}$$

To these formulas the construction depicted in Fig. 5.3 corresponds. (The plane wave incident from the right to the left is reflected at the point ξ according to the rule "the angle of incidence equals the angle of reflection." The phase is $\theta = -\cos\varphi + \tau$.) The construction can be performed until $-\pi/2 < \varphi < \pi/2$, i.e., in the illuminated domain. In the domain $3\pi/2 > \varphi > \pi/2$ (Fig. 5.4) the vector z first enters the disc as τ varies from 0, then crosses the circle in z_0 at $\tau = \tau_0$ and goes out. The constructed function θ satisfies the equation outside the disc but not on the circle: at $z = z_0$ it does not satisfy any more the boundary value. This is due to the fact that in the shaded domain characteristics are first directed inside the disc.

Now let us solve (5.6.3) in the illuminated domain. A solution is sought in the form

$$A = A_0 + \varepsilon A_1 + \cdots, \quad A_0|_{\bar{S}} = 1, \quad A_k|_{\bar{S}} = 0 \quad (k = 1, 2, \ldots) \tag{5.6.5}$$

and A_k are defined from equations

$$2i\frac{\partial A_k}{\partial \tau} + \frac{2i}{\delta}A_k + \left(\frac{\partial^2}{\partial x^2} + \frac{\partial^2}{\partial y^2}\right)A_{k-1} = 0, \tag{5.6.5$'$}$$

$$\delta = \delta(\tau, \varphi) = \cos\varphi + 2\tau, \quad A_0 = \sqrt{\delta(0,\varphi)/\delta(\tau,\varphi)} = \frac{\sqrt{\cos\varphi}}{\sqrt{\cos\varphi + 2\tau}}. \tag{5.6.5$''$}$$

Note that in accordance with the said earlier $\delta = |\partial(x,y)/\partial(\tau,\varphi)|$, where $-\pi/2 \le \varphi \le \pi/2, \tau \ge 0$. Presently A_k have singularities on the variety $\delta = 0$. This variety in the considered domain consists of two points $a_\pm = (\varphi = \pm\pi/2, \tau = 0)$. Since points $a_\pm$ belong to the boundary of the domain, the conditions when the Maslov method is applicable are not satisfied. The problem should be reconstructed and a new leading operator in a neighborhood of $a_\pm$ must be chosen. For this let us make use of some general considerations

$$x = \varphi + \tau$$

formulated in the subsection "Reconstruction."

In a neighborhood of $\delta = 0$, A_k's are of the form

$$A_k = \frac{B_k(\varphi,\tau)}{\delta^{3k+1/2}}, \qquad B_k(\varphi, 0) \ne 0, \qquad B_k(\varphi,\tau) = 0(1). \tag{5.6.6}$$

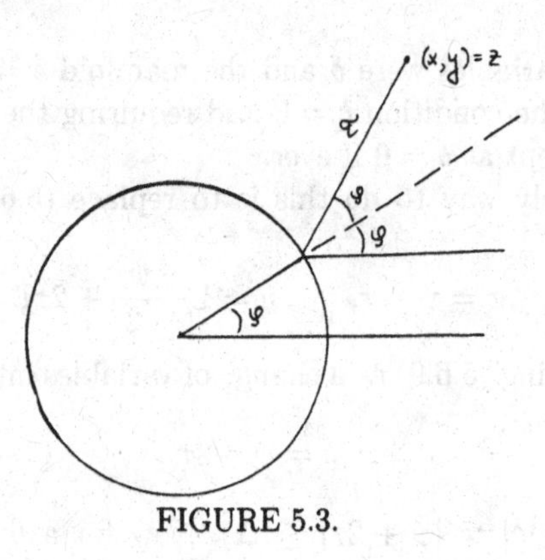

FIGURE 5.3.

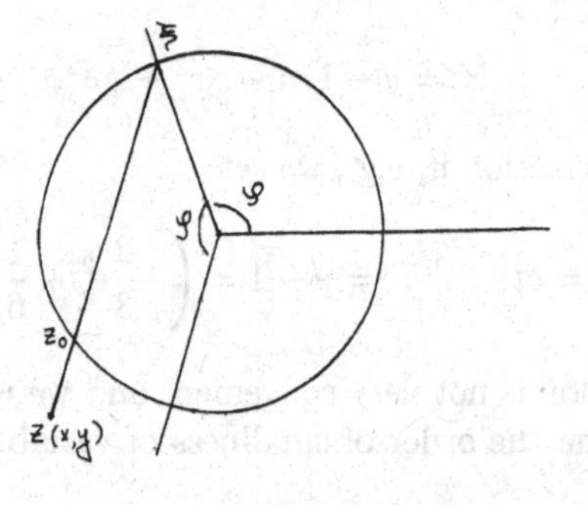

FIGURE 5.4.

[The simplest way to verify (5.6.6) is to substitute (5.6.6) in (5.6.5′). Then Δ being applied to A_{k-1} increases the order of singularity of A_{k-1} by 4 and $(2i(\partial/\partial\tau)+2i/\delta)$ being applied to A_k increases the order of singularity of A_k by 1 proving the validity of (5.6.6).]

Therefore considering only a neighborhood of a_+, the asymptotics (5.6.5) can be used in the domain

$$\mathcal{D}_\Delta : (0 \le \varphi \le \pi/2, \delta \ge \Delta, \varepsilon/\Delta^3 \ll 1, \Delta \ll 1) \tag{5.6.7}$$

with $\varepsilon^k A_k \sim (\varepsilon/\Delta^3)^k (1/\Delta^{1/2})$ if the *whole* ray connecting ξ and z (Fig. 5.3) is contained in $\mathcal{D}_\Delta$.

Let us pass to the study of the problem in a neighborhood of $\delta = 0$. The variety $\delta = 0$ (in the domain $\varphi \ge 0$, $\tau \ge 0$) is the point $(x = 0, y = 1)$. Setting $\varphi = \pi/2 - \psi$ we get

$$\begin{aligned} x &= \sin\psi - \tau\cos 2\psi, & y &= \cos\psi + \tau \sin 2\psi, \\ \delta &= \sin\psi + 2\tau, & \theta &= -\sin\psi + \tau. \end{aligned} \tag{5.6.8}$$

Taking what had been said in the subsection "Reconstruction" into account let us parametrize the neighborhood of (0,1) defined by $\delta < \Delta$ so that one of the variables were δ and the manifold $\delta = 0$ in new variables were defined by the condition $\delta = 0$ and requiring the change of variables not to vanish except at $\delta = 0$ if ever.

The most simply way to do this is to replace (5.6.8) by approximate formulas

$$x = \psi - \tau, \qquad y = 1 - \frac{\psi^2}{2} + 2\tau\psi \tag{5.6.9}$$

and then considering (5.6.9) as a change of variables introduce the function

$$\delta = \psi + 2\tau \tag{5.6.10}$$

(and the domain $|\delta| = |\psi + 2\tau| \leq \Delta$). Now for a fixed δ (5.6.10) is an equation of parabola P_δ

$$Y = y - 1 = -\tfrac{2}{3}x^2 + \tfrac{1}{6}\delta^2. \tag{5.6.11}$$

We get a parametrization if, e.g., we set

$$x = \delta t, \qquad Y = y - 1 = \left(-\frac{2}{3}t^2 + \frac{1}{6}\right)\delta^2.$$

This parametrization is not very convenient and we will use it for the time being only to define the order of smallness of variables in a neighborhood of $\delta = 0$. Clearly

$$x \sim \delta, \qquad Y = y - 1 \sim \delta^2. \tag{5.6.12}$$

To get the main part of an equation governing the problem in a neighborhood of $(0, 1)$ set

$$x = \tilde{\varepsilon}\xi, \qquad Y = y - 1 = \tilde{\varepsilon}^2 y, \tag{5.6.12$'$}$$

where

$$\tilde{\varepsilon}^3 = \varepsilon. \tag{5.6.13}$$

($\tilde{\varepsilon}$ is a "smallest" value of $\Delta : \varepsilon/\Delta^3 \ll 1$.)

Before we proceed let us make several remarks.

Remark 5.6.1 In order to use (5.6.5) in $\mathcal{D}_\Delta$ we need that the ray that joins ξ and z (Fig. 5.3) was contained in $\mathcal{D}_\Delta$. In terms of parabolas P_δ this can be formulated as follows: choose Δ and let us construct the parabola P_Δ and the ray S_Δ. It is easy to see that the whole ray S_Δ is situated outside of P_Δ. Therefore if $\varepsilon/\Delta^3 \ll 1$, then in the shaded domain we can make use of (5.6.5).

Remark 5.6.2 Formulas (5.6.5) are applicable in a considerably wider domain if A_k, where $k = 0, 1, \ldots$, are determined from (5.6.5) and (5.6.5$'$) but with their values given not on $\bar{S}$ but on P_Δ.

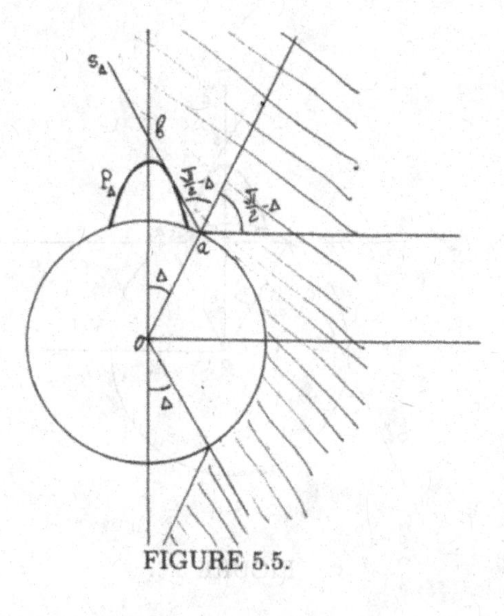

FIGURE 5.5.

In Figure 5.6 this possibility is shown. The domain where (5.6.5) is applicable is bounded by a part of the plane situated over S_Δ outside of P_Δ. In what follows we intend to find these values on P_Δ making use of the reconstructed operator.

Remark 5.6.3 This purely "technical" remark concerns the parametrization of a neighborhood of $(0,1)$. A successful choice of parametrization might simplify calculations considerably. General recipes clearly cannot be given though it is advisable to parametrize the greater part of the domain.

In our case the solution is known on a part of a circle that belongs to the interior of P_Δ (it is defined by initial values) and on S_Δ [computed with the help of (5.6.5)].

It seems natural to choose as new coordinates the distance h along the normal from (x,y) to the circle and the angle ν between the vector (x,y) and the y axis. Then

$$x = (1+h)\sin\nu, \qquad y = (1+h)\cos\nu. \tag{5.6.14}$$

The equation of the circle is $h = 0$. Taking into account that $x \sim \tilde{\varepsilon}$, $Y \sim \tilde{\varepsilon}^2$ implies $\nu \sim \tilde{\varepsilon}$, $h \sim \tilde{\varepsilon}^2$, set (factors are chosen from the convenience considerations)

$$\nu = \tilde{\varepsilon}\sigma 2^{1/3}, \qquad h = \tilde{\varepsilon}^2\gamma 2^{-1/3}. \tag{5.6.15}$$

However there is one more possibility to simplify calculations. We are going to find a solution u in the domain O_Δ defined by $|\delta| \leq \Delta$, $\varepsilon/\Delta \ll 1, \Delta \ll 1$. It means that variables σ and γ take values bounded by the condition

$$\sigma = O(\Delta/\tilde{\varepsilon}), \gamma = O(\Delta^2/\tilde{\varepsilon}^2) \tag{5.6.16}$$

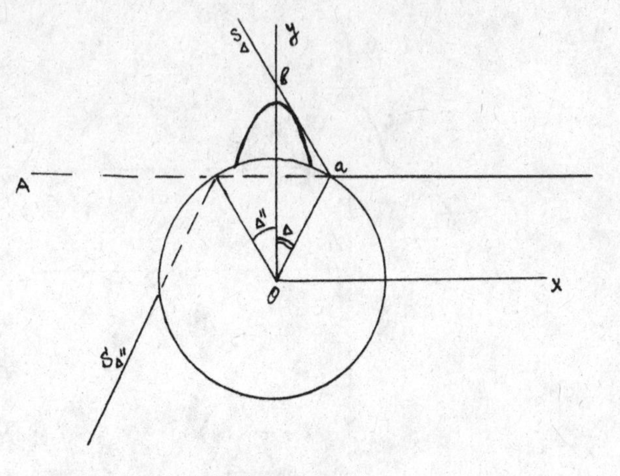

FIGURE 5.6.

and actually may take values of order $\Delta/\tilde{\varepsilon}$ and $(\Delta/\tilde{\varepsilon})^2$, i.e., very large. But it follows from (5.6.8) that

$$\frac{\theta}{\varepsilon} = \frac{-x + 2\tau \sin^2 \psi}{\varepsilon}$$
$$= -\frac{\nu}{\varepsilon} + \frac{1}{\varepsilon}\left(-h\nu + \frac{\nu^3}{3} + 2\tau\psi^2\right) + [\,]_5/\varepsilon + [\,]_7/\varepsilon + \cdots, \tag{5.6.17}$$

where $[\quad]_k$ stands for the k-th order term. (ν, τ, and ψ are considered to be of order 1 with respect to $\tilde{\varepsilon}$ and h of the power 2.)

Formula (5.6.17) shows that in the "principal" part of O_Δ, where σ and γ are finite, functions $e^{i\theta/\varepsilon}$ and $e^{-i\nu/\varepsilon}$ change with the same speed, i.e., in this domain we may put

$$u = e^{-i\nu/\varepsilon} B(\gamma, \sigma) \tag{5.6.18}$$

instead of $u = e^{i\theta/\varepsilon} A$ and find B. [Third order terms entering (5.6.17) are already of order ε, since $\tilde{\varepsilon}^3 = \varepsilon$, hence they contribute a nonoscillating factor in $u = e^{i\theta/\varepsilon} A$ which is now ascribed to B.] Besides let us put the additional constraint

$$\Delta^5/\tilde{\varepsilon}^3 \ll 1 \tag{5.6.19}$$

on Δ. This condition guarantees the smallness of $[\quad]_5/\varepsilon$, $[\quad]_7/\varepsilon$, ... in (5.6.17) in the whole domain of σ, γ admissible by (5.6.16).

Now, to calculations. Substituting (5.6.18) in equation which follows from $\varepsilon^2 \Delta u + u = 0$

$$\varepsilon^2 \left(\frac{\partial^2 u}{\partial h^2} + \frac{1}{1+h}\frac{\partial u}{\partial h} + \frac{1}{(1+h)^2}\frac{\partial^2 u}{\partial \nu^2}\right) + u = 0$$

and making the change (5.6.15) we get

$$\frac{\partial^2 B}{\partial \gamma^2} + \gamma B - i\frac{\partial B}{\partial \sigma} - i\left(\frac{1}{1+\tilde{\varepsilon}^2 2^{-1/3}\gamma} - 1\right)\frac{\partial B}{\partial \sigma}$$
$$+ 2^{-2/3}\left(\frac{-3\tilde{\varepsilon}^2 2^{-2/3}\gamma - \tilde{\varepsilon}^4\gamma^3}{(1+\tilde{\varepsilon}^2 2^{-1/3}\gamma)^2}\right) B + \frac{\tilde{\varepsilon}^2 2^{-1/3}}{(1+\tilde{\varepsilon}^2 2^{1/3}\gamma^2)^2}\frac{\partial B}{\partial \gamma}$$
$$+ \frac{2^{-4/3}\tilde{\varepsilon}^2}{(1+2^{-1/3}\tilde{\varepsilon}^2\gamma)^2}\frac{\partial^2 B}{\partial \sigma^2} = 0. \tag{5.6.20}$$

Note that the domain T of values of σ and γ, where the solution of (5.6.20) should be sought for, is defined by (5.6.16).

Due to (5.6.19) coefficients of (5.6.20) except the first three are small in T. Hence, in T the leading equation is the Schroedinger equation

$$i\frac{\partial B}{\partial \sigma} = \frac{\partial^2 B}{\partial \gamma^2} + \gamma B. \tag{5.6.21}$$

Let us formulate a boundary value problem for (5.6.21). We know that $B(0,\sigma) = e^{i\nu/\varepsilon} u|_{\gamma=0}$. But $u|_{\gamma=0} = e^{-ix/\varepsilon}|_{\gamma=0}$ hence $B(0,\sigma) = e^{i(\nu-x)/\varepsilon}|_{\gamma=0}$. Making use of (5.6.14) we get

$$(\nu - x)\big|_{\gamma=0} = \nu - \sin\nu = \frac{\nu^3}{3!} - \frac{\nu^5}{5!} + \cdots$$

or

$$B(0,\sigma) = e^{i\sigma^3/3} e^{iO(\Delta^5/\tilde{\varepsilon}^3)} = e^{i\sigma^3/3}(1 + O(\Delta^5/\tilde{\varepsilon}^3)). \tag{5.6.22}$$

Besides, B is known on the line (b,a) (Fig. 5.5). In order not to compute too much let us evaluate B only at b. [We will see that it suffices to know B (the asymptotics of B) in an arbitrary point of the line (a,b) in order to formulate a correct problem.] First of all $B = e^{i\nu/\varepsilon} e^{i\theta/\varepsilon} A$, hence $B|_b = e^{i\theta/\varepsilon}|_b A|_b$ (since $\nu = 0$ at b). Besides, see (5.6.8):

$$\tau = \frac{\sin\psi}{\cos 2\psi},$$
$$h = \tau\sin 2\psi - (1 - \cos 2\psi),$$
$$\theta = \tau - \sin\psi = \sin\psi\left(\frac{1}{\cos 2\psi} - 1\right)$$

at b. Taking $h = \tilde{\varepsilon}^2\gamma 2^{-1/3}$ into account we get after simple calculations

$$\frac{\theta}{\varepsilon} = \frac{4}{3^{3/2}}\gamma^{3/2} + O\left(\frac{\Delta^5}{\tilde{\varepsilon}^3}\right). \tag{5.6.23}$$

Let us resume: the order of h at the point b equals Δ^2 and $\gamma \sim (\Delta/\tilde{\varepsilon})^2$ is large. Since there were no restrictions on Δ except $\Delta \gg \tilde{\varepsilon}$ and $\Delta^5/\tilde{\varepsilon}^3 \ll 1$, then (5.6.23) can be read simply as a formula for the *principal* part of the asymptotics of the phase on the line $\sigma = 0$ when $\gamma \gg 1$. Small terms being ignored we come to the following boundary value problem for (5.6.21): Find a solution of (5.6.21) if it is known that [our notations differ from those in (4) by the sign of γ]:

$$B(0,\sigma) = e^{i\sigma^3/3} \qquad (-\infty \le \sigma \le +\infty),$$
$$B(\gamma,0) = e^{4i\gamma^{3/2}/3^{3/2}} O(1), \qquad \gamma \gg 1 \tag{5.6.24}$$

(let us stress, as it had been already done, that γ can take large values).

Let us show that (5.6.24) suffices to define B uniquely.

It is natural to seek (5.6.21) in the form

$$B = \int_{-\infty}^{+\infty} \mathcal{E}_\alpha(\gamma+\alpha) e^{i\alpha\sigma} f(\alpha)\, d\alpha \tag{5.6.25}$$

where $\mathcal{E}_\alpha(\gamma)$ is an Airy function

$$\mathcal{E}_\alpha''(\gamma) + \gamma \mathcal{E}_\alpha(\gamma) = 0. \tag{5.6.26}$$

Recall that (5.6.26) has solutions $w_1(\gamma)$, $w_2(\gamma)$, and $v(\gamma)$ with the following properties [$A|_b = O(1)$ as follows from (5.6.5″) for $\varphi = \pi/2 - \Delta$, $\tau \sim \Delta$; further $O(1)$ is replaced by 1]:

$$w_1(\gamma)_{\gamma\to\infty} \sim \frac{1}{\sqrt[4]{\gamma}} \exp\left(\frac{2}{3}\gamma^{3/2} + \frac{\pi}{4}\right) i,$$
$$w_2(\gamma)_{\gamma\to\infty} \sim \frac{1}{\sqrt[4]{\gamma}} \exp\left(-\left(\frac{2}{3}\gamma^{3/2} + \frac{\pi}{4}\right) i\right) \tag{5.6.27}$$
$$w_1(\gamma) \sim w_2(\gamma)_{\gamma\to-\infty} \sim (-\gamma)^{-1/4} \exp\frac{2}{3}(-\gamma)^{3/2} \tag{5.6.27'}$$

and

$$v(\gamma) = \frac{w_1(\gamma) - w_2(\gamma)}{2i} = \frac{1}{2\sqrt{\pi}} \int_{-\infty}^{+\infty} e^{i(-\gamma y + y^3/3)}\, dy,$$
$$v \sim \frac{1}{\sqrt[4]{-\gamma}} e^{-2/3(-\gamma^{3/2})} \qquad \gamma\to-\infty. \tag{5.6.28}$$

Putting $\gamma = 0$ in (5.6.25) and making use of (5.6.24) we get

$$\int_{-\infty}^{+\infty} \mathcal{E}_\alpha(\alpha) f(\alpha) e^{i\alpha\sigma}\, d\alpha = e^{i\sigma^3/3}, \tag{5.6.28'}$$

whence inverting (5.6.28′) and comparing with (5.6.24) we have

$$\mathcal{E}_\alpha(\alpha) f(\alpha) = \frac{1}{2\pi} \int e^{i\sigma^3/3} e^{-i\alpha\sigma}\, d\sigma = \frac{1}{\sqrt{\pi}} v(\alpha).$$

Hence (5.6.25) yields

$$B(\gamma, \sigma) = \frac{1}{\sqrt{\pi}} \int_{-\infty}^{+\infty} \frac{\mathcal{E}_\alpha(\gamma+\alpha)}{\mathcal{E}_\alpha(\alpha)} e^{i\alpha\sigma} v(\alpha)\, d\alpha. \tag{5.6.29}$$

Let us define the form of $\mathcal{E}_\alpha$. Clearly, $\mathcal{E}_\alpha(\gamma) = c_1(\alpha)w_1(\gamma) + c_2(\alpha)w_2(\gamma)$. Now we should apply the second condition in (5.6.24), i.e., the asymptotics of $B(0,\gamma)$. If putting $\sigma = 0$ in (5.6.29) we replace w_1 and w_2 by the asymptotics (5.6.27) we will see that (5.6.24) is satisfied only if $c_2 = 0$. We will only verify that the second condition is satisfied at $c_2 = 0$.

Thus we should find the asymptotics of

$$\mathcal{T}(\gamma) = \int_{-\infty}^{+\infty} \frac{w_1(\gamma+\alpha)}{w_1(\alpha)} v(\alpha)\, d\alpha, \qquad \gamma \gg 1. \tag{5.6.30}$$

Choose $0 < \delta < 1$, $0 < q < 1$ and present $\mathcal{T}(\gamma)$ in the form

$$\mathcal{T}(\gamma) = \int_{-\infty}^{-\delta\gamma} + \int_{-\delta\gamma}^{\gamma^q} + \int_{\gamma^q}^{+\infty} v(\alpha) \frac{w_1(\gamma+\alpha)}{w_1(\alpha)}\, d\alpha = \mathcal{T}_1 + \mathcal{T}_2 + \mathcal{T}_3.$$

Computing $\mathcal{T}_1$ we may clearly apply (5.6.27′) and (5.6.28) and show that $\mathcal{T}_1$ exponentially decreases. In $\mathcal{T}_2$ we may replace $w_1(\gamma+\alpha)$ by its asymptotics since $\gamma + \alpha \gg 1$ if $-\delta\gamma \leq \alpha \leq \gamma^q$ and get

$$\mathcal{T}_2 \simeq e^{i\pi/4} \int_{-\delta\gamma}^{\gamma^q} \frac{v(\alpha)}{w_1(\alpha)} \frac{1}{i} \frac{1}{(\gamma+\alpha)^{1/4+1/2}}\, d e^{i\frac{2}{3}(\gamma+\alpha)^{3/2}}.$$

Integrating this by parts and replacing $v(\alpha)/w_1(\alpha)$ at limit points by asymptotics we see that the contribution of $-\delta\gamma$ exponentially decreases while that of γ^q can be ignored due to the arbitrariness of q (or it will cancel with the corresponding contribution of $\mathcal{T}_3$). Concerning $\mathcal{T}_3$ we can apply the asymptotics

$$\mathcal{T}_3(\gamma) \simeq \int_{\gamma^q}^{+\infty} \frac{e^{(2/3)(\gamma+\alpha)^{3/2} i}}{e^{(2/3)\alpha^{2/3} i}(\gamma+\alpha)^{1/4}} \times \frac{1}{2i}\left(e^{(2/3)(\alpha^{3/2}+\pi/4)i} - e^{-(2/3)i\alpha^{3/2}-(\pi/4)i}\right) d\alpha.$$

The first summand of this integrand decreases. The second summand is up to a factor equal to

$$\int_{\gamma^q}^{+\infty} \frac{e^{(2i/3)(\gamma+\alpha)^{3/2} i - (4/3)\alpha^{3/2} i}}{(\gamma+\alpha)^{1/4}}\, d\alpha = \int_{\gamma^{q-1}}^{+\infty} \gamma^{3/4} \frac{e^{(2/3(1+\alpha)^{3/2} - (4/3)\alpha^{3/2}) i \gamma^{3/2}}}{(1+\alpha)^{1/4}}\, d\alpha.$$

Applying the stationary phase method and ignoring the contribution of the lower limit of integration we get an expression for $\mathcal{T}_3$ coinciding with (5.6.24).

Now let us consider the whole problem. In (5.6.20) we will seek a partial solution in the form $e^{i\sigma\alpha}\mathcal{E}_\alpha(\gamma)$. Then for $\mathcal{E}_\alpha$ we get the equation

$$\frac{d^2\mathcal{E}}{d\gamma^2} + (\gamma+\alpha)\mathcal{E} + \alpha f_1\mathcal{E} + f_2\mathcal{E} + f_3\frac{\partial\mathcal{E}}{\partial\gamma} + f_4\alpha^2\mathcal{E} = 0 \qquad (5.6.31)$$

where f_i originate from the last four terms in (5.6.20) and all of them are small, if $\gamma \le O((\Delta/\tilde{\varepsilon})^2)$ or if (5.6.19) holds. Our goal should be the construction of two solutions of (5.6.31), $\mathcal{E}_1(\gamma,\alpha)$ and $\mathcal{E}_2(\gamma,\alpha)$, such that asymptotically they were like $w_1(\gamma+\alpha)$ and $w_2(\gamma+\alpha)$, and then search for B in the form

$$B = \int_{-\infty}^{+\infty} \mathcal{E}(\gamma,\alpha)e^{i\alpha\sigma}\psi(\alpha)\,d\alpha. \qquad (5.6.32)$$

In this form the problem is clearly impossible to carry on since in (5.6.32) the variable α can take arbitrary values and for an arbitrary large α there are simply no solutions of (5.6.31) with the needed asymptotics. The way out is the following consideration: though γ takes arbitrary values in (5.6.31), still $\gamma \le (O(\Delta/\tilde{\varepsilon})^2)$. If we can find $\psi(\alpha)$ and $r(\Delta,\tilde{\varepsilon})$ such that $\psi(\alpha) = 0$, when $|\alpha| \ge r(\Delta,\tilde{\varepsilon})$, if we can construct solutions $\mathcal{E}_1(\gamma,\alpha)$, $\mathcal{E}_2(\gamma,\alpha)$ of the needed form, when $|\alpha| \le r(\Delta,\tilde{\varepsilon})$, and if at the same time we can find ψ such that $B(\gamma,\sigma)$ defined by (5.6.32) has the given total asymptotics (5.6.22) at $\gamma = 0$ and the behavior of $B(\gamma,0)$ coincides with the prescribed one on the infinity, then the problem is solved. Let us try to fulfill this program.

Boundary values at $\gamma = 0$ are

$$B(0,\sigma) = e^{i(\sigma^3/3)}e^{(-i\nu^5/5!\varepsilon)}e^{(i\nu^7/7!\varepsilon)}\ldots = h(\sigma)(\nu = \tilde{\varepsilon}\sigma 2^{1/3},\ |\nu| \le \Delta). \qquad (5.6.33)$$

Since in what follows we will only need the value σ satisfying $|\sigma| \le (\Delta/\tilde{\varepsilon})2^{-1/3} = \rho$, we replace $h(\sigma)$ by $\hat{h}(\sigma) = h(\sigma)$ for $|\sigma| \le 2\rho$ and $\hat{h}(\sigma) = 0$ for $|\sigma| > 2\rho$. Choose a function $\varphi_n(\mu)$ with $3n+2$ continuous derivatives when $-\infty \le \mu \le +\infty$ and such that

$$\varphi_n(\mu) = 0, \qquad |\mu| \ge 1, \qquad \sqrt{2\pi}\varphi_n(0) = 1, \qquad \varphi_n^{(k)}(0) = 0,$$

where $k = 1,\ldots,n-1$. Put

$$g_n(\nu) = \frac{1}{2\pi}\int_{-\infty}^{+\infty} e^{i\nu\mu}\varphi_n(\mu)\,d\mu,$$

$$h_n(\sigma) = \frac{1}{\delta}\int_{-\infty}^{+\infty} g_n\left(\frac{\sigma-\nu}{\delta}\right)\hat{h}(\nu)\,d\nu.$$

where δ is a parameter. The inversion formula implies that

$$\int_{-\infty}^{+\infty} g_n(\nu)\, d\nu = 1 \quad \text{and} \quad \int_{-\infty}^{+\infty} g_n(\nu)\nu^k\, d\nu = 0,$$

when $k = 1, \ldots, n-1$. Therefore

$$\begin{aligned} h_n(\sigma) - \hat{h}(\sigma) &= \int_{-\infty}^{+\infty} \frac{1}{\delta} g_n\left(\frac{\sigma-\nu}{\delta}\right)(\hat{h}(\nu) - \hat{h})(\sigma)))\, d\nu \\ &= \int_{-\infty}^{+\infty} g_n(-\nu)(\hat{h}(\sigma+\delta\nu) - \hat{h}(\sigma))\, d\nu \\ &= \int_{-\rho}^{\rho} g_n(-\nu)(\hat{h}(\sigma+\delta\nu) - \hat{h}(\sigma))\, d\nu \\ &\quad + \int_{\rho}^{+\infty} + \int_{-\infty}^{-\rho} g_n(-\nu)(\hat{h}(\sigma+\delta\nu) - \hat{h}(\sigma))\, d\nu \\ &= T_1 + T_2 + T_3. \end{aligned}$$

But (assuming $|\delta| \ll 1$, $|\sigma| \le \rho$) we may replace $\hat{h}$ by h in T_1. Then

$$T_1 = \int_{-\rho}^{\rho} g_n(-\nu)\left(\sum_{1\le k\le n-1} \frac{\delta^k\nu^k}{k!} h^{(k)}(\sigma) + \frac{\delta^n\nu^n}{n!} h^{(n)}(\sigma + c\delta\nu)\right) d\nu$$

($|c| \le 1$). It is easy to verify that $|h^{(k)}(\sigma)| = O(|\sigma|^{2k})$ and $|g_n(\nu)| = O(1/|\nu|^{3n+2})$ as $|\nu| \to \infty$. This immediately implies

$$\begin{aligned} \left|\delta^k \int_{-\rho}^{\rho} g_n(-\nu)\nu^k h^{(k)}(\sigma)\, d\nu\right| &= \delta^k \left|\int_{\rho}^{+\infty} + \int_{-\infty}^{-\rho} g_k(-\nu)\nu^k\, d\nu\right| \\ &= O(\delta^k(\tilde{\varepsilon}/\Delta)^{3n+1-k}(\Delta/\tilde{\varepsilon})^{2k}), \end{aligned} \tag{5.6.34}$$

where $k = 1, 2, \ldots, n-1$. Further

$$\begin{aligned} &\left|\int_{-\rho}^{\rho} g_n(-\nu)\delta^n\nu^n h^{(n)}(\sigma + c\delta\nu)\, d\nu\right| \\ &\qquad \le \delta^n \int |g_n(-\nu)||\nu|^n\sigma^{2n}\, d\nu \cdot O(1) = O(\delta^n\sigma^{2n}). \end{aligned} \tag{5.6.35}$$

Finally, it is clear that

$$|T_{2,3}| = O(\tilde{\varepsilon}/\Delta)^{3n+1}. \tag{5.6.36}$$

Inequalities (5.6.34)–(5.6.36) show that if

$$\delta\rho^2 = \delta(\Delta/\tilde{\varepsilon})^2 2^{-2/3} \ll 1, \tag{5.6.37}$$

then $\hat{h}(\sigma)$ can be approximated however good by $h_n(\sigma)$ for sufficiently large n (assuming $|\sigma| \leq \rho$) which allows one to replace $\hat{h}$ by $\hat{h}_n$ in the conditions of the problem.

Setting $\gamma = 0$ in (5.6.32) we get

$$\int_{-\infty}^{+\infty} \mathcal{E}(0,\alpha)e^{i\alpha\sigma}\psi(\alpha)\,d\alpha = h_n(\sigma) = \frac{1}{\delta}\int_{-\infty}^{+\infty} g_n\left(\frac{\sigma-\nu}{\delta}\right)\hat{h}(\nu)\,d\nu. \tag{5.6.38}$$

Inverting (5.6.38) we have

$$\begin{aligned}\mathcal{E}(0,\alpha)\psi(\alpha) &= \varphi_n(\alpha\delta)\cdot\frac{1}{\sqrt{2\pi}}\int_{-\infty}^{+\infty} e^{-i\nu\alpha}\hat{h}(\nu)\,d\nu\\ &= \varphi_n(\alpha\delta)\hat{H}(\alpha),\end{aligned}$$

where

$$\hat{H}(\alpha) = \frac{1}{\sqrt{2\pi}}\int_{-\infty}^{+\infty} e^{-i\nu\alpha}\hat{h}(\nu)\,d\nu.$$

Thus

$$B(\gamma,\sigma) = \int_{-\infty}^{+\infty}\frac{\mathcal{E}(\gamma,\alpha)}{\mathcal{E}(0,\alpha)}\varphi_n(\alpha\delta)\hat{H}(\alpha)e^{i\alpha\sigma}\,d\alpha \tag{5.6.39}$$

and $\varphi_n(\alpha\delta) = 0$ for $|\alpha| \geq 1/\delta$. The last fact allows one to consider only values α that satisfy

$$|\alpha| \leq 1/\delta. \tag{5.6.40}$$

To conclude the consideration of the general case we should study the behavior of coefficients in (5.6.31). First of all we easily get that

$$\begin{aligned}|f_1| &= O(\Delta^2),\\ f_2 &= O\left(\max\left(\Delta^2, \frac{\varepsilon^4\Delta^6}{\varepsilon^6}\right)\right) = O\left(\max\left(\frac{\Delta^5}{\varepsilon^3}\Delta\varepsilon\right), \Delta^2\right) = O(\Delta^2),\\ |f_3| &= O(\tilde{\varepsilon}^2),\\ f_4 &= O(\tilde{\varepsilon}^2).\end{aligned} \tag{5.6.41}$$

Besides $\tilde{\varepsilon}$, Δ, and δ satisfy

$$\delta(\Delta^2/\tilde{\varepsilon})^2 \ll 1, \qquad \tilde{\varepsilon}/\Delta \ll 1, \qquad \Delta^5/\tilde{\varepsilon}^3 \ll 1. \tag{5.6.42}$$

Since $|\alpha| \leq 1/\delta$, then all coefficients in (5.6.31) are small if

$$\Delta^2 \ll \delta, \qquad \tilde{\varepsilon} \ll \delta. \tag{5.6.43}$$

But $\Delta^5/\tilde{\varepsilon}^3 = (\Delta^2/\tilde{\varepsilon})^2\Delta/\tilde{\varepsilon} \ll 1$ implies

$$\Delta^2 \ll \tilde{\varepsilon}. \tag{5.6.44}$$

Therefore it suffices to choose δ satisfying the condition

$$\tilde{\varepsilon} \ll \delta \ll (\tilde{\varepsilon}/\Delta)^2 \tag{5.6.45}$$

which due to (5.6.44) is not self-contradictory.

Now note that

$$|\gamma + \alpha| \leq |\gamma| + |\alpha| \leq \left(\frac{\Delta}{\varepsilon}\right)^2 + \frac{1}{\delta} = \frac{1}{\delta}\left(\delta\left(\frac{\Delta}{\tilde{\varepsilon}}\right)^2 + 1\right) = O\left(\frac{1}{\delta}\right).$$

Therefore we can put $\gamma + \alpha = s$ in (5.6.31) and seek a solution of (5.6.31) with asymptotics

$$s^{-1/4} \exp i\left(\frac{2}{3}s^{3/2} + \frac{\pi}{4}\right)$$

assuming $|s| \leq 1/\delta$. Clearly this very solution should be inserted in (5.6.39). Since we only intended to present main principles, we will not dwell on the construction of a needed solution, $\mathcal{E}$. It is a well-known process.

Diffraction of Short Waves on a Circle Shade

As had been mentioned earlier, it is impossible to define θ in a shaded domain so that boundary values were satisfied since characteristics grow into the exterior part of the domain (i.e., inside of the circle). However in the shade the setting of the problem clearly should be modified: first, splitting of a solution onto the incident wave plus the reflected one becomes meaningless; hence, in the shade we should seek the solution itself and accordingly its boundary value will be vanishing. Further, since the wave goes along characteristics, the physical intuition hints that in the shaded domain a "flow about" the boundary must take place, i.e., in this domain, characteristics should be tangent to the boundary; in other words the shaded part of the boundary must be a caustic. The latter requirement corresponds also to the mathematical intuition: the requirement for characteristics to be tangent is a compromise between requirements that characteristic should be reflected by a boundary and that it should go out of the domain.

Let us try to solve the eikonal equation under this condition. We should solve the system

$$\frac{dx}{dt} = 2p_x, \qquad \frac{dy}{dt} = 2p_y, \qquad \frac{dp_x}{dt} = 0, \qquad \frac{dp_y}{dt} = 0. \tag{5.6.46}$$

On the circle $x = \sin\psi$, $y = \cos\psi$ the tangent condition is $(dx/dt)n_x + (dy/dt)n_y = 0$, where (n_x, n_y) is the normal vector to the circle, or $x\bar{p}_x + y\bar{p}_y = 0$, where $(\bar{p}_x, \bar{p}_y)$ is the value of (p_x, p_y) at the point (x, y) belonging to a circle. Since (p_x, p_y) are constants on the characteristic due to (5.6.46), then $p_x = \bar{p}_x = -\alpha(\psi)\cos\psi$, $p_y = \bar{p}_y = \alpha(\psi)\sin\psi$ and the eikonal equation yields that we can put $\alpha(\psi) = 1$. Then integrating (5.6.46) we get for $\tau = 2t$

$$x = \sin\psi - \tau\cos\psi, \quad y = \cos\psi + \tau\sin\psi, \quad p_x = -\cos\psi, \quad p_y = \sin\psi. \tag{5.6.47}$$

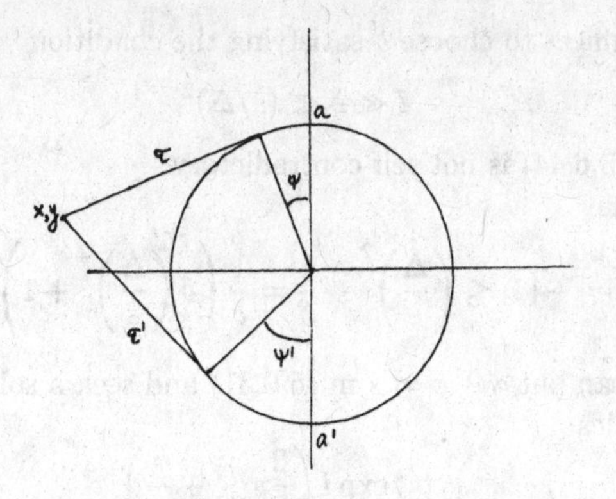

FIGURE 5.7.

Now we can find θ, which is natural to seek in coordinates ψ, τ

$$\frac{d\theta}{d\tau} = \frac{\partial\theta}{\partial x}\frac{dx}{d\tau} + \frac{\partial\theta}{\partial y}\frac{dy}{d\tau} = -(\cos^2\psi + \sin^2\psi) = -1$$

or $\theta(x,\tau) = \theta(\psi,0) - \tau$. But since characteristics are tangent to the circle we can define $\theta(\psi,0)$. In fact,

$$\frac{\partial\theta(\psi,0)}{\partial\psi} = \frac{\partial\theta(\psi,0)}{\partial x}\frac{\partial x}{\partial\psi} + \frac{\partial\theta(\psi,0)}{\partial y}\frac{\partial y}{\partial\psi} = -(\cos^2\psi + \sin^2\psi) = -1$$

or $\theta(\psi,0) = -\psi + \text{const}$. Choosing the constant equal to 0 we get

$$\theta = -\psi - \tau. \tag{5.6.48}$$

Thus, the condition that characteristics are tangent to the circle defines, actually, θ uniquely. (In the three-dimensional case we get for θ on the surface the eikonal equation.)

Formulas (5.6.47) define the coordinate system depicted in Fig. 5.7. Note that θ defined by (5.6.48) is a many-valued function in (x,y).

As earlier, let us define a solution in the form

$$u = e^{i\theta/\varepsilon}A. \tag{5.6.49}$$

From the beginning let us make an important, though trivial, remark. Usually when a solution is sought for in the form (5.6.49) or in an analogous form corresponding to the nonlinear problem, they say that A should vary slowly meaning that partial derivatives of A should be bounded. In fact, one must only require that A should vary slower than $e^{i\theta/\varepsilon}$, i.e., $|\varepsilon \operatorname{grad} A|_{\varepsilon\to 0} \to 0$. Now we will encounter this very situation.

The equation for A in coordinates (ψ, τ) is

$$2iA_\tau + \frac{iA}{\tau} + \varepsilon\left(A_{\tau\tau}\frac{1+\tau^2}{\tau^2} + \frac{2A_{\tau\psi}}{\tau^2} + \frac{A_{\psi\psi}}{\tau^2}\right.$$
$$\left. + A_\tau\frac{\tau^2 h(\psi) - 1}{\tau^3} + A_\psi(-1/\tau^3)\right) = 0 \tag{5.6.50}$$

where $h(\psi) = 1 - 2\cos^2\psi$. The following formulas follow from (6.47):

$$\psi_y = \frac{\cos\psi}{\tau}, \qquad \tau_y = \frac{\cos\psi + \tau\sin\psi}{\tau}, \qquad \psi_x = \frac{\sin\psi}{\tau},$$
$$\tau_x = \frac{\sin\psi - \tau\cos\psi}{\tau}, \qquad \frac{\partial(x,y)}{\partial(\psi,\tau)} = \tau, \qquad \Delta\theta = \frac{1}{\tau}.$$

Choose $A|_{\hat{S}} = 0$ as boundary values for A. We should learn to "glue" this solution with the one obtained earlier in the semi shade, i.e., for small τ. For finite and "large" τ the solution of (5.6.50) can be chosen in the form

$$A = A_0 + \varepsilon A_1 + \varepsilon^2 A_2 + \cdots. \tag{5.6.51}$$

If $|\tau| \ll 1$, then $A_0 \sim 1/\sqrt{\tau}$ and $A_1 \sim 1/\tau^{3+1/2}$. Thus, the same arguments as in the study of the semi shade yield that (5.6.51) is the asymptotic series if $|\tau| \geq \Delta$ and $c/\Delta^3 \ll 1$.

This shows that to find an equation which governs the problem for $|\tau| \ll 1$ we should set

$$\tau = \tilde{\varepsilon}s, \qquad \tilde{\varepsilon}^3 = \varepsilon. \tag{5.6.52}$$

However, let us stress that in our arguments we have implicitly assumed that differentiation along ψ does not "spoil" the boundedness of derivatives along ψ. For the present we have no reasons for such an assumption. Hence we should keep the most possible generality and consider the problem of constructing (5.6.51) in more detail. Equations for A_k are

$$i\left(2(A_0)_\tau + \frac{A_0}{\tau}\right) = 0, \qquad i\left((2A_k)_\tau + \frac{A_k}{\tau}\right) + LA_{k-1} = 0, \tag{5.6.53}$$

where $k = 1, 2, \ldots$, and L is the operator in square brackets in (5.6.50). Then $A_0 = a_0(\psi)\tau^{-1/2}$. For εA_1, we have

$$\varepsilon A_1 = \tilde{\varepsilon}^3\frac{a_0(\psi)}{\tau^{1/2+3}}\gamma_1 + \frac{\tilde{\varepsilon}^3 a_0(\psi)}{\tau^{1/2+1}}\gamma_2 + \frac{\tilde{\varepsilon}^3}{\tau^{1/2+2}}a_0'(\psi)\gamma_3$$
$$+ \frac{\tilde{\varepsilon}^3}{\tau^{1/2+1}}a_0''(\psi)\gamma_4 + \frac{\tilde{\varepsilon}^3 a_0(\psi)h(\psi)}{\tau^{1/2+1}}\gamma_5,$$

where γ_k's are some numbers. Now εA_1 is small if $|\tau| \geq \Delta$ and $\tilde{\varepsilon}/\Delta \ll 1$ and if at the same time $\tilde{\varepsilon}a_0'(\psi)$ and $\tilde{\varepsilon}^2 a_0''(\psi)$ are bounded.

The last condition means that we can assume $a_0(\psi) = b_0(\psi/\varepsilon)$, where $b_0(\eta)$ is a smooth function in η.

The trivial consideration of other equations (5.6.53) shows that, in general, one should seek a solution in the form

$$A = A_0\left(\tau, \frac{\psi}{\tilde{\varepsilon}}\right) + \varepsilon A_1\left(\tau, \frac{\psi}{\tilde{\varepsilon}}\right) + \cdots \tag{5.6.54}$$

indicating the necessity of additional change

$$\psi = \varepsilon\eta. \tag{5.6.55}$$

Final deduction: the master equation for the problem in the neighborhood of $\tau = 0$, $\psi = 0$ is obtained from (5.6.50) after the change of variables (5.6.52), (5.6.54).

Remark 5.6.4 The necessity of studying separately the domain of small parameter ψ follows also from the fact that small values of ψ correspond to the already solved problem in the semi shade, hence should define a further step in constructing of a solution.

The indicated change being performed, we derive from (5.6.50)

$$2iA_s + \frac{iA_s}{s} + \frac{A_{ss} + 2A_{s\eta} + A_{\eta\eta}}{s^2} - \frac{A_s + A_\eta}{s^3} + \tilde{\varepsilon}^2\left(A_{ss} - \frac{A_s}{s}\cos 2\tilde{\varepsilon}\eta\right) = 0. \tag{5.6.56}$$

The leading equation is again "parabolic":

$$2iA_s + \frac{iA}{s} + \frac{1}{s^2}\left(\frac{\partial}{\partial s} + \frac{\partial}{\partial \eta}\right)^2 A - \frac{1}{s^3}\left(\frac{\partial}{\partial s} + \frac{\partial}{\partial \eta}\right) A = 0. \tag{5.6.57}$$

The change $s = s$, $s - \eta = p$ gives an equation

$$2i\frac{\partial A}{\partial s} + 2i\frac{\partial A}{\partial p} + i\frac{A}{s} + \frac{1}{s^2}\frac{\partial A}{\partial s^2} - \frac{1}{s^3}\frac{\partial A}{\partial s} = 0.$$

The change

$$s = 2^{1/3}\sqrt{h}, \qquad -\eta + s = p, \qquad A = e^{-ih^{3/2}\cdot 2/3}B \tag{5.6.58}$$

gives the equation for B:

$$2^{1/3}i\frac{\partial B}{\partial p} + hB + \frac{\partial^2 B}{\partial h^2} = 0. \tag{5.6.59}$$

After the change $s = s$, $s - \eta = p$, Equation (5.6.57) takes the form

$$2i\frac{\partial A}{\partial s} + 2i\frac{\partial A}{\partial p} + i\frac{A}{s} + \frac{1}{s^2}\frac{\partial^2 A}{\partial s^2} - \frac{1}{s^3}\frac{\partial A}{\partial s} = 0.$$

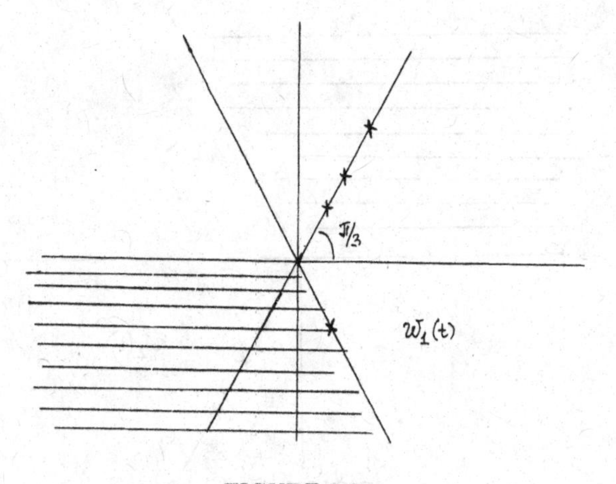

FIGURE 5.8.

[It is advisable to kill terms containing $\partial A/\partial s$. The imaginary summand is killed by the change $A = e^{i\psi(s)}B$ and the real one is killed afterwards by the change $s = s(h)$. The result is given by the change (5.6.58).]

In our problem negative values of ψ are needed. Therefore, we can assume further that $p > 0$. Let us find a solution of (5.6.59) in the form

$$B = \exp\left(2^{-1/3}ip\alpha\right)\tilde{B}(h). \tag{5.6.60}$$

Then $\tilde{B} = w(\alpha - h)$, where w is a solution of the Airy equation $d^2\mathcal{E}/dt^2 - t\mathcal{E} = 0$. If B is in the indicated form, then $\tilde{B}(0) = w(\alpha) = 0$. In what follows we will denote Airy functions as in [4]. Figures 5.8 and 5.9 are taken from [4] also. In the shaded domains function w_i exponentially decrease. Crosses mark zeros of corresponding functions. Thus α must be a zero of an Airy function.

But (5.6.60) implies that $\operatorname{Re}\alpha > 0$, since $p > 0$, and can be large. Therefore taking Figures 5.8 and 5.9 into account we should put

$$w(t) = cw_1(t). \tag{5.6.61}$$

For α one should choose a zero of $w_1(t)$ with the minimal real part. In fact, the solution B is in the long run a function in $(\tau/\tilde{\varepsilon}, \psi/\tilde{\varepsilon})$, whereas both ψ and τ can take finite values, or one can say that p and h can take large values.

Formula (5.6.61) shows that B exponentially decreases as p grows and therefore, though, generally speaking, we might have looked for a solution in the form of the sum of summands of the form (5.6.60) corresponding to different α, zeros with large real parts give no contribution to asymptotics. In this connection we should make the following remark: we can count the

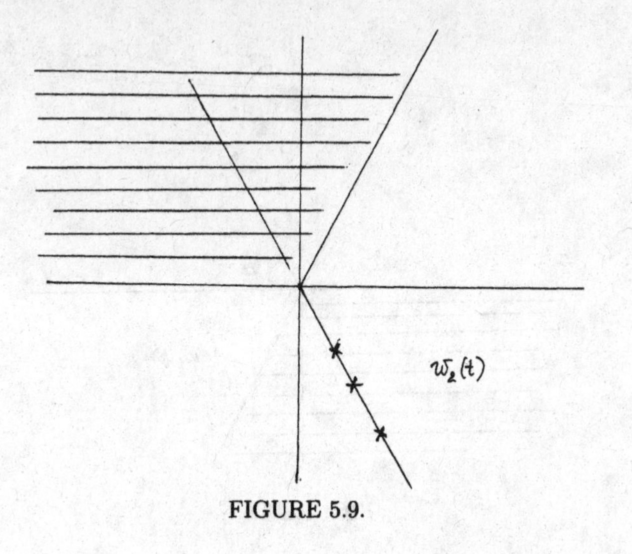

FIGURE 5.9.

angle ψ from the lines $(0, a')$, see Fig. 5.7. A new coordinate system (ψ', τ') would arise. Then only the greatest summands are left in the asymptotics corresponding to the new system.

Let us show how to define the constant c in (5.6.61).

We know a solution of the problem on the line (A, a), see Fig. 5.6, the case of semi shade. It is defined by coordinates σ and γ from (5.6.15) and $\sigma^2 \sim \gamma$ on (A, a).

Variables σ and γ can then take large values and $x \sim \nu = \tilde{\varepsilon}\sigma 2^{1/3}$. On the other hand in coordinates (5.6.58) (in the shade) we have $p = s = \eta = 0$ and $x = \tilde{\varepsilon}s$ on (A, a), implying $s \sim \sigma 2^{1/3}$.

Now making use of the solution of the Shroedinger equation for the semi shade on (A, a) and comparing it with $cw_1(\alpha - h)$ for large s, i.e., computing asymptotics of both formulas, we define c.

We will confine ourselves to the principal term of the asymptotics constructed this way. Examination of the complete equation (5.6.56) in which we should again perform the change (5.6.58) enables one to make corrections. Then as it had been done in the case of semi shade, we obtain a solution for large, but not too large, p and h and then pass to solving (5.6.53).

In the following problem as in the Section 5.5, arguments from Section 5.3 are used but in the "boundary layer" situation. Unlike in Section 5.5, we will not use the canonical reduction of the corresponding system to the normal form and will seek a solution directly in the form (5.3.11). The method of constructing the asymptotics is similar to the one used in the simple example, Section 3.9.

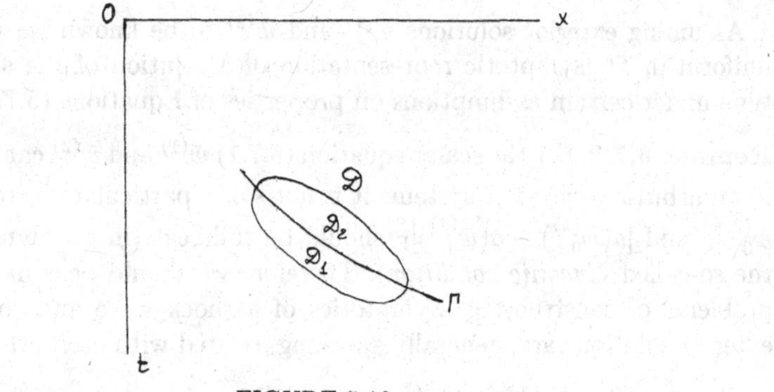

FIGURE 5.10.

5.7 One-Dimensional Shock Wave [7]

Introduction

Consider the system of differential equations

$$\frac{\partial w}{\partial t} + \frac{\partial}{\partial x}\left(a(w) - \varepsilon\frac{\partial b(w)}{\partial x}\right) = 0 \qquad (5.7.1)$$

where w is a vector, $a(w)$ and $b(w)$ are vector functions $x, t \in \mathcal{D}$, where $\mathcal{D}$ is a domain on the plane (x, t).

An *exterior solution* of Equations (5.7.1) is the formal series in

$$\bar{w} = \bar{w}_0 + \varepsilon\bar{w}_1 + \cdots, \qquad \bar{w}_j = \bar{w}_j(t, x), \qquad (5.7.2)$$

where $j = 0, 1, \ldots$, satisfying (5.7.1) in each order in powers of ε, i.e.,

$$\frac{\partial \bar{w}_0}{\partial t} + \frac{\partial}{\partial x}a(\bar{w}_0) = 0,$$

$$\frac{\partial \bar{w}_1}{\partial t} + \frac{\partial}{\partial x}\left((\bar{w}_1\nabla)a(\bar{w}_0) - \frac{\partial b(\bar{w}_0)}{\partial x}\right) = 0 \qquad (5.7.3)$$

[here and in what follows $(p\nabla)c(q)$ means componentwise application of $\sum_k p_k(\partial/\partial q_k)$ to $c(q)$.]

A solution $w(t, x; \varepsilon)$ of (5.7.1) in $\mathcal{D}$ is a *shock wave* if there is a smooth curve Γ separating $\mathcal{D}$ into two parts $\mathcal{D}_1$ and $\mathcal{D}_2$ and two exterior solutions $\bar{w}^{(1)}$ and $\bar{w}^{(2)}$ which do not coincide in the principal order on Γ, i.e., $[\bar{w}_0^{(1)} - \bar{w}_0^{(2)}]_\Gamma \neq 0$ and such that $w(t, x; \varepsilon)$ asymptotically coincides with $\bar{w}^{(i)}$ in each interior subdomain $\mathcal{D}_i$ $(i = 1, 2)$. Thus in the limit $w = \bar{w}_0^{(i)}$ in $\mathcal{D}_i$ at $\varepsilon = 0$ and $w(t, x; 0)$ has a first-type discontinuity on Γ.

Remark 5.7.1 We will not strictly follow the given definition conceding without proof that the formal asymptotics (of the boundary-layer type) which satisfies the definition is the asymptotics of $w(t, x, \varepsilon)$.

Assuming exterior solutions $\bar{w}^{(1)}$ and $\bar{w}^{(2)}$ to be known we will seek a uniform in $\mathcal{D}$ asymptotic representation of a solution of the shock wave type under certain assumptions on properties of Equations (5.7.1).

Remark 5.7.2 For the scalar equation (5.7.1) $\bar{w}^{(1)}$ and $\bar{w}^{(2)}$ can be defined in an arbitrary way. For systems it is not so; in particular, vectors $[\bar{w}_0^{(1)} - \bar{w}_0^{(2)}]_\Gamma$ and $[a(\bar{w}_0^{(1)}) - a(\bar{w}_0^{(2)})]_\Gamma$ should be collinear (in gas dynamics it is the so-called *Gugonic condition*). Therefore we should bear in mind that problems of constructing asymptotics of a shock wave and constructing exterior solutions are, generally speaking, related with each other.

First, recall several known (see, e.g., [36]) considerations on the structure of a shock wave.

The shock line Γ ; $x = x_0(t)$ being known, putting $y = (1/\varepsilon)(x - x_0(t))$ and passing from t, x to t, y we get $\varepsilon(\partial w/\partial t) + \partial/\partial y[-x_0'(t)w + a(w) - \partial b(w)/\partial y] = 0$. Assuming that $\partial w/\partial t$ is bounded as $\varepsilon \to 0$ we see that the principal term of the asymptotics in a neighborhood of Γ is defined by equations $(\partial/\partial y)\,[-x_0'(t)w + a(w) - \partial b(w)/\partial y] = 0$ with boundary values $w(t, -\infty) = w_0^{-(1)}|_\Gamma$, $w(t, \infty) = \bar{w}_0^{(2)}|_\Gamma$. Integrating once with respect to y we get $-x_0'(t)w + a(w) - \partial b(w)/\partial y = m(t)$, where due to the boundary values we have, on the one hand, $m(t) = [-x_0'(t)\bar{w}_0^{(1)} + a(\bar{w}_0^{(1)})]_\Gamma$ and, on the other hand, $m(t) = [-x_0'(t)\bar{w}_0^{(2)} + a(\bar{w}_0^{(2)})]_\Gamma$. Thus, the so-called conditions on the shock line should be satisfied,

$$[\bar{w}_0^{(1)} - \bar{w}_0^{(2)}]_\Gamma x_0'(t) = [a(\bar{w}_0^{(1)}) - a(\bar{w}_0^{(2)})]_\Gamma; \tag{5.7.4}$$

see Remark 5.7.2, which can be also obtained integrating both parts of (5.7.1) with respect to x along the segment containing $x = x_0(t)$ and passing to the limit as $\varepsilon \to 0$. Conditions on the shock (5.7.4) give the equation for the shock line $\Gamma : x_0'(t) = F(t, x_0(t))$ [and constraints for exterior solutions on Γ if (5.7.1) is a system] which allow one to find $x_0(t)$ if one of the points of Γ is known, e.g., $x_0(t_0)$, where t_0 is the moment when the shock wave is born.

Further, to obtain the principal term of the asymptotics we should solve the ordinary (in y) system of equations

$$-x_0'(t)w + a(w) - \frac{\partial b(w_0)}{\partial y} = m(t)$$

with the above-mentioned boundary value. If the phase trajectory of this system, connecting points $\bar{w}_0^{(1)}|_\Gamma$ and $\bar{w}_0^{(2)}|_\Gamma$ exists and is unique (assumption of this kind will be important in what follows) then w (the principal term) as function in t, y can be found up to the change of y by $y - c(t)$, where $c(t)$ is arbitrary.

Thus these arguments enable one to find the form of the shock wave in the principal order up to a shift along y-axis by $c(t)$ or up to values of w on Γ. It remains also unclear how to construct higher approximations.

In what follows we indicate a regular process of recovering a uniform in $\mathcal{D}$ asymptotics from known exterior solutions with accuracy up to perhaps one *constant* defined by values of w and playing the same role as $x_0(t_0)$ (see above).

Remark 5.7.3 For *systems* the knowledge of w in one of points of Γ is often termed to be formally unnecessary, e.g., $x_0(t)$ can be often found from (5.7.4) excluding $x_0'(t)$ if $\bar{w}^{(1)}$ and $\bar{w}^{(2)}$ are known; see Remark 5.7.2.

ASYMPTOTICS IN $\mathcal{D}_i \cup \Gamma$

An essential feature of this problem is that the desired asymptotics is represented by different uniform developments $w^{(1)}$ and $w^{(2)}$ in $\mathcal{D}_1 \cup \Gamma$ and $\mathcal{D}_2 \cup \Gamma$. Cases admitting a "uniform" asymptotic development, e.g., the known Burger's equation, are very rare exceptions.

In formulas of this section the index i, where $i = 1, 2$, indicating the domain $\mathcal{D}_i \cup \Gamma$ is omitted for brevity's sake but will be restored later. The line Γ is assumed to be known. [The shock line Γ can be found from (5.7.4) which will also follow from further constructions.]

Asymptotics of $w(t, x, \varepsilon)$ will be sought in the form

$$w(t,x,\varepsilon) \simeq w_0(t,x,\zeta) + \varepsilon w_1(t,x,\zeta) + \cdots, \tag{5.7.5}$$

where $\zeta = e^{\theta(t,s)/\varepsilon}$ under the following assumptions:

1. $\theta|_\Gamma = 0$, $\theta_x \neq 0$ ($\theta < 0$ outside of Γ).
2. $w_j(t,x,\zeta) = \bar{w}_j(t,x) + \zeta\hat{w}_j(t,x) + O(\zeta^2)$ as $\zeta \to 0$ ($|\hat{w}_0| \neq 0$).
$$\tag{5.7.6}$$
3. There exist $\lim_{\zeta\to+\infty} w_0 < \infty$, $\lim_{\zeta\to+\infty} \zeta(\partial w_0/\partial\zeta) = 0$ and w_j's do not grow faster than $O(\ln\zeta)^j)$ for fixed t, x as $\zeta \to +\infty$.
4. The series (5.7.3) formally satisfies equations (5.7.1) in the following sense. Functions $w_j(t,x,\zeta)$ are differentiated according to the rule

$$(w_j)_t = \frac{\theta_t}{\varepsilon}\zeta\frac{\partial w_j}{\partial\zeta} + \frac{\partial w_j}{\partial t}, \qquad (w_j)_x = \frac{\theta_x}{\varepsilon}\zeta\frac{\partial w_j}{\partial\zeta} + \frac{\partial w_j}{\partial x},$$

$$(w_j)_{xx} = ((w_j)_x)_x$$

and Equation (5.7.1), which is $(w)_t + (a(w))_x - \varepsilon(b(w))_{xx} = 0$, is satisfied in each order in powers of ε identically in t, x, ζ.

Let us elucidate these assumptions. If (3) in (5.7.6) is not mentioned then we are evidently speaking of a direct generalization of the method indicated in Section 3.9 of constructing boundary-layer type asymptotics. Regularity conditions (2) in (5.7.6) will lead, as in Section 3.9, first, to the fact that $\bar{w} = \bar{w}_0 + \varepsilon\bar{w}_1 + \cdots$ is actually an exterior solution of (5.7.2) satisfying (5.7.3) which gives w "far" ($\zeta \to 0$) from Γ and, second, will give equations for θ (eikonal equation) and $\hat{w}_j$ [since the condition (1) $\theta|_\Gamma = 0$ gives a boundary value for θ ; boundary values for $\hat{w}_j$ will be discussed later].

As to condition (3) in (5.7.6), it deals with the possibility of extending (5.7.5) across Γ into *another* domain $\mathcal{D}_i$, where $\theta > 0$, but only for "small" distances of order ε from Γ [since $\ln\zeta = \theta/\varepsilon$, then (5.7.5) can be applied for a positive $\theta \sim \varepsilon^k$, $k \geq 0$, according to (3) in (5.7.6); the critical value of k is 0].

Remark 5.7.4 Condition (3) in (5.7.6) as will be clear in what follows is essential: it will follow from formulas themselves that one should expect that $w_j = O((\ln\zeta)^j)$ as $\zeta \to +\infty$.

Though the general scheme of computations should be clear from Section 3.9 nevertheless we will dwell on equations for $w_0, w_1, \ldots$. Inserting (5.7.5) in (5.7.1) and taking (4) in (5.7.6) into account and equating coefficients of ε^{-1} to 0 we get

$$\omega\zeta\frac{\partial}{\partial\zeta}\left(-kw_0 + a(w_0) - \omega\zeta\frac{\partial b(x_0)}{\partial\zeta}\right) = 0, \tag{5.7.7}$$

where

$$\omega = \partial\theta/\partial x, \qquad -\omega k = \partial\theta/\partial t \tag{5.7.8}$$

[in fact (5.7.7) is already known from the Introduction to this example—equation for the principal term of the asymptotics].

First, let us make use of the regularity condition (2) in (5.7.6): $w_0 = \bar{w}_0 + \zeta\hat{w}_0 + O(\zeta^2)$ as $\zeta \to 0$. Since (5.7.7) turns into a trivial identity at $\zeta = 0$, then equating to zero coefficients of ζ we get only one corollary (recall that $\omega = \theta_x \neq 0$):

$$(\hat{w}_0\nabla)(-k\bar{w}_0 + a(\bar{w}_0) - \omega b(\bar{w}_0)) = 0 \tag{5.7.9}$$

[here, as had been said in the remark to (5.7.3) ∇ is the grad-operation with respect to $\bar{w}_0$; in particular $(\hat{w}_0\nabla)\bar{w}_0 = \hat{w}_0$].

Considering (5.7.9) as a system of linear equations for $\hat{w}_0$ let us write it in the form

$$T(\omega, k, \bar{w}_0)\hat{w}_0 = 0, \tag{5.7.9'}$$

where $T(\omega, k, \bar{w}_0)$ is the corresponding matrix.

Since we have assumed that $|\hat{w}_0| \neq 0$ we should have

$$\det T(\omega, k, \bar{w}_0) = 0. \tag{5.7.10}$$

Let us postpone for the time being the discussion of this (important) equation and turn to (5.7.7). Dividing it by $\omega\zeta$ let us integrate it from 0 to ζ. We will get

$$-kw_0 + a(w_0) - \omega\zeta\frac{\partial b(w_0)}{\partial\zeta} = -k\bar{w}_0 + a(\bar{w}_0). \qquad (5.7.11)$$

Put $w_0^* = \lim_{\zeta\to+\infty} w_0$ which exists due to (3). Then

$$-k\bar{w}_0 + a(\bar{w}_0) = -kw_0^* + a(w_0^*). \qquad (5.7.12)$$

Further, we should make several assumptions, one of which is the existence of a solution $w_0^* = w_0^*(k, \bar{w}_0)$ of (5.7.12) different from $\bar{w}_0$. For the sake of simplicity we will assume that the solution $w_0^* \neq \bar{w}_0$ of (5.7.12) is unique.

Remark 5.7.5 Cases of nonuniqueness of w_0^* are possible though we do not give such examples.

Now considering (5.7.11) as a system of ordinary equations [k, ω, and $\bar{w}_0$ are parameters related by (5.7.10)] which should be solved with initial values $w_0|_{\zeta=0} = \bar{w}_0$, $w_0|_{\zeta=+\infty} = w_0^*$. Let us call the following statement the Main Theorem: *There exists the unique phase trajectory of the system (5.7.11) (in the space of w_0) passing through points $\bar{w}_0$ and w_0^* for any admissible values of parameters.*

If the Main Theorem is valid, which will be assumed in what follows, then it is possible to recover w_0 from (5.7.11) as a function in $M_0(t,x)\zeta$ and in parameters k, ω, and $\bar{w}_0$, where $M_0(t,x)$ is arbitrary [the change of ζ by $M_0\zeta$ does not affect (5.7.11)].

Parameters k, ω, $\bar{w}_0$ should be related in a quite definite manner: to real roots $\omega(k, \bar{w}_0)$ of (5.7.10) directions $\hat{w}_0/|\hat{w}_0|$ in the space w_0 tangent to phase trajectories of (5.7.11) passing through $\bar{w}_0$ correspond and one can only pick the root

$$\omega = \Omega(k, \bar{w}_0) \qquad (5.7.13)$$

corresponding to the direction $\hat{w}_0/|\hat{w}_0|$ tangent to the trajectory connecting $\bar{w}_0$ and w_0^*. We will assume that such a root $\Omega(k, \bar{w}_0)$ is unique and smoothly depends on k, $\bar{w}_0$. Further we will also assume that $\operatorname{rank} T(\Omega; k, \bar{w}_0) = N - 1$, where N is the dimension of the space of w's.

Before we turn to other equations let us indicate what is and what is not known about w_0 (in a fixed domain $\mathcal{D}_i$). Until now we have not established that $\bar{w}_0$ is in fact the corresponding exterior solution satisfying the first of Equations (5.7.3). However, it is clear that it is so and it will easily follow from regularity conditions in the next order (recall Section 3.9). Further, we have assumed the shock line Γ to be known, i.e., have not as yet derived (5.7.4). Below it will be obtained from continuity considerations for the desired solution on Γ but it is already clear from the introduction that (5.7.4) takes place.

Remark 5.7.6 It does not follow from (5.7.4) and (5.7.12) that w_0^* is another exterior solution.

Now note that (5.7.13) is, if we take (5.7.8) into account, an (eikonal) equation for θ in partial derivatives of the first order (i.e., actually, ordinary differential equation) with the known boundary value $\theta|_\Gamma = 0$. Hence, θ can be assumed to be known.

Thus a true unknown as yet function in w_0 is the factor $M_0(t,x)$ which plays the same role as the shift $y(t)$ in a nonuniform approach described earlier. The search for $M_0(t,x)$ is now the main problem in constructing the principal term of asymptotics.

Now consider equations arising from terms with ε^0 in (5.7.1) after substituting (5.7.5). They are of the form

$$\omega\zeta\frac{\partial}{\partial\zeta}\left(-kw_1 + (w_1\nabla)a(w_0) - \frac{\partial b(w_0)}{\partial x} - \omega\zeta\frac{\partial}{\partial\zeta}((w_1\nabla)b(w_0))\right)$$
$$= -\frac{\partial w_0}{\partial t} - \frac{\partial}{\partial x}\left(a(w_0) - \omega\zeta\frac{\partial b(w_0)}{\partial\zeta}\right). \tag{5.7.14}$$

Just as above let us first make use of the regularity conditions (2) in (5.7.6). Comparison terms containing ζ^0 yields

$$\frac{\partial\bar{w}_0}{\partial t} + \frac{\partial}{\partial x}a(\bar{w}_0) = 0, \tag{5.7.15}$$

i.e., the first of Equations (5.7.3), as had to be expected [similarly, next equations imply the second equation of (5.7.3), etc., i.e., $\bar{w}$ is actually the exterior solution].

Further, comparing coefficients of ζ^1 we get

$$T(\omega; k, \bar{w}_0)\hat{w}_1 = R_1 \tag{5.7.16}$$

and $T(\omega; k, \bar{w}_0)$ is the already known matrix and the right-hand side of (5.7.16) can be written in the form

$$R_1 = -(\bar{w}_1\nabla)[(\hat{w}_0\nabla)a(\bar{w}_0) - \omega(\hat{w}_0\nabla)b(\bar{w}_0)] + \frac{\partial}{\partial x}[(\hat{w}_0\nabla)b(\bar{w}_0)]$$
$$- \frac{1}{\omega}\left(\frac{\partial\hat{w}_0}{\partial t} + \frac{\partial}{\partial x}(k\hat{w}_0)\right)$$

if we make use of (5.7.8) to simplify terms arising from the right-hand side of (5.7.14). To avoid a misunderstanding, note that here ∇ is grad with respect to $\bar{w}_0$ and the first square bracket should not be replaced by $k\hat{w}_0$ and also substitute ω from (5.7.13) in it before differentiation is performed.

Now note that in R_1 only one unknown function actually enters. It is the factor $M_0(t,x)$ up to which $\hat{w}_0$ had been determined above and R_1 can be written in the form

$$R_1 = A_1\frac{\partial M_0}{\partial t} + B_1\frac{\partial M_0}{\partial x} + c_1 M_0,$$

where A_1, B_1, and C_1 are known vectors.

The linear system (5.7.16) is compatible if and only if

$$(QR_1) = 0, \tag{5.7.17}$$

where $(T^*Q) = 0$, if T^* is the transposed of T [for $\omega = \Omega(k, \bar{w}_0)$]. Due to our assumption on the rank of $T(\omega; k, \bar{w}_0)$ the vector Q is unique up to a multiple.

Thus we get for M_0 a first order linear differential equation

$$(Q, A_1)\frac{\partial M_0}{\partial t} + (Q, B_1)\frac{\partial M_0}{\partial x} + (Q, C_1)M_0 = 0,$$

i.e., (5.7.17).

To find M_0 we now should need $M_0|_\Gamma$ (boundary value) and it is for the time being the unique arbitrariness in w_0.

Let us return to (5.7.14) and dividing it by $\omega\zeta$ integrate from 0 to ζ. Inserting $\omega\zeta(\partial b(w_0)/\partial\zeta)$ from (5.7.11) in the right-hand side and making use of (5.7.15), we will get

$$\begin{aligned} -kw_1 + (w_1\nabla)a(w_0) - \omega\zeta\frac{\partial}{\partial\zeta}((w_1\nabla)b(w_0)) - \frac{\partial b(w_0)}{\partial x} \\ = -k\bar{w}_1 + (\bar{w}_1\nabla)a(\bar{w}_0) - \frac{\partial b(\bar{w}_0)}{\partial x} \\ -\frac{1}{\omega}\int_0^\zeta \frac{1}{\zeta}\left\{\frac{\partial(w_0 - \bar{w}_0)}{\partial t} + \frac{\partial}{\partial x}[k(w_0 - \bar{w}_0)]\right\} d\zeta. \end{aligned} \tag{5.7.18}$$

Like (5.7.11) for w_0, this is ordinary and now linear system for w_1. Note that the integral in the right-hand side grows, generally speaking, like $\ln\zeta$ as $\zeta \to +\infty$.

Let us assume the validity of the main-theorem-type assumption about (5.7.18): the solution w_1 satisfying $w_1|_{\zeta=0} = \bar{w}_1$ at $\zeta = 0$ and $|w_1| = O(\ln\zeta)$ as $\zeta \to +\infty$ exists and is unique up to a scalar multiple $M_1(t, x)$ [it is clear from (5.7.9') and (5.7.16) that $\hat{w}_1$ is defined up to a summand $M_1\hat{w}_0$].

Thus, in w_1 only one arbitrary function $M_1(t, x)$ remains for which in the next order we will find the equation $(Q, R_2) = 0$ of the form (5.7.17), etc. Further arguments (we hope) are clear.

Thus we have indicated the process of construction the asymptotics in a fixed domain $\mathcal{D}_i \cup \Gamma$ with accuracy up to values $M_0|_\Gamma$ and $M_1|_\Gamma$ which are to be found and (5.7.4) is to be derived.

Matching Conditions. Conservation Laws

Now we will endow the quantities introduced above with the index $i = 1, 2$, indicating the domain $\mathcal{D}_i \cup \Gamma$.

Matching conditions consist first of all in the continuity of the obtained solution on Γ. Since $\theta^{(1)} = \theta^{(2)} = 0$ on Γ, i.e., $\zeta^{(1)} = \zeta^{(2)} = 1$, the continuity of $w(t, x; \varepsilon)$ on Γ is the condition of coincidence of $w_j^{(1)}(t, x, \zeta^{(1)})$ and $w_j^{(2)}(t, x, \zeta^{(2)})$:

$$w_j^{(1)}(t, x_0(t), 1) = w_j^{(2)}(t, x_0(t), 1), \tag{5.7.19}$$

where $j = 0, 1, \ldots$, or in short $[w_j^{(i)}(t, x, \zeta^{(i)})]x = x_0(t)]_{(1)}^{(2)} = 0$. These conditions give *one* relation between $M_j^{(1)}|_\Gamma$ and $M_j^{(2)}|_\Gamma$, where $j = 0, 1, 2, \ldots$.

Since

$$(b(w))_x = \frac{1}{\varepsilon}\omega\zeta\frac{\partial b(w_0)}{\partial \zeta} + \omega\zeta\frac{\partial}{\partial \zeta}((w_1\nabla)b(w_0)) + \frac{\partial b(w_0)}{\partial x} + \cdots,$$

the continuity condition of $(b(w))_x$ takes the form

$$\left[\omega^{(i)}\zeta^{(i)}\frac{\partial b(w_0^{(i)})}{\partial \zeta^{(i)}}\bigg|_{\substack{x=x_0(t)\\ \zeta^{(i)}=1}}\right]_{(1)}^{(2)} = 0,$$

$$\left[\left(\omega^{(i)}\zeta^{(i)}\frac{\partial}{\partial \zeta^{(i)}}((w_1^{(i)}\nabla)b(w_0^{(i)})) + \frac{\partial b(w_0^{(i)})}{\partial x}\right)\bigg|_{\substack{x=x_0(t)\\ \zeta^{(i)}=1}}\right]_{(1)}^{(2)} = 0. \tag{5.7.20}$$

The first of these conditions, as is clear from (5.7.11), (5.7.19) and $k^{(1)}|_\Gamma = k^{(2)}|_\Gamma = x_0'(t)$, is the condition on the shock line (5.7.4) and Γ is now known.

Other conditions will give the second relation between $M_j^{(1)}|_\Gamma$ and $M_j^{(2)}|_\Gamma$, where $j = 0, 1, 2, \ldots$.

Conditions (5.7.20) can be considered as conservations laws on the shock line. Let us illustrate this by the example of the first two of these conditions.

For the exterior solution, let us introduce the notion of a vector potential $U = U_0 + \varepsilon U_1 + \cdots$, where $U_j = U_j(t, x)$, setting

$$\bar{w} = \frac{\partial U}{\partial x}, \qquad a(\bar{w}) - \varepsilon\frac{\partial b(\bar{w})}{\partial x} = \frac{\partial U}{\partial t} \tag{5.7.21}$$

which is possible due to the divergent form of (5.7.1). Then the first of conditions (5.7.20), i.e., the condition on the shock line (5.7.4) can be written in the form

$$\left[(-k^{(i)}\bar{w}_0^{(i)} + a(\bar{w}_0^{(i)}))|_{x=x_0(t)}\right]_{(1)}^{(2)}$$

$$= \left[\left(x_0'(t)\frac{\partial U_0^{(i)}}{\partial x} + \frac{\partial U_0^{(i)}}{\partial t}\right)\bigg|_{x=x_0(t)}\right]_{(1)}^{(2)} = \left(\frac{\partial U_0^{(i)}|_\Gamma}{dt}\right)_{(1)}^{(2)} = 0,$$

i.e., in the form of the conservation law for U_0, i.e., $[U_0^{(i)}|_\Gamma]_{(1)}^{(2)} = 0$ (the constant of integration is included in U).

Further, the second of the conditions (5.7.20) is of the form

$$\Bigg[\Bigg(-\frac{\partial b(\bar{w}_0^{(i)})}{\partial x} - k^{(i)}\bar{w}_1^{(i)} + (\bar{w}_1^{(i)}\nabla)a(\bar{w}_0^{(i)}) - \frac{1}{\omega^{(i)}}\int_0^1 \frac{1}{\zeta}\left\{\frac{\partial(w_0^{(i)} - \bar{w}_0^{(i)})}{\partial t} + \frac{\partial}{\partial x}\left(k^{(i)}(w_0^{(i)} - \bar{w}_0^{(i)})\right)\right\} d\zeta\Bigg)\Bigg|_{x=x_0(t)}\Bigg]_{(1)}^{(2)} = 0 \tag{5.7.22}$$

due to (5.7.18) and (5.7.19). Set

$$T_1^{(i)} = \int_0^1 \frac{w_0^{(i)} - \bar{w}_0^{(i)}}{\omega^{(i)}\zeta}\, d\zeta. \tag{5.7.23}$$

Then

$$\frac{1}{\omega^{(i)}}\left(\frac{\partial}{\partial t} + \frac{\partial}{\partial x}k^{(i)}\right)\int_0^1 \frac{w_0^{(i)} - \bar{w}_0^{(i)}}{\zeta}\, d\zeta = \left(\frac{\partial}{\partial t} + k^{(i)}\frac{\partial}{\partial x}\right)T_1^{(i)}$$

since $\omega_t + (k\omega)_x = 0$ due to (5.7.8).

Now, according to (5.7.21), formula (5.7.22) takes the form

$$\left(\left(\frac{\partial}{\partial t} + k^{(i)}\frac{\partial}{\partial x}\right)(U_1^{(i)} - T_1^{(i)})\Big|_\Gamma\right)_{(1)}^{(2)} = 0$$

and since $(\partial/\partial t) + k(\partial/\partial x) = d/dt$ on Γ, then integrating and including the constant of integration in U we get the conservation law

$$\left((U_1^{(i)} - T_1^{(i)})|_\Gamma\right)_{(1)}^{(2)} = 0. \tag{5.7.24}$$

Note that $M_0^{(i)}|_\Gamma$ enter explicitly in (5.7.24), i.e., there are no derivatives of M_0 in (5.7.24). (M_0 only enters in T_1.)

Similar conservation laws take place also in all orders.

Thus (5.7.20) produces the second relation between $M_j^{(1)}|_\Gamma$ and $M_j^{(2)}|_\Gamma$, where $j = 0, 1, 2, \ldots$, and we can now terminate the constructing of asymptotics with accuracy up to a constant (potentials, and therefore $M_j^{(i)}|_\Gamma$, are defined with accuracy up to a constant).

Notice also that (5.7.20) implies new conditions on the shock line (one might say conditions of the second order) relating $\bar{w}_1^{(1)}$ and $\bar{w}_1^{(2)}$.

In conclusion, let us give several examples.

EXAMPLE I

The generalized Burger's equation $u_t + h(u)u_x = \varepsilon u_{xx}$, where $h(u)$ is a monotonous (increasing) function. Equation (5.7.11) is

$$-ku_0 + Q(u_0) - \omega\zeta\frac{\partial u_0}{\partial \zeta} = -k\bar{u}_0 + Q(\bar{u}_0)Q(u) = \int_0^u h(\eta)\, d\eta.$$

Equation (5.7.12), i.e., $-k\bar{u}_0 + Q(\bar{u}_0) = -ku_0^* + Q(u_0^*)$, has the unique root $u_0^* \neq \bar{u}_0$ since $h(u)$ is monotonous. Further, (5.7.9') is of the form $T\hat{u}_0 \equiv [-k+h(\bar{u}_0)-\omega]\hat{u}_0 = 0$ so that (5.7.10) has the unique root $\omega = \Omega = h(\bar{u}_0)-k$. This is the equation for θ. In more details $\theta_x^{(i)} = h(\bar{u}_0^{(i)})+\theta_t^{(i)}/\theta_x^{(i)}$, where $i = 1, 2$. The equation for the shock line is

$$x_0'(t) = [Q(\bar{u}_0^{(1)}) - Q(\bar{u}_0^{(2)})]/[\bar{u}_0^{(1)} - \bar{u}_0^{(2)}]|_{x=x_0(t)}.$$

The solution of the equation for u_0, where $\omega = h(\bar{u}_0) - k$; $u_0|_{\zeta=0} = \bar{u}_0, u_0|_{\zeta=\infty} = u_0^*$, i.e., of $(u_0 - \bar{u}_0)G(u_0; \bar{u}_0, k) = M_0(t,x)\zeta$ is

$$G(u_0; \bar{u}_0, k) = \exp\int_{\bar{u}_0}^{u_0}\left(\frac{h(\bar{u}_0) - k}{V(\eta, \bar{u}_0, k)} - \frac{1}{\eta - \bar{u}_0}\right) d\eta,$$

where $V(u_0; \bar{u}_0, k) = Q(u_0) - Q(\bar{u}_0) - k(u_0 - \bar{u}_0)$.

Functions $M_0^{(i)}$, where $i = 1, 2$, are related on the shock line Γ first by (5.7.19): $u_0^{(1)}|_\Gamma = u_0^{(2)}|_\Gamma = u_0|_\Gamma$ (a common value), i.e.,

$$(u_0|_\Gamma - \bar{u}_0^{(i)}|_\Gamma)G(u_0|_\Gamma = u_0^{(i)}|_\Gamma, \qquad x_0'(t)) = M_0^{(i)}|_\Gamma \qquad (i = 1, 2).$$

Further, since we consider scalar equations, Equation (5.7.17) is of the form $R_1 = 0$ $(\hat{u}_0 = M_0)$ and we have

$$-\bar{u}_1^{(i)} M_0^{(i)} h'(\bar{u}_0^{(i)}) + \frac{\partial M_0^{(i)}}{\partial x} - \frac{1}{\omega^{(i)}}\left(\frac{\partial M_0^{(i)}}{\partial t} + \frac{\partial}{\partial x}(k^{(i)} M_0^{(i)})\right) = 0,$$

where $h'(u) = dh/du$.

The boundary value for this equation is given by the preceding formula if $u_0|_\Gamma$ is known. We find this quantity from (5.7.22) in the integrated form; we do not introduce a potential:

$$\left(\int_{u_0^{(i)}|_\Gamma}^{\bar{u}_0|_\Gamma} \frac{\eta - \bar{u}_0^{(i)}|_\Gamma}{V(\eta; \bar{u}_0^{(i)}|_\Gamma, x_0'(t))}\, d\eta - \int_{t_0}^{t}\left(\omega^{(i)}\bar{u}_1^{(i)} - \frac{\partial \bar{u}_0^{(i)}}{\partial x}\right)\bigg|_\Gamma dt\right)_{(1)}^{(2)} = c(t_0),$$

where $c(t_0)$ is defined from the value of $u_0|_\Gamma$ at the fixed point of Γ.

Remark 5.7.7 For the classical Burgers equation $h(u) = u$ all computations can be carried out explicitly $[c(t_0) = 0$, where t_0 is the moment when the shock wave is born] and obtain the asymptotics of the exact Hopf–Cole solution. In this case we get "uniform" asymptotics: in $\mathcal{D}$ a single function θ can be introduced so that $\theta^{(1)} = \theta$, $\theta^{(2)} = -\theta$ and $M_0^{(1)} = 1/M_0^{(2)} = M_0$.

The essential feature of the case $h(u) \neq u$ is that there can be no "uniform" asymptotics. It is clear from the formula $\omega^{(i)} = h(\bar{u}_0^{(i)}) - k^{(i)}$ and the equation for the shock line

$$x_0' = [Q(\bar{u}_0^{(1)}) - Q(\bar{u}_0^{(2)})]_\Gamma / [\bar{u}_0^{(1)} - \bar{u}_0^{(2)}]_\Gamma.$$

If we would have $\theta^{(1)} = -\theta^{(2)} = \theta$, then simultaneously we would have $\omega = h(\bar{u}_0^{(1)}) - k$, $-\omega = h(\bar{u}_0^{(2)}) - k$, i.e., $k = \frac{1}{2}h(\bar{u}_0^{(1)}) + h(\bar{u}_0^{(2)})$ which is incorrect when $h(u) \neq u$ even on Γ [and holds for $h(u) = u$].

Example II

(See [42]).

$$\frac{\partial u}{\partial t} - \frac{\partial v}{\partial x} = 0, \qquad \frac{\partial v}{\partial t} - u\frac{\partial u}{\partial x} + \varepsilon\frac{\partial^2 u}{\partial x^2} = 0.$$

The corresponding system (5.7.11) is

$$-ku_0 - v_0 = -k\bar{u}_0 - \bar{v}_0, \qquad -kv_0 - \frac{1}{2}u_0^2 + \omega\zeta\frac{\partial u_0}{\partial \zeta} = -k\bar{v}_0 - \frac{1}{2}\bar{u}_0^2.$$

The first of these equations is a "ready-made" equation of the trajectory passing through $(\bar{u}_0, \bar{v}_0)$ and (u_0^*, v_0^*), where $u_0^* = 2k^2 - \bar{u}_0$, $v_0^* = \bar{v}_0 + 2k(\bar{u}_0 - k^2)$ is a solution of (5.7.12). The formula (5.7.9') is

$$T\hat{w}_0 = \left\| \begin{matrix} -k & -1 \\ -\bar{u}_0 + \omega & -k \end{matrix} \right\| \begin{pmatrix} \hat{u}_0 \\ \hat{v}_0 \end{pmatrix} = 0.$$

Equation (5.7.10) has the unique root $\omega = \bar{u}_0 - k^2$ giving the equation for θ. Conditions on the shock (5.7.4) are

$$x_0'(t) = -\left.\frac{\bar{v}_0^{(1)} - \bar{v}_0^{(2)}}{\bar{u}_0^{(1)} - \bar{u}_0^{(2)}}\right|_{x=x_0(t)} = -\frac{1}{2}\left.\frac{\left(\bar{u}_0^{(1)}\right)^2 - \left(\bar{u}_0^{(2)}\right)^2}{\bar{v}_0^{(1)} - \bar{v}_0^{(2)}}\right|_{x=x_0(t)}$$

[they can be obtained changing $\bar{w}_0$ and w_0^* by $\bar{w}_0^{(1)}|_\Gamma$ and $\bar{w}_0^{(2)}|_\Gamma$ and k by $x_0'(t)$ in (5.7.12)]. Solutions of (5.7.11) are $u_0 = \bar{u}_0 + u_0^* M_0\zeta/1 + M_0\zeta$ and $v_0 = \bar{v}_0 + v_0^* M_0\zeta/1 + M_0\zeta$ where $\hat{w}_0 = \{\hat{u}_0, \hat{v}_0\} = -2\omega M_0\{1, -k\}$.

Further, we easily find $Q = \{k, -1\}$ and (5.7.17) takes the form

$$2k^{(i)}\left(\frac{\partial M_0(i)}{\partial t} + k^{(i)}\frac{\partial M_0(i)}{\partial x}\right) - \frac{\partial}{\partial x}(\omega^{(i)} M_0^{(i)})$$
$$+ M_0^{(i)}\left(\omega^{(i)}\bar{u}_1^{(i)} + \frac{\partial k(i)}{\partial t} + k^{(i)}\frac{\partial k(i)}{\partial x}\right) = 0 \qquad \text{for } i = 1, 2.$$

To find values of $M_0^{(i)}|_\Gamma$ we require first the coincidence of $u_0^{(1)}$ and $v_0^{(1)}$ with $u_0^{(2)}$ and $v_0^{(2)}$ on (5.7.19). Due to the formulas for u_0, v_0, u_0^*, and v_0^* and conditions on the shock (see above) we find one condition $M_0^{(1)}|_\Gamma = 1/M_0^{(2)}|_\Gamma = \mu(t)$ and should also find $\mu(t)$..

As we have seen $M_0|_\Gamma$, hence $\mu(t)$, enters only $T_1^{(i)}|_\Gamma$. We find $T_1^{(i)}|_\Gamma$. We find $T_1^{(i)}|_\Gamma = -2\ln(1 + M_0^{(i)}|_\Gamma)\{1, -x_0'(t)\}$.

Formulas (5.7.22) (we do not integrate them) after certain simplifications take the form

$$\frac{4}{1+\mu}\frac{d\mu}{dt} = x_0'(t)[\bar{u}_1^{(1)} - \bar{u}_1^{(2)}]_\Gamma + [\bar{v}_1^{(1)} - \bar{v}_1^{(2)}]_\Gamma,$$

$$-\frac{8x_0'(t)}{1+\mu}\frac{d\mu}{dt} - 2x_0''(t)\ln\mu$$
$$= \frac{4}{[\bar{u}_0^{(1)} - \bar{u}_0^{(2)}]_\Gamma}\left(2x_0'^2(t)x_0''(t) - \frac{1}{2}\frac{d}{dt}[\bar{v}_0^{(1)} + \bar{v}_0^{(2)}]_\Gamma\right)$$
$$+ \frac{1}{2}[(\bar{u}_0^{(1)} - \bar{u}_0^{(2)})(\bar{u}_1^{(1)} + \bar{u}_1^{(2)})]_\Gamma$$

[the value of $\partial b(\bar{w}_0)/\partial x|_\Gamma$ can be found expanding $\bar{w}_0$ in powers of $x - x_0(t)$ with the help of (5.7.3)].

This shows that if the shock line is not a straight line, i.e., $x_0''(t) \neq 0$, then $\mu(t)$ can be recovered from $\bar{u}$ and $\bar{v}$ without integration (see once more Remarks 5.7.3 and 5.7.2).

In [43] the last conditions are absent. The author's solution corresponds to the case $\mu = 1$.

Example III. One-Dimensional Gas Flow

Equations of one-dimensional motion of a heat-conducting perfect gas with viscosity, without mass forces and supply of heat can be written in the form (5.7.1) where

$$w = \{\rho, \sigma, \tau\},$$
$$b(w) = \left\{0, \frac{\sigma}{\rho}, \beta\frac{\tau}{\rho} - \frac{\beta - 1}{2}\frac{\sigma^2}{\rho^2}\right\},$$
$$a(w) = \left\{\sigma, (\gamma - 1)\tau + \frac{3-\gamma}{2}\frac{\sigma^2}{\rho}, \gamma\frac{\tau\sigma}{\rho} - \frac{\gamma - 1}{2}\frac{\sigma^3}{\rho^2}\right\}.$$

Here $\sigma = \rho u$, ρ is the density and u the velocity, $\tau = \rho(c_v T + \frac{1}{2}u^2)$, T is temperature, and C_v is heat capacity at constant volume; $\varepsilon = 4/3\mu$, where μ is the viscosity coefficient, $\gamma = c_p/c_v$, c_p is heat capacity at constant pressure, $\beta = 3\gamma/4p_r$, p_r is the Prandtle number. Let us assume parameters γ, β, and ε to be constant and $\varepsilon \ll 1$ and equations will be considered as dimensionless.

The main theorem for equations of one-dimensional gas flow had been proved by R. Mizes under certain restrictions (see [41], pp. 155–176, 483–488).

For the case $p_r = 0,\ 75$ ($\beta = \gamma$) the phase trajectory had been obtained in analytical form by Becker (see, i.e., p. 168). Below we will discuss only this case.

Setting $-k\bar{w}_0 + a(\bar{w}_0) = \{e_1, e_2, e_3\}$ so that $e_1 = -k\bar{\rho}_0 + \bar{\sigma}_0$, etc., we find that (5.7.12) has the unique solution $w_0^* \neq \bar{w}_0$. In particular (we give only main formulas)

$$\rho_0^* = \frac{(\gamma + 1/2\gamma)e_1^2}{e_2 - ke_1 - \frac{\gamma+1}{2\gamma}\frac{e_1^2}{\bar{\rho}_0}} = \frac{\gamma+1}{\gamma-1}\frac{e_1^3/\bar{\rho}_0}{2e_3 + k^2 e_1 - 2ke_2}.$$

The phase trajectory of (5.7.11) connecting $\bar{w}_0$ and w_0^* exists and is unique as had been already said and condition (5.7.4) in our notations are of the form $e_j^{(1)}|_\Gamma = e_j^{(2)}|_\Gamma$, where $j = 1, 2, 3$ [recall that they are obtained from (5.7.12) changing $\bar{w}_0$, w_0^* and k by $\bar{w}_0^{(1)}|_\Gamma$, $\bar{w}_0^{(2)}$ and $x_0'(t)$ respectively].

The Becker trajectory is

$$\tau_0 = \frac{\gamma-1}{2\gamma}\frac{\sigma_0^2}{\rho_0} + \frac{k}{\gamma}\sigma_0 + \frac{e_3 - ke_2}{\gamma e_1}, \qquad \sigma_0 = k\rho_0 + e_1.$$

The system (5.7.11) is reduced to the equation

$$\frac{-2\gamma\bar{\rho}_0\rho_0^*}{(\gamma+1)e_1}\omega\zeta\frac{\partial\rho_0}{\partial\zeta} = \rho_0(\rho_0 - \bar{\rho}_0)(\rho_0 - \rho_0^*).$$

Equation (5.7.10) turns out to be a quadratic equation in ω so that the choice of a root in this problem is essential (rank T is 2 for both roots). Equation (5.7.10) is of the form ($\beta = \gamma$)

$$(\omega - e_1)\left(\omega - \frac{\gamma+1}{\gamma}e_1 + \frac{e_2 - ke_1}{e_1}\bar{\rho}_0\right) = 0.$$

It turns out that only for

$$\omega = \Omega = \frac{\gamma+1}{\gamma}e_1 - \frac{e_2 - ke_1}{e_1}\bar{\rho}_0$$

the vector $\hat{w}_0$ defined by (5.7.9') is parallel to the tangent to the Becker curve at the point $(\bar{\rho}_0, \bar{\sigma}_0, \bar{\tau}_0)$.

Inserting $\omega = \Omega$ and making use of the expression for ρ_0^* we find

$$(\rho_0 - \bar{\rho}_0)\frac{\bar{\rho}_0}{\rho_0}\left(\frac{1 - \rho_0^*/\bar{\rho}_0}{1 - \rho_0^*/\rho_0}\right)^{\bar{\rho}_0/\rho_0^*} = M_0(t, x)\zeta.$$

This equation gives the principal term of the asymptotics of ρ_0.

How to search for $M_0(t,x)$ should be clear from previous examples: the equation for M_0 is given by (5.7.17) and we get boundary values from (5.7.19) via $\rho_0|_\Gamma$. As to $\rho_0|_\Gamma$ we will find it from (5.7.22). Namely

$$T_1 = \frac{2\gamma\bar{\rho}_0}{(\gamma+1)e_1} \ln\left(\frac{\rho_0|_\Gamma}{\bar{\rho}_0}\frac{\bar{\rho}_0-\rho_0^*}{\rho_0|_\Gamma-\rho_0^*}\right)\left\{1, k, \frac{(3-\gamma)(e_3-ke_2)+2e_2k^2e_1}{(\gamma+1)e_1}\right\}$$
$$+\left\{0, 0, \frac{\gamma-1}{\gamma+1}e_1\left(\frac{1}{\rho_0|_\Gamma}-\frac{1}{\bar{\rho}_0}\right)\right\}.$$

This makes it clear that since k and e_i are preserved when the shock line is crossed then $\rho_0|_\Gamma$ enter $(T_1^{(1)} - T_1^{(2)})_\Gamma$ only in the form

$$\bar{\rho}_0^{(1)} \ln\left(\frac{\rho_0|_\Gamma}{\rho_0^{(1)}|_\Gamma}, \frac{\rho_0^{(1)}|_\Gamma - \rho_0^{(2)}|_\Gamma}{\rho_0|_\Gamma - \rho_0^{(2)}|_\Gamma}\right) - \bar{\rho}_0^{(2)}|_\Gamma \ln\left(\frac{\rho_0|_\Gamma}{\rho_0^{(2)}|_\Gamma}, \frac{\rho_0^{(2)}|_\Gamma - \rho_0^{(1)}|_\Gamma}{\rho_0|_\Gamma - \rho_0^{(1)}|_\Gamma}\right).$$

If $\bar{w}_0$ and $\bar{w}_1$ are known, this difference can be computed and $\rho_0|_\Gamma$ can be found.

Recall once more that problems of constructing exterior solutions and shock wave are related with each other, see, e.g., the case of gas before piston in a tube.

References

[1] Alfven, H. and Fälthammer, G. Cosmical electrodynamics, Oxford Univ. Press, Oxford, 1963.

[2] Appel', P. Theoretical mechanics, v. II. Physmathgiz, Moscow, 1960 (Russian).

[3] Arnol'd, V.I., Proof of A.N. Kolmogorov's theorem on conservartion of conditionally-periodic motions at small variation of Hamiltonian function. Russian Math. Surveys, v. 18, n. 5, 1963 (Russian).

[4] Babich, V.M., Buldyrev, V.S., Asymptotic methods in short wave diffraction problems. Nauka, Moscow, 1982 (Russian).

[5] Bogaevski, V.N., Drift equations and currents. In: "Questions of mathematical physics and functional analysis". Proceedings of Scientific Seminars of Phys. Tec. Int. of Low Temp. (Kharkov), Naukova Dumka, Kiev, 1966 (Russian).

[6] Bogaevski, V.N., Ostrer, L.A. On fast rotation of a mass solid body with fixed point, Problems of Math. & Mech., v. 44, n. 6, 1980 (Russian).

[7] Bogaevski, V.N., Povzner, A.Ya. On one-dimensional shock waves. Int. J. Non-Linear Mechanics, v. 13, 1979, 337–349.

[8] Bogaevski, V.N., Povzner, A.Ya. Nonlineary generalization of a shearing transformation. Funct. Anal. Appl. v. 16, n. 3, 1982, 45–46 (Russian).

[9] Bogaevski, V.N., Povzner, A.Ya. Linear methods in nonlinear problems with a small parameter. Lecture Notes in Mathematics, Vol. 985, 1983, 431–449.

[10] Bogolyubov, N.N., Mitrolopsky, Yu.A. Asymptotical methods in nonlinear oscillations theory. Physmathgiz, Moscow, 1958 (Russian),

[11] Braginski, S.I. Ukranian Math. J. v. 8, 1956, 119– (Russian).

[12] Caplun, S. The role of coordinate systems in boundary-layer theory. Z. Angew. Math. Phys. v. 5, 1954, 111-135.

[13] Cole, D. Perturbation methods in applied mathematics. Balisdell Publ. Company, Waltham-Massachusetts-Toronto-London, 1968.

[14] Courant, R. and Hilbert, D. Methods of mathematical physics, New York-London, 1962.

[15] Dorodnitzyn, A.A. Asymptotic of van der Pol equation, 11, 1947, 313–318 (Russian)

[16] Dyke, M. Van, Perturbation methods in fluid mechanics, Academic Press, New York-London, 1964.

[17] Fröman, N. Fröman, Per Olaf, SWKB-Approximation. North-Holland, 1965.

[18] Gelfand, I.M., Minlos, R.A., Shaprio, Z.Ya. Representation of the rotation group and the Lorentz group and their applications. Gostekhizdat, Moscow, 1953.

[19] Giacaglia, G.E.O. Perturbation methods in non-linear systems, Springer-Verlag, New York-Heidelberg-Berlin, 1972.

[20] Golubev, V.V. Lectures on integration of motion of a mass solid body with fixed point. GITTL, Moscow, 1953 (Russian).

[21] Grad, H. Asymptotic of the Botzmann equation, The Physics of Fluids, 6(2), 1963, 147–181.

[22] Iooss, G., Joseph, D. Elementary stability and bifurcation theory. Springer, 1980.

[23] Joseph, D. Stability of fluid motions. I and II. Springer, Tracts in Natural Philosophy, v. 27 and 28, Berlin-Heidelberg-New York, 1976.

[24] Kamke, E. Differentialgleichungen. Losungsmethoden und Lösungen. I. Govohnliche differential Gleichungen, Leipzig, 1959.

[25] Kapitza, P.L. Dynamical stability of pendulum with oscillating hanging point. JETPh, v. 21, n. 5, 1951 (Russian).

[26] Khazin, L.G. Remarks on Pyapunov's paper "A special case of stability of motion problem". M.V. Keldysh, Int. Appl. Math. USSR Acad. Sci. Preprint n. 9, 1980 (Russian).

[27] Kirchgraber, U., Stiefel, E. Methoden der analytischen Störungsrechnung und ihre Audwendungen, B.G. Tenbuer, Stuttgart, 1978.

[28] Langer, R. The asymptotic solutions of certain linear ordinary differential equations of the second order. Trans. Math. Soc., v. 36, 1934.

[29] Lepert, B. Dynamics of changed particles,

[30] Lyapunov, A.M. General stability of motion problem. Collected Papers, v. I, USSR Acad. Sci., Moscow-Leningrad, 1956 (Russian).

[31] Lyapunov, A.M. A special case of stability of motion problem. Collected Papers, v. II, USSR Acad. Sci., Moscow-Leningrad, 1956 (Russian).

[32] Marsden-McCracken, M. The Hopf-bifurcation and its applications. Lecture Notes in Applied Mathematical Sciences, v. 18, 1976.

[33] Maslov, V.P., Fedoryuk, M.V.[1] Quasiclassical approximations for equations of quantum mechanics. Nauka, Moscow, 1970 (Russian).

[34] Mises, R. Mathematical theory of compressible fluid flow. Academic Press, New York, 1958.

[35] Mischenko, E.F., Rozov, N.Kh. Differential equation with small parameter and relaxation oscillations. Nauka, Moscow, 1975 (Russian).

[36] Moiseev, N.N. Asymptotic methods of nonlinear mechanics. Nauka, Moscow, 1981 (Russian).

[37] Morozov, A.I., Soloviev, L.S. Drift of charged particles in electromagnetic fields. Plasma theory questions, Gosatumisdat, Moscow, 1963 (Russian).

[38] Nayfeh, A.H. Perturbation methods. J. Wiley, New York.

[39] Neu, Y.C. The method of near-identity transformations and its applications. SIAM J. Appl., v. 38, n. 2, 1980.

[40] Povzner, A. Linear methods in problems of non-linear differential equations with a small parameter. Intern. Journal of Nonlinear Mech., v. 9, 1974, 279–323.

[41] Povzner, A.Ya. Stokes constants for the Schroedinger equation with polynomial potential. Theor. Math. Phys., v. 51, n. 1, 1982 (Russian).

[42] Szaniawksi, A. The asymptotic structure of weak shock waves, Acta Mechanica, v. 5, n. 2, 1968.

[43] Smirnov, Treatize on higher mathematics, v. 3, part 2. Gostechizdat, 1949 (Russian).

[44] Stokes, J. Water waves, J. Wiley, New York, 1957.

[45] Wainberg, M.M., Trenogin, V.A. Branching theory for solutions of nonlinear equations. Nauka, Moscow, 1969 (Russian).

---

[1] According to Russian alphabetical order

[46] Wasow, W. Asymptotic expansions for ordinary differential equations, Wiley, New York, 1965.

[47] Whitham, B. Linear and nonlinear waves, J. Wiley, New York, 1974.

Index

Applied Mathematical Sciences

cont. from page ii

52. *Chipot:* Variational Inequalities and Flow in Porous Media.
53. *Majda:* Compressible Fluid Flow and System of Coservation Laws in Several Space Variables.
54. *Wasow:* Linear Turning Point Theory.
55. *Yosida:* Operational Calculus: A Theory of Hyperfunctions.
56. *Chang/Howes:* Nonlinear Singular Perturbation Phenomena: Theory and Applications.
57. *Reinhardt:* Analysis of Approximation Methods for Differential and Integral Equations.
58. *Dwoyer/Hussaini/Voigt (eds):* Theoretical Approaches to Turbulence.
59. *Sanders/Verhulst:* Averaging Methods in Nonlinear Dynamical Systems.
60. *Ghil/Childress:* Topics in Geophysical Dynamics: Atmospheric Dynamics, Dynamo Theory and Climate Dynamics.
61. *Sattinger/Weaver:* Lie Groups and Algebras with Applications to Physics, Geometry, and Mechanics.
62. *LaSalle:* The Stability and Control of Discrete Processes.
63. *Grasman:* Asymptotic Methods of Relaxation Oscillations and Applications.
64. *Hsu:* Cell-to-Cell Mapping: A Method of Global Analysis for Nonlinear Systems.
65. *Rand/Armbruster:* Perturbation Methods, Bifurcation Theory and Computer Algebra.
66. *Hlavácek/Haslinger/Necasl/Lovísek:* Solution of Variational Inequalities in Mechanics.
67. *Cercignani:* The Boltzmann Equation and Its Applications.
68. *Temam:* Infinite Dimensional Dynamical System in Mechanics and Physics.
69. *Golubitsky/Stewart/Schaeffer:* Singularities and Groups in Bifurcation Theory, Vol. II.
70. *Constantin/Foias/Nicolaenko/Temam:* Integral Manifolds and Inertial Manifolds for Dissipative Partial Differential Equations.
71. *Catlin:* Estimation, Control, and the Discrete Kalman Filter.
72. *Lochak/Meunier:* Multiphase Averaging for Classical Systems.
73. *Wiggins:* Global Bifurcations and Chaos.
74. *Mawhin/Willem:* Critical Point Theory and Hamiltonian Systems.
75. *Abraham/Marsden/Ratiu:* Manifolds, Tensor Analysis, and Applications, 2nd ed.
76. *Lagerstrom:* Matched Asymptotic Expansions: Ideas and Techniques.
77. *Aldous:* Probability Approximations via the Poisson Clumping Heuristic.
78. *Dacorogna:* Direct Methods in the Calculus of Variations.
79. *Hernández-Lerma:* Adaptive Markov Processes.
80. *Lawden:* Elliptic Functions and Applications.
81. *Bluman/Kumei:* Symmetries and Differential Equations.
82. *Kress:* Linear Integral Equations.
83. *Bebernes/Eberly:* Mathematical Problems from Combustion Theory.
84. *Joseph:* Fluid Dynamics of Viscoelastic Fluids.
85. *Yang:* Wave Packets and Their Bifurcations in Geophysical Fluid Dynamics.
86. *Dendrinos/Sonis:* Chaos and Socio-Spatial Dynamics.
87. *Weder:* Spectral and Scattering Theory for Wave Propagation in Preturbed Stratified Media.
88. *Bogaevski/Povzner:* Algebraic Methods in Nonlinear Perturbation Theory.